# Television Receivers

## Second Edition

**K. F. Ibrahim**

Senior Lecturer, Willesden College of Technology

LONGMAN

**Addison Wesley Longman Limited**
Edinburgh Gate
Harlow
Essex CM20 2JE
England
*and Associated Companies throughout the world*

*Visit Addison Wesley Longman on the world wide web at:*
*http://www.awl-he.com*

First edition published 1992
Second impression 1994
Third Impression 1995
Fourth Impression 1997
This edition published 1999

ISBN 0 582 35631 8

**British Library Cataloguing-in-Publication Data**
A catalogue record for this book is available from the British Library

**Library of Congress Cataloging-in-Publication Data**
A catalog record for this book is available from the Library of Congress

Typeset by 35 in 10/12pt Times
Produced by Addison Wesley Longman Singapore (Pte) Ltd.,
Printed in Singapore

To Valerie

00036267

# Contents

# Preface to the second edition

Television receivers have in the past few years undergone a rapid transformation in terms of the processing techniques employed and the circuitry used. The use of integrated circuits is only one aspect which has dominated the development of TV receiver design. More fundamental has been the use of digital techniques previously confined to data communication. The development of digital television and the use of microprocessors in TV receivers are other recent innovations that are now widespread throughout the industry.

For the TV engineer or student, a knowledge of highly sophisticated and complex digital processing techniques, including microprocessor applications, is essential to the understanding of modern TV receiver and decoder circuitry. Over a third of the book is devoted to these new developments.

I have attempted to describe both the system of television transmission, analogue and digital, and the circuitry of a TV receiver/decoder without undue reliance on previous technical or mathematical knowledge on the part of the reader.

The principles of mono and colour television transmission and reception are described in chapters 1 and 2. This is then followed by a detailed analysis of each section of an analogue television receiver (chapters 3 to 14). Circuits using discrete components as well as IC chips are fully covered with several practical examples as used by various manufacturers. Digital processing, including NICAM, the use of computers and remote control, is covered in chapters 15 to 18. Recent developments in satellite and digital television transmission and reception are covered in chapters 20 to 23. Testing and servicing digital receiver/decoder boxes is dealt with in chapter 24.

My thanks go to the British TV receiver manufacturers for giving me permission to use their circuits. Particular thanks must go to Ferguson, Philips and Pace for their special help and assistance.

K. F. Ibrahim
1998

# 1 Principles of monochrome television

At the television studio, the scene to be transmitted is projected on a photosensitive plate located inside the TV camera. The scene is repeatedly scanned by a very fast electron beam which ensures that consecutive images differ only very slightly. At the receiving end, a cathode-ray tube (c.r.t.) is used to recreate the picture by an identical process of scanning a screen by an electron beam. The phenomenon of persistence of vision then gives the impression of a moving picture in the same way as a cine film does. In the UK, 25 complete pictures are scanned every second.

## Scanning

In order to explore the scene in detail, the brightness of each element is examined line by line as shown in Fig. 1.1. The electron beam sweeps across the scene from left to right, *the sweep*, returns back very quickly, *the flyback*, to begin scanning the next line, and so on. A very large number of lines are employed to give adequate representation of the contents of the picture. In the UK, 625 lines are used while the USA uses 525 lines. The waveform that provides the scanning movement of the electron beam is the sawtooth waveform shown in Fig. 1.2. At the end of each complete scan the electron beam moves back to the top of the scene and the sequence is repeated. In the UK, 25 complete pictures are scanned every second with each picture containing 625 lines. This gives a line frequency of $25 \times 625 = 15\,625$ Hz or 15.625 kHz.

**Fig. 1.1**  *Television line scanning*

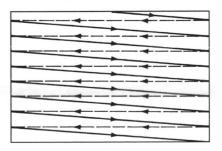

**Fig. 1.2**  *Scanning sawtooth waveform*

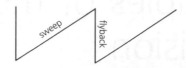

# Interlacing

Normal sequential scanning, i.e. scanning complete pictures (625 lines) at one time followed by another complete picture scan, introduces unacceptable flicker. This is avoided by the simple technique known as interlacing. Interlace scanning involves scanning the 'odd' lines 1, 3, 5, etc., followed by the 'even' lines 2, 4, 6, etc. Only one-half of the picture, known as a *field*, is scanned each time. A complete picture therefore consists of two fields, odd and even, resulting in a field frequency of $2 \times 25 = 50$ Hz.

At the end of each field the electron beam is deflected rapidly back to the beginning of the next scan. To ensure the same flyback time for both fields, the even flyback is started halfway along the last line of the even field (point B in Fig. 1.3) to take the beam to the start of the following odd field (point C) halfway along line 1. And the odd flyback starts at the end of the last line of the odd field (point D) to take the beam to the start of the first line of the even field (point A). As can be seen from Fig. 1.3, the beam is made to move through the same vertical distance, hence the travel time is the same for both fields. Since the line scan continues to move the electron beam across the screen during the field flyback, the path traced by the beam during the flyback is as shown in Fig. 1.4.

With half a line included in each field, the total number of lines must be odd, hence the UK's 625 lines and the USA's 525 lines.

In the absence of picture information, scanning produces what is known as a *raster*.

**Fig. 1.3**  *Field flyback*

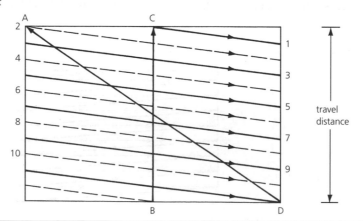

**Fig. 1.4** *Field flyback path: (a) end of odd field to start of even field; (b) end of even field to start of odd field*

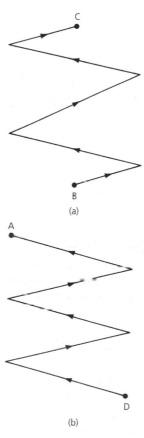

(a)

(b)

## Sync. pulses

For faithful reproduction of the picture by the cathode-ray tube, the scanning at the receiving end must follow the scanning at the transmitting end, line by line and field by field. To make sure this takes place, *synchronising pulses* are introduced at the end of each line to initiate the line flyback at the receiver; these pulses are called the *line sync*. Another synchronising pulse is introduced at the end of a field to initiate the start of the field flyback; this is called the *field sync.*

## Composite video waveform

The picture information and the sync. pulses constitute the composite or complete video waveform. Such a waveform for a one-line scan is shown in Fig. 1.5. The picture information is represented by the waveform between the two line sync. pulses and thus may acquire any shape, depending on the varying picture brightness along the line. The

**Fig. 1.5** *One-line composite video showing the relative video amplitudes*

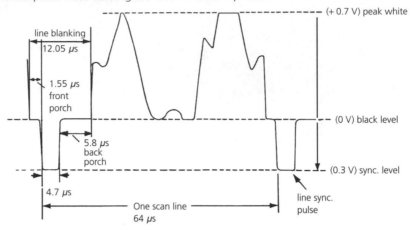

waveform shown represents a line that starts at half-peak white then gradually increases in brightness to peak white (maximum brightness); it then falls to black before rising again to peak white, and finally it returns to half-peak white.

The total available voltage is divided into two regions:

- *Below black level region* 0 to −0.3 V reserved for the sync. pulses (line and field)
- *Above black level region* 0 to 0.7 V (peak white), used for the video or picture information

Before and after every sync. pulse the voltage is held at the black level for short periods of time, respectively known as the front porch and the back porch. The *front porch* has a duration of 1.55 $\mu$s; it ensures the video information is brought down to the black level before the sync. pulse is applied. The *back porch* has a longer duration of 5.8 $\mu$s; it provides time for the flyback to occur before the application of the video information. The back porch is also used for black level clamping. As can be seen, the front porch, the sync. pulse and the back porch are at or below the black level. During this time, a total of 12.05 $\mu$s, the video information is completely suppressed; this is known as the line blanking period.

The duration of one complete line of a composite video waveform may be calculated from the line frequency:

$$\text{line duration} = \frac{1}{\text{line frequency}} = \frac{1}{15.625\,\text{kHz}} = 64\,\mu\text{s}$$

## Video bandwidth

The frequency of the video waveform is determined by the change in the brightness of the electron beam as it scans the screen line by line. Maximum video frequency is obtained when adjacent bits or elements are alternately black and white (Fig. 1.6); this represents

Fig. 1.6

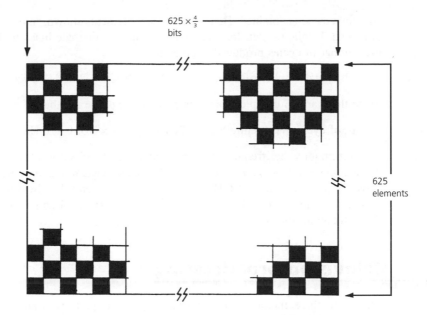

$625 \times \frac{4}{3}$
bits

625
elements

the maximum definition of a TV image. Along a vertical line there are a maximum of 625 alternating black and white elements. For equal definition along a horizontal line, the separation between the black and white bits must be the same as the separation along a vertical line. For a perfectly square TV screen, an equal number of bits would be obtained in both directions. However, the TV screen has an *aspect ratio* (ratio of width to height) of 4:3. (Digital TV broadcasting uses an aspect ratio of 4:5.) This increases the number of bits along a horizontal line to $625 \times 4/3 = 833.3$ elements per line giving a picture total of

$$625 \times 625 \times 4/3 = 520\ 833 \text{ bits or elements}$$

When an electron beam scans a line containing alternate black and white elements, the video waveform is that shown in Fig. 1.7, representing the variation of brightness along the line. As can be seen, for any adjacent pair of black and white elements, one

**Fig. 1.7** *Video waveform for alternate black and white elements*

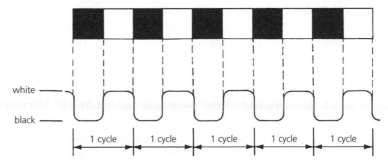

white

black

1 cycle | 1 cycle | 1 cycle | 1 cycle | 1 cycle

complete cycle is obtained. Hence, for the ten elements shown, five complete cycles are produced. It follows that, for a complete picture of alternate black and white elements, the number of cycles produced is given by

$$1/2 \times \text{total number of elements} = 1/2 \times 520\,833 = 260\,417 \text{ cycles per picture}$$

Since there are 25 complete pictures every second, the number of cycles per second is

$$\text{no. of cycles per picture} \times 25 = 260\,417 \times 25 = 6\,510\,416 \text{ Hz} = 6.5 \text{ MHz}$$

The minimum video frequency is obtained when the electron beam scans elements of unchanging brightness. This corresponds to an unchanging amplitude of the video wave-form, a frequency of 0 Hz or d.c., giving an overall bandwidth of 0 to 6.5 MHz. In practice, however, a bandwidth of 0 to 5.5 MHz is found to be satisfactory for domestic applications.

## Television broadcasting

There are three methods of television broadcasting: terrestrial, satellite and cable. Terrestrial is the traditional method of broadcasting television signals to the home employing UHF radio frequencies. Satellite broadcasting involves two stages. In the first place, the TV signals are sent to a satellite circulating the earth at a distance of 35 765 km. The satellite sends these signals back to earth (on a different frequency) and they may be received using a simple satellite dish aerial. Cable broadcasting, as the name implies, uses a transmission cable to send TV signals to home subscribers. Unless otherwise stated, terrestrial broadcasting is assumed throughout the book.

## Modulation

TV transmission uses amplitude modulation for the video information. Ordinary amplitude modulation gives rise to two sets of sidebands on either side of the carrier, thus doubling the bandwidth requirement for the transmission. However, since each sideband contains all the video information, it is possible to suppress one sideband completely, employing what is known as single-sideband (SSB) transmission. However, pure single-sideband transmission demands a more complicated synchronous detector at the receiving end, making the receiver more expensive. The simple and cheap diode detector introduces a distortion known as *quadrature distortion*, caused mainly by the lower end of the video frequency spectrum. To avoid this and still use the diode detector, *vestigial sideband transmission* is employed in which double-sideband transmission is used for low video frequencies and single-sideband transmission for higher video frequencies. The frequency response of vestigial sideband transmission used in the UK is shown in Fig. 1.8, in which part of the lower sideband, up to 1.25 MHz, is transmitted with the unsuppressed upper sideband.

As well as the composite video, it is also necessary to transmit a sound signal. Unlike the video information, sound is frequency modulated on a separate carrier with a

**Fig. 1.8** *Frequency response for UK TV transmission*

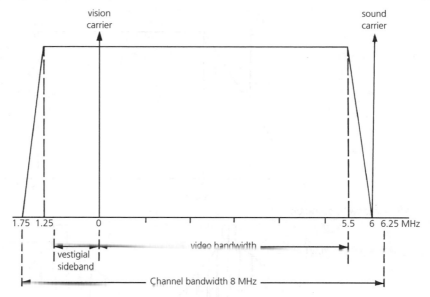

bandwidth of 100 kHz. The sound carrier is chosen to be 6 MHz away from the vision carrier, so it falls just outside the highest transmitted video frequency (Fig. 1.8). For a vision carrier of 510 MHz the sound carrier is 510 + 6 = 516 MHz.

Figure 1.8 shows that the response remains constant over the range of video frequencies up to 5.5 MHz on the upper sideband and 1.25 MHz on the lower sideband. Above 5.5 MHz a sharp but gradual attenuation takes place to ensure that no video information remains beyond 6 MHz; this prevents any overlap with the sound information. An additional 0.25 MHz is needed to accommodate the sound bandwidth and to provide a buffer space for the adjacent channel. Similar attenuation is necessary for frequencies extending beyond 1.25 MHz on the lower sideband to ensure that no video information extends beyond 1.75 MHz, thus preventing any overlap with an adjacent channel. Gradual attenuation is necessary since it is not possible to have filters with instantaneous cut-off characteristics. An 8 MHz (1.75 + 6.25) bandwidth is therefore allocated for each TV channel.

As a consequence of vestigial sideband transmission, video frequencies up to 1.25 MHz are present in both sidebands and frequencies above 1.25 MHz are present in one sideband only. When detected by a simple diode detector, the frequencies below 1.25 MHz will produce twice the output of the frequencies above 1.25 MHz. To compensate for this, it is necessary to shape the frequency response of the receiver so that frequencies that are present in both sidebands are afforded less amplification than those present in one sideband only. Such a response curve is shown in Fig. 1.9.

The video signal may be used to modulate the vision carrier in either a positive or negative way. Negative modulation (Fig. 1.10) is used in the UK, with peak white corresponding to a minimum voltage. Positive modulation (Fig. 1.11) was used in the old 405-line British system. Here peak white corresponds to a maximum voltage.

**Fig. 1.9** *Television receiver frequency response*

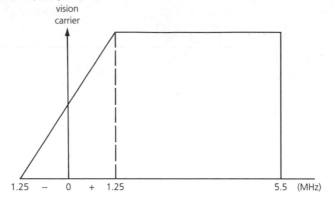

**Fig. 1.10** *Negative modulation*

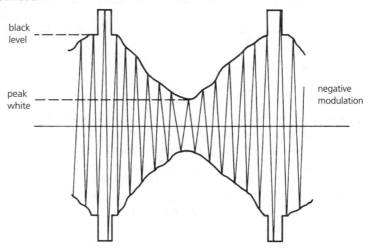

**Fig. 1.11** *Positive modulation*

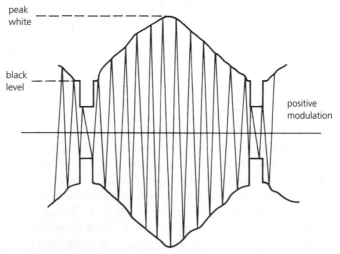

# Channel allocation

In order to cover a large area, a number of transmitting stations are used. Each transmitting station is assigned a number (4 or 5) of channels and each channel occupies a bandwidth of 8 MHz. The transmitting frequencies of all the channels must fall within a specified bandwidth for the transmitting station. Only half of the available channels may be used for analogue TV broadcasting. This minimises the inter-channel interference by avoiding the use of adjacent channels. However, the remaining channels may be used for digital television broadcasting. The UK is divided into several areas each served by a station transmitting UHF frequencies in bands IV and V. For example, Crystal Palace serves the Greater London area and uses the following analogue channels:

| ITV | Channel 23 | 486–494 MHz |
| BBC1 | Channel 26 | 510–518 MHz |
| C4 | Channel 30 | 542–550 MHz |
| BBC2 | Channel 33 | 566–574 MHz. |

Channel 5 transmits at channel 37 (600–608 MHz) from Croydon.

# TV receiver (terrestrial)

Figure 1.12 shows a block diagram for a monochrome (black and white) TV receiver. The tuner selects an appropriate carrier frequency and converts it to an *intermediate frequency* of 39.5 MHz (mixer–oscillator stage). The modulated i.f. is then amplified through several stages and demodulated to reproduce the original composite video signal. The composite TV signal is then separated into its three component parts. The 6 MHz *sound inter-carrier* is taken off at the emitter follower stage (or video driver stage) following the detector. The f.m. sound signal is then detected, amplified and fed into the loudspeaker. The sync. pulses are clipped from the video information at the video output stage, separated into line and field and taken to the appropriate timebase. After amplification, line and field pulses are used to deflect the electron beam in the horizontal and vertical directions via a pair of scan coils. The video information itself is amplified by the video output amplifier and fed into the cathode of the cathode-ray tube. *Automatic gain control* (*a.g.c.*) is employed to ensure the output of the i.f. stage remains steady irrespective of changes in the strength of the received signal. *Automatic frequency control* (*a.f.c.*) is sometimes used to keep the intermediate frequency stable at 39.5 MHz. Apart from providing the drive for the line scan coils, the line output stage also provides the *extra high tension* (*e.h.t.*) for the c.r.t. by the use of an overwind at the line output transformer. The line output stage is also used to provide other stabilised d.c. supplies for the receiver.

# The cathode-ray tube

The cathode-ray tube (c.r.t.) used for television picture display operates on the same principle as the old thermionic valve. A negatively charged, hot cathode emits electrons which are attracted to and collected by a positively charged anode.

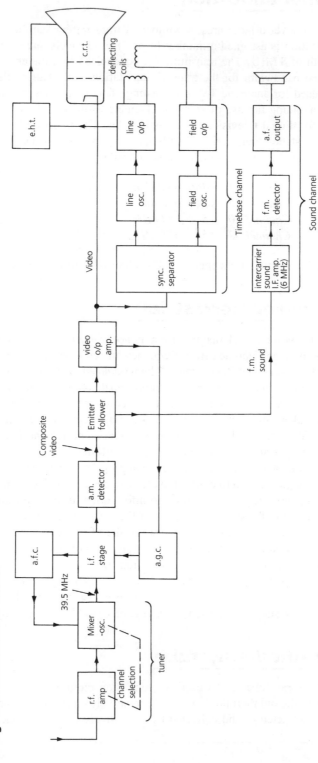

**Fig. 1.12** *Monochrome TV receiver*

**Fig. 1.13** *Monochrome TV cathode-ray tube*

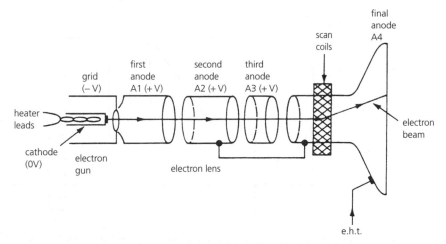

In the c.r.t., high-speed electrons are emitted by an electron gun. They are focused and accelerated by an electron lens and directed towards a screen which acts as the positively charged anode (Fig. 1.13). The screen is a glass faceplate coated on the inside with a fluorescent powder or phosphor which gives a visible glow when hit by high-speed electrons. The colour of the emitted light is determined by the type of the phosphor.

The electron beam generated by the electron gun gives a stationary dot on the screen. In order to produce a display, the c.r.t. must have the capacity to deflect the beam in both the horizontal (line) and vertical (field) directions. Electromagnetic deflection is employed using two sets of coils, known as *scan coils* (line and field), placed along the neck of the tube in order to deflect the beam horizontally and vertically.

The monochrome display tube consists of a single electron gun, an anode assembly acting as the electron lens, and a viewing surface. The function of the electron gun is to produce a high-velocity concentrated beam of electrons which strike the phosphor-coated screen.

The electrons are attracted and accelerated by a positively charged first anode known as the accelerating anode (A1 in Fig. 1.13). The number of electrons leaving the cathode is controlled by the grid. Because it has a negative potential with respect to the cathode, the grid controls the emission of electrons, hence the luminance or brightness of the display. Electron beam suppression, i.e. tube blanking, may thus be obtained by applying a suitable negative-going pulse to the grid. The final anode voltage, known as the *extra high tension* (*e.h.t.*), is in the region of 15–20 kV. It is produced using an overwind on the line output transformer.

---

### QUESTIONS

1. For the British UHF television broadcasting system, state
   (a) the number of lines that make up one field
   (b) the number of fields that make up one complete picture

---

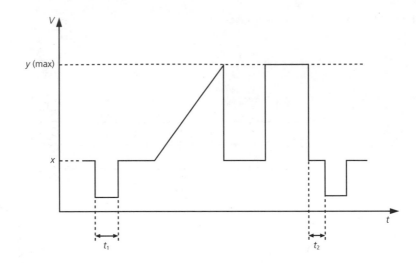

(c) the number of complete pictures per second
(d) the line frequency
(e) the field frequency

2. Refer to Fig. Q1.1.
   (a) State the duration of period $t_1$.
   (b) Name the voltage level $x$.
   (c) Name the voltage level $y$.
   (d) Name the part of waveform shown as $t_2$.
   (e) Describe the kind of display that would be produced on a monochrome receiver if the waveform were repeated for each line.

3. For British UHF TV broadcasting, state
   (a) the video bandwidth
   (b) the bandwidth of one channel
   (c) the frequency spacing between the vision carrier of channels 59 and 60

# 2 Colour transmission

The first problem facing the transmission of colour TV signals is compatibility with the existing monochrome transmission. Colour TV signals must be capable of producing a normal black and white image on a monochrome receiver without any modification to the television set. Conversely, a colour receiver must be capable of producing a black and white image from a monochrome signal. A colour transmission system must therefore retain the monochrome information, sync. pulses and the sound intercarrier in the same form as those of normal monochrome transmission. The additional colour information has to be included without interfering with the composite monochrome signal. Furthermore, the colour signal must occupy the same bandwidth as that allocated for monochrome transmission. To understand how this may be done, we must isolate the component part of visible light that stimulates the sensation of colour in the human eye.

## Visible light

Visible light is an electromagnetic wave similar to radio waves, X-rays, and so on. It forms the narrow band of the electromagnetic spectrum shown in Fig. 2.1. Light waves falling on the eye pass through the pupil, which focuses the image on the retina at the back of the eye (Fig. 2.2). The retina is sensitive to electromagnetic waves within the visible band; it can therefore translate the electromagnetic energy into suitable information, which is then passed to the brain via numerous optic nerve fibres.

**Fig. 2.1** *Electromagnetic wave spectrum*

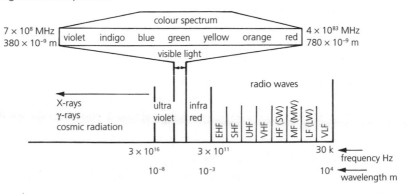

**Fig. 2.2** *The human eye*

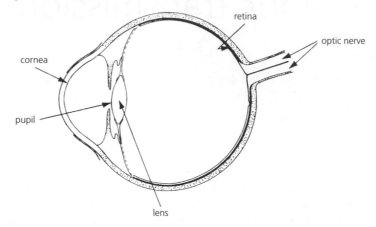

The retina contains a large number of light-sensitive cells. Cells known as *rods* are sensitive to brightness (or luminance) only; cells known as *cones* are sensitive to colour (or chrominance) only. Rods outnumber cones by a factor of 20 and they are 10 000 times more sensitive. The eye therefore reacts predominantly to the luminance of a colour picture, much more than to its chrominance. With high video frequencies, for example, fine picture details are perceived in black and white only.

## Primary colours

Cones themselves are of three different types. One is energised by red, another by green and another by blue. These colours are known as *primary colours*. Colours other than the three primary colours are perceived through energising two or more types of cones simultaneously. For example, the sensation of yellow is produced by energising the red and green cones simultaneously. Other colours may be produced by different mixtures of colours. In general all colours may be produced by the addition of appropriate quantities of the three primary colours red R, green G and blue B. This is known as *additive mixing*. For example:

$$yellow = R + G$$
$$magenta = R + B$$
$$cyan = B + G$$
$$white = R + G + B$$

Yellow, magenta and cyan are known as *complementary colours*, complementary to blue, green and red respectively. A complementary colour produces white when added to its corresponding primary colour. For example, if yellow is added to blue, then

$$yellow + blue = red + green + blue = white$$

Colours may also be produced by a process of *subtractive mixing*. For example, yellow may be produced by subtracting blue from white. Since $W = R + G + B$, then

$$W - B = (R + G + B) - B = R + G = \text{yellow}$$

Similarly,

$$W - G = R + B = \text{magenta}$$
$$W - R = G + B = \text{cyan}$$
$$W - R - G - B = \text{black (absence of colour)}$$

## The colour triangle

A colour triangle (Fig. 2.3) may be used to represent the chrominance content of a colour picture. Pure white is represented by a point W at the centre of the triangle; other colours are represented by phasors (or vectors) extending from the centre W to a point on or inside the triangle. Phasors going to the three corners of the triangle WR, WG and WB represent the primary colours, red, green and blue. Other colours are represented by appropriate phasors. For instance, yellow is represented by phasor WY with point Y falling between its two primary components, red and green. Phasors for cyan (WC) and magenta (WM) are constructed in a similar way.

**Fig. 2.3**  *Colour triangle*

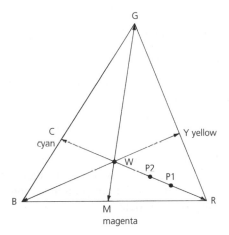

## Saturation and hue

Phasor WR in Fig. 2.3 represents a pure red with no trace of any other colour present. It is said to be *fully saturated*. Desaturation is obtained when white is added to a pure colour. For instance, if white is added to red, desaturated red or pink is produced. On the colour triangle, this is represented by moving along phasor WR away from R (pure red) and towards W (pure white). Point P1 thus represents desaturated red, or pink, with WP1 as its phasor representation. A greater desaturation produces a shorter phasor; WP2 represents pale pink, and so on. The phasor length represents *saturation*; the phasor direction or angle represents *hue*. Hue denotes the principal primary component of a colour.

For instance, pink has red as its principal primary colour, hence phasor WP is in phase with pure red WR. Yellow, on the other hand, has two primary components, red and green, hence phasor WY falls between pure red WR and pure green WG. Other phasors, such as for cyan and magenta, are constructed in a similar way. It follows that to represent the chrominance component of a colour picture, two qualities have to be ascertained:

- *Hue* its place on the colour spectrum, e.g. red, lemon, moss green and purple
- *Saturation* e.g. pink, pale green, dark blue and other pastel colours

## Principles of colour transmission

Colour transmission involves the simultaneous transmission of the luminance and chrominance components of a colour picture. The luminance signal Y is transmitted directly in the same way as a monochrome system. As for the chrominance component, it is first 'purified' by removing the luminance component from each primary colour, resulting in what is known as *colour difference signals*:

R − Y
G − Y
B − Y

Since the luminance signal Y = R + G + B, only two colour difference signals need to be transmitted, R − Y and B − Y. The third colour difference signal (G − Y) is recovered at the receiving end from the three transmitted components Y, R − Y and B − Y. Start with Y = R + G + B, then

R = (R − Y) + Y
B = (B − Y) + Y and
G = Y − R − B

The problem that remains to be solved is the manner in which this additional information, R − Y and B − Y, may be added to the monochrome signal without causing it any interference. To do this, quadrature amplitude modulation (QAM) is used on a separate carrier frequency of 4.43 MHz.

## Quadrature amplitude modulation

Two 4.43 MHz carriers, OV and OU, are arranged at right angles (quadrature) to each other (Fig. 2.4). The two colour difference signals are then used to modulate the two carriers with R − Y modulating OV and B − Y modulating OU. As with ordinary amplitude modulation, each modulated carrier produces two bands of side frequencies, one on each side of the carrier. $VR_1$, $VR_2$ and $UB_1$, $UB_2$ in Fig. 2.4 represent pairs of side frequencies for the red and blue colour difference signals respectively. The carriers themselves contain no information and are therefore suppressed, leaving the side frequencies only (Fig. 2.5). The pairs of side frequencies $VR_1$, $VR_2$ and $UB_1$, $UB_2$ produce resultant colour difference phasors E(R − Y) and E(B − Y) respectively. The two colour

**Fig. 2.4** *Quadrature amplitude modulation (QAM)*

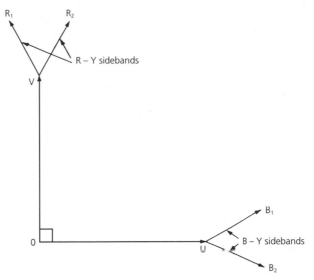

**Fig. 2.5** *Suppressed carrier (QAM)*

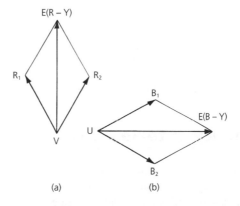

difference phasors retain the 90° or quadrature angle difference. This is because a common frequency of 4.43 MHz is used for both carriers. The two colour difference signals, R − Y and B − Y, themselves produce a resultant chrominance phasor OC (Fig. 2.6). Although the carrier itself is suppressed, the resultant phasor has the same frequency as the suppressed carrier. This chrominance phasor corresponds to the phasor associated with the colour triangle with its length (or amplitude) representing saturation and its angle (or phase) $\theta$ representing hue.

The bandwidth of the chrominance information is limited to approximately 1 MHz on each side of the colour carrier, which is now known as a *subcarrier* since it falls within the TV transmission frequency spectrum (Fig. 2.7). The relatively narrow bandwidth allocated to the chrominance is quite sufficient for an adequate reproduction of a colour image at the receiving end. This is because the eye perceives high video frequencies in black and white only.

**Fig. 2.6** *Colour difference phasors and their resultant*

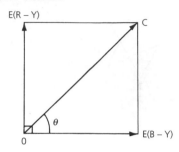

**Fig. 2.7** *Frequency spectrum for colour TV transmission*

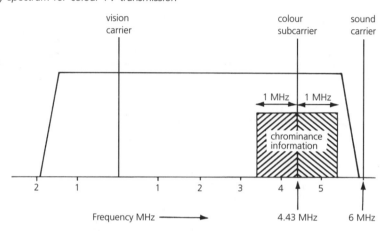

# Frequency interleaving

The colour subcarrier is chosen to fall within the monochrome frequency spectrum so as not to increase the bandwidth of TV transmission. The resulting overlap between chrominance and luminance signals would create patterning on a monochrome receiver tuned to a colour transmission, but this is avoided by frequency interleaving.

When the frequency spectrum of a TV signal is examined in detail, it is found that the distribution of frequencies is not uniform. Frequencies tend to gather in bunches centred around the harmonics of the line frequency (Fig. 2.8). This is because part of the composite video signal modulating the vision carrier includes the line sync. pulses. Square in shape and with unequal mark and space, the line sync. pulses contain an infinite number of harmonics; these harmonics produce side frequencies on either side of the carrier around which the video information clusters. Furthermore, the amplitude of these side frequencies gets progressively smaller as we move away from the vision carrier. It follows that, for minimum interference with the monochrome signal, the subcarrier must fall between two bunches. The subcarrier is itself amplitude modulated by the chromin-ance signal and thus produces side frequencies of its own in similar bunches on each side of the subcarrier centred around harmonics of the line frequency. By choosing a

**Fig. 2.8**  *Side frequency bunching*

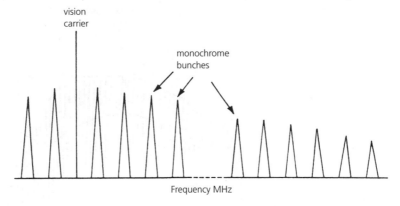

**Fig. 2.9**  *Frequency interleaving, half-line offset*

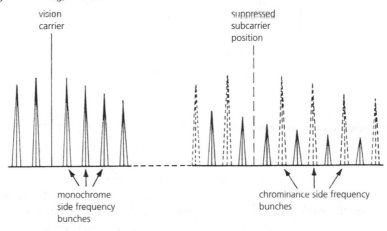

subcarrier to fall between two monochrome bunches, the chrominance bunches will then fall in the spaces between the bunches produced by the monochrome signal (Fig. 2.9). This is known as *frequency interleaving* or *interlacing*.

The subcarrier must therefore be a multiple of half the line frequency, known as half-line offset. However, to avoid any possible dot pattern on a monochrome receiver, quarter-line offset is used; the subcarrier is a multiple of one-quarter the line frequency. This is further modified by adding the half-field frequency, giving a subcarrier frequency of 4.433 618 75 MHz.

## Composite colour signal

The modulated subcarrier is added to the luminance (monochrome) signal to form the composite colour signal (Fig. 2.10). The modulated subcarrier appears as a sine

**Fig. 2.10** *One line of composite colour signal*

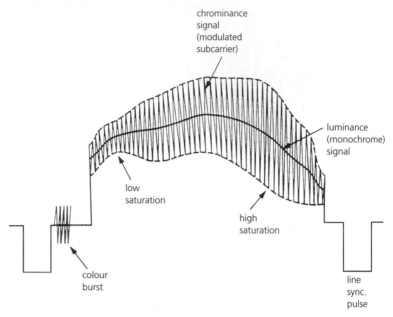

wave superimposed on the monochrome signal; it changes in amplitude and phase. The amplitude of the subcarrier represents saturation. Thus a fully saturated colour is represented by maximum subcarrier amplitude; black and white is represented by zero subcarrier amplitude. Hue, on the other hand, is represented by the phase angle of the subcarrier. To ascertain the phase angle, a 'burst' of about 10 cycles of the original sub-carrier is transmitted for use as a reference at the receiver. This *colour burst* is mounted on the back porch of the line sync. The phase of the modulated colour signal is then compared with the phase of the regenerated subcarrier to provide a measure of the phase angle and therefore hue. The absence of a colour burst indicates a black and white transmission.

## Gamma-correction

Receiver cathode-ray tubes use the voltage at the grid to control the beam current, hence the brightness of the display on the screen. However, the relationship between the two suffers from a non-linearity that must be corrected to avoid severe deterioration in the quality of the picture. This non-linearity at the receiving end is compensated by the intro-duction of an equal and opposite non-linearity at the transmitting end; it is known as gamma-correction ($\gamma$-correction). The voltage $E$ from the camera is raised to a power of $1/\gamma$ (i.e. $E^{1/\gamma}$), where for UK TV transmission $\gamma = 2.2$ and $1/\gamma = 0.45$. Gamma-corrected signals are indicated using a prime, e.g. Y', R' and gamma-corrected colour difference signals are indicated as Y' − R' and Y' − B'.

# Weighting factors

The subcarrier has a maximum amplitude when 100% or fully saturated colour is transmitted. Since the subcarrier is added to the luminance signal, the amplitude of the composite colour signal may exceed the maximum possible voltage. To avoid this, the peak amplitude of the chrominance signal, i.e. the peak amplitude of the colour difference signals, is reduced by a factor known as the weighting factor. The new weighted components of the chrominance signal are called $U$ and $V$ where

$$U = 0.493(B' - Y')$$
$$V = 0.877(R' - Y')$$

The resultant chrominance phasor is now produced by the phasor sum of $U$ and $V$.

# PAL colour system

There are three main systems of colour transmission: NTSC, PAL and SECAM. All three systems split the colour picture into luminance and chrominance; all three use colour difference signals to transmit the chrominance information. The difference between the systems lies in the way in which the subcarrier is modulated by the colour difference signals. SECAM (used in France) transmits the colour difference signals $U$ and $V$ on alternate lines: $U$ on one line, $V$ on the next, and so on. The other two systems, NTSC (used in the USA) and PAL (used in the UK), transmit both chrominance components simultaneously using quadrature amplitude modulation. However, it is found that errors in hue may occur as a result of phase errors (delay or advance) of the chrominance phasor (Fig. 2.11). Such errors are caused either by the receiver itself or by the way the signal is propagated. They are almost completely corrected by the PAL system.

In the PAL (phase alternate line) system, the $V$ signal is reversed on successive lines, $V$ on one line followed by $-V$ on the next, and so on . The first line is called an NTSC line and the second line is called a PAL line. Phase errors are thus reversed from one line to the next. At the receiving end, a process of averaging consecutive lines by the eye cancels out the errors to reproduce the correct hue. This simple method is known as PAL-S. A more accurate method is the use of a delay line to allow for consecutive lines (NTSC and PAL) to be added to each other, thus cancelling phase errors. This is known as PAL-D.

**Fig. 2.11**

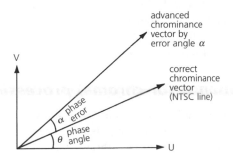

In order to distinguish between the two types of lines, the colour burst is made to swing by approximately 180° as an NTSC line is followed by a PAL line and back again.

# Colour TV receiver

Figure 2.12 shows a generalised block diagram of the colour processing parts of a PAL TV receiver.

After detection the composite colour signal is fed into three processing sections: burst, luminance and chrominance.

## Burst processing

The burst section provides the necessary reference subcarrier frequency, colour kill and burst blanking signals. The burst gate amplifier is opened by a line sync. pulse from the line timebase (Fig. 2.12) allowing only the colour burst to go through. The burst is fed into a phase discriminator and a reactance stage in order to phase lock the crystal-controlled reference oscillator to that of the original suppressed subcarrier. Two quadrature (90°) 4.43 MHz subcarriers are necessary for the two colour difference demodulators. For this reason one output of the oscillator is made to suffer a 90° phase change before going into the B' – Y' (U component) demodulator. The other output is fed directly to the R' – Y' (V component) demodulator. However, since the phase of the V component is reversed on alternate lines, a PAL switch is introduced as shown to ensure that the phase of the V subcarrier changes by 180° in step with the transmitted signal.

The burst which is present on colour transmission only changes phase from one line to the next. This change in phase produces a half-line frequency at the phase discriminator stage, i.e. 7.8 kHz; it is known as the *ident signal*. The absence of the ident signal indicates a monochrome-only transmission. The *colour kill* is then activated to shut the chrominance processing channel. This is essential in order to prevent information from the luminance signal which falls within the chrominance bandwidth from getting through the chrominance channel and causing colour interference on the screen.

## Luminance processing

After amplification the signal is delayed to compensate for the delay introduced to the chrominance component by the relatively narrowband chrominance amplifier. The *delay line* ensures that both the luminance and chrominance signals reach the c.r.t. at the same time. The modulated colour subcarrier, which contains the colour information, is then removed by a notch filter, leaving the luminance signal only. This is then fed into the output amplifier before going to the matrix and the c.r.t.

## Chrominance (or chroma) processing

For colour transmission the colour killer allows the signal through to the chrominance amplifier. The chrominance amplifier is preceded by a bandpass filter which allows only colour information to pass through. The PAL decoder includes two synchronous

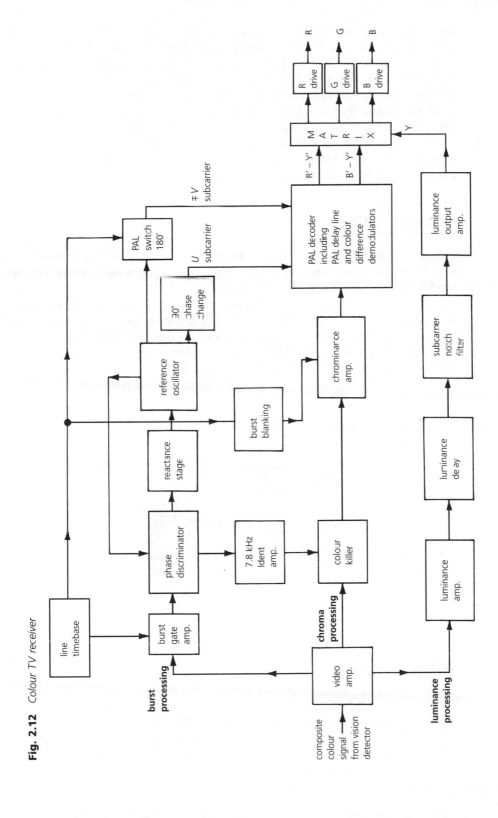

**Fig. 2.12** *Colour TV receiver*

demodulators, one for the B′ − Y′ component and the other for the R′ − Y′ component of the colour signal. The B′ − Y′ demodulator is fed with the 4.43 MHz subcarrier directly from the reference oscillator, whereas the R′ − Y′ demodulator receives the subcarrier signal via a PAL switch. In PAL-D receivers a chrominance delay line is incorporated in the decoder unit which stores one line of chrominance information for a one-line period of 64 $\mu$s. This is then added to the chrominance component of the following line to cancel out errors in hue introduced by incorrect pulse angles.

The chrominance amplifier is turned off for the duration of the burst by a blanking pulse from the line timebase. Known as *burst blanking*, it prevents the burst from creating colour interference on the screen.

## Colour tube

Cathode-ray tubes for colour display contain three separate guns that bombard a screen coated with phosphors arranged in triads. Each triad contains three different phosphors, one for each primary colour. Placed behind the coated screen, a steel *shadow mask* allows the three electron beams to converge and pass through slots before they strike their respective phosphors on the screen (Fig. 2.13). As the deflection (scan) coils scan the electron beams across the screen, the shadow mask ensures that each beam strikes its own particular phosphor and no other (blue gun for blue phosphor, etc.). Three primary colours are thus produced which, because they are very close to each other, are not resolved individually by the human eye. Instead an additive mixture is perceived, giving the sensation of colour.

**Fig. 2.13** *Colour tube colour production*

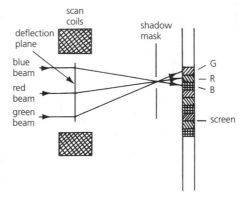

## Tube drive

There are two colour tube drive techniques: RGB and colour difference. In the RGB technique, the output Y of the luminance section and the two colour difference outputs, B′ − Y′ and R′ − Y′, from the chrominance section are fed into a matrix to reproduce the three primary colours R, G and B; these signals are then amplified and used to drive the cathodes of the colour tube. Alternatively the colour tube may be driven directly by the colour difference signals, R − Y, G − Y and B − Y, with luminance Y going into the grid to perform the subtraction process. The RGB technique is almost universally used by manufacturers.

## QUESTIONS

1.  Explain the purpose of gamma-correction.

2.  With reference to PAL colour transmission, state
    (a) the frequency of the colour subcarrier
    (b) the frequency of the sound subcarrier
    (c) the type of modulation used for the chrominance component of the picture information
    (d) the type of modulation used for the sound component of the transmission

3.  Sketch one single line of a composite video signal showing
    (a) the front and back porches
    (b) the colour subcarrier
    (c) the sync. pulse

4.  (a) State the function of the colour burst.
    (b) State the purpose of luminance delay in a PAL colour TV receiver.

5.  What would be the effect on a television display if
    (a) the colour burst were absent
    (b) the delay line were
        (i) short-circuited
        (ii) open-circuited

# 3 The UHF tuner

The function of the tuner is to select a TV channel frequency, amplify it and convert it into an intermediate frequency for further amplification by the i.f. stage (Fig. 3.1). The tuner must be capable of selecting any channel from bands IV and V, and it must provide sufficient r.f. amplification with good signal-to-noise ratio and minimal frequency drift. One or more stages of r.f. amplification are therefore necessary before the mixer–oscillator stage. A high-pass filter is normally used at the input to the r.f. amplifier to produce a correctly shaped response curve. The mixer–oscillator changes the tuned r.f. to a common *intermediate frequency* (*i.f.*) of 39.5 MHz. The tuner unit is built inside a metal case to screen it from outside r.f. interference. Further screening is also provided between various stages of the unit by metal walls. These inner walls which form part of the tuned circuits prevent unwanted coupling between one compartment and another. But when a coupling capacitance is needed, a small hole or slot may be cut in the dividing wall.

**Fig. 3.1** *UHF tuner*

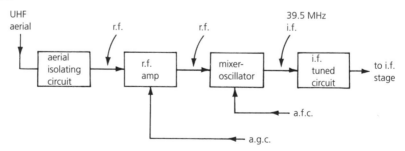

## UHF tuned circuits

The resonant frequency of a tuned circuit is given as

$$f = \frac{1}{2\pi\sqrt{LC}}$$

Due to the high carrier frequencies employed in TV transmission (370–862 MHz) the values of $L$ and $C$ are very small, with the result that normal inductors and capacitors cannot be used. Instead, conductors known as tuned or *lecher lines* are employed. These are short transmission lines, normally a quarter of a wavelength long; in combination with a tuning capacitor, they resonate at the required frequency. Coupling between stages is achieved by simple loops of wire or by tapping the tuned line.

## Aerial isolation

The aerial is usually connected to the tuner via an isolating circuit. This is essential in receivers operating from the mains without the use of an isolating transformer. If the receiver is fed directly from the mains, the chassis can become live; and if a live chassis were connected directly to the aerial, it would make the aerial live as well.

A commonly used aerial isolation circuit is shown in Fig. 3.2. Capacitors C1 and C2 are large enough to give adequate coupling for radio frequencies, but of high enough impedance at the mains frequency of 50 Hz to effectively isolate the aerial from the mains supply. Resistors R1 and R2 prevent static charge from building up on the aerial.

**Fig. 3.2** *Aerial isolation circuit*

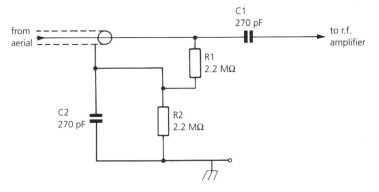

## R.f. oscillator

At the frequencies of operation within the UHF range, feedback in the oscillator is obtained by mutual inductance, employing tuned lines to form a Hartley or a Colpitts oscillator. Use may also be made of the interelectrode capacitors of the transistor (Fig. 3.3). The circuit shows a common base Colpitts oscillator in which interelectrode capacitors $C_{ce}$ (between collector and emitter) and $C_{be}$ (between emitter and base) provide the necessary feedback for sustained oscillation. $C_{ce}$ in series with $C_{be}$ effectively fall across the output developed between the collector and base. Part of this output, that across $C_{be}$, is fed back into the input between the emitter and base. R1 is the emitter resistor, R2/R3 is the base bias chain with C2 as the base or bias decoupling capacitor, TL1 is the output tuned line resonating with variable capacitor C3.

**Fig. 3.3** *R.f. oscillator*

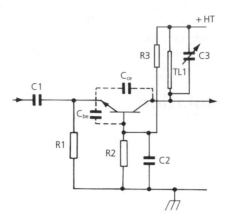

# R.f. amplifier

A typical r.f. amplifier using npn transistor TR1 in the common base configuration is shown in Fig. 3.4, in which TL1, TL2 and TL3 are tuned lines. The base of TR1 is decoupled to the chassis by capacitor C2 with potential divider R2/R3 providing the base bias. R1 is the emitter resistor for d.c. stability and L3 is an r.f. choke. Loops L1/L2 provide the input coupling with the output developing across resonant circuit TL1/C4 and trimmer C5. The signal is then coupled to the second tuned line TL2 through the small gaps in the screening between the two stages. TL2, which is tuned by C6 and trimmed by C7, acts as a bandpass coupling element between TL1 and TL3, which feeds the mixer–oscillator stage. The high-pass filter at the input together with bandpass element TL2/C6/C7 ensures the correct shape for the frequency response of the tuner.

**Fig. 3.4** *Tuner r.f. amplifier*

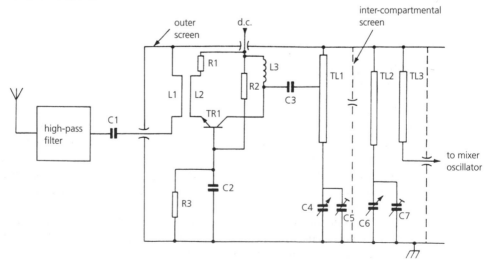

## Mixer–oscillator

Frequency changing may be achieved either through multiplication or addition. Addition is the preferred method for TV tuners. A single transistor is made to oscillate at frequency $f_0$, which is 39.5 MHz above the selected channel frequency $f_c$. The non-linear part of the transistor characteristic is then used to produce the sum $(f_0 + f_c)$ and the difference $(f_0 - f_c)$ of the two frequencies together with the two original frequencies, $f_0$ and $f_c$. A tuned circuit at the output is then made to select the frequency difference, $f_0 - f_c = 39.5$ MHz.

## Complete UHF tuner

A typical tuner used for a monochrome TV receiver employing mechanical push-button tuning is shown in Fig. 3.5. VT351 and VT352 are the r.f. amplifier and the mixer–oscillator respectively. Both are connected in the common-base configuration. The signal from the aerial is fed through an isolating circuit via coupling capacitor C351 into the emitter of VT351 with R351 as the emitter resistor. The output of the r.f. amplifier across tuned circuit L352/C354/C355 is transformer-coupled (via bandpass coupling circuit L353/C356/C357) to coupling loop L355 at the input of the mixer oscillator transistor VT352. The output is taken at the collector of VT352 via filter components L357/C366. The i.f. is produced across L358, which is tuned to 39.5 MHz. L359/L360 together with the input circuit of the first i.f. amplifier form a bandpass filter allowing the video bandwidth to go through.

The gain of the r.f. amplifier may be varied by local distance gain control R74, which controls the base voltage of VT351 and hence its gain. Resistor R74 is normally set for maximum gain. However, the r.f. amplifier may saturate in strong reception areas, causing clipping and hence distortion of the signal. In such cases the gain of the amplifier is reduced.

## Automatic frequency control

The purpose of automatic frequency control (a.f.c.) is to ensure the stability of the intermediate frequency produced by the mixer–oscillator. A drift in the i.f. results in a slight deterioration in the monochrome content of the picture as a consequence of the loss of a part of the bandwidth. For colour reception such a loss in bandwidth attenuates the chrominance information positioned at the upper end of the frequency spectrum, which results in colour desaturation and ultimately in loss of colour altogether if the drift is large enough. For this reason, a.f.c. is an essential feature for colour receivers.

Deviation of the i.f. from its correct value of 39.5 MHz is normally caused by a drift in the frequency of the oscillator at the tuner. To overcome this, an a.f.c. loop is used (Fig. 3.6). The frequency of the i.f. oscillator is monitored at the i.f. amplification stage. A change in the i.f. is amplified and fed into a frequency discriminator, which produces a d.c. correcting voltage of a level proportional to the change in frequency. Following amplification, the correcting voltage is fed into the tuner oscillator via a varicap diode (varactor) stage. The varicap diode then changes the oscillator frequency by an amount determined by the d.c. correcting voltage to keep the frequency constant.

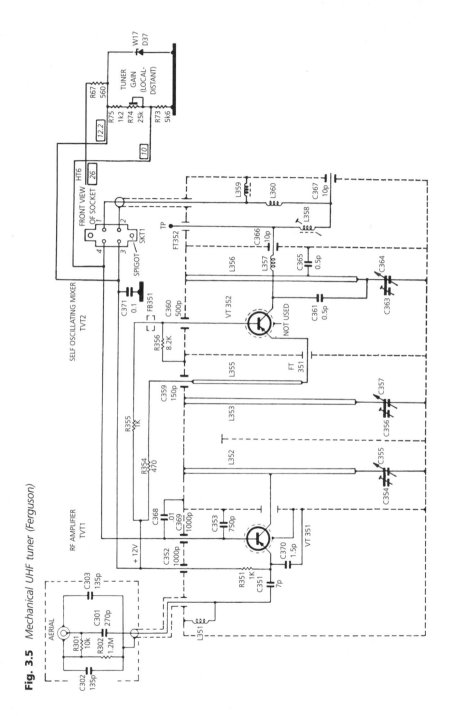

**Fig. 3.5** *Mechanical UHF tuner (Ferguson)*

**Fig. 3.6** *Automatic frequency control (a.f.c.) loop*

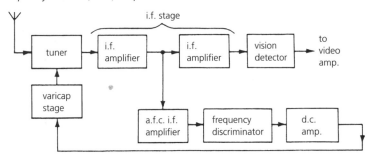

An a.f.c. switch is normally incorporated in order to disable (i.e. disconnect) the a.f.c. loop when the receiver is being tuned. The disabling switch, sometimes known as an *a.f.c. mute*, allows tuning to take place independently of the loop, thus ensuring that the frequency discriminator is operating at the centre point of its characteristics in order that it may be able to correct frequency deviations on either side of the i.f. An a.f.c. mute is sometimes used for changing channels to prevent what is known as *lock-out* of the signal. This is caused by the inability of the a.f.c. to change as quickly as the channel selection system. A.f.c. disabling switches are either operated manually or automatically when a channel is selected.

## Varicap tuners

Varicap tuners are widely used in modern TV receivers. Compared with the ganged-capacitor type, varicap tuners are more reliable; they have no moving parts hence they are less prone to frequency drift; they are more suitable for touch-sensitive tuning and remote control; and they may be located in the most suitable place on the chassis with only the control panel having to be on the front cabinet of the receiver.

A typical switching arrangement for a varicap tuner is shown in Fig. 3.7. Varicap diode W1 acts as the tuning capacitor for lecher line L1. The reverse voltage for W1 is

**Fig. 3.7** *Varicap switching arrangement: channel 1 selected*

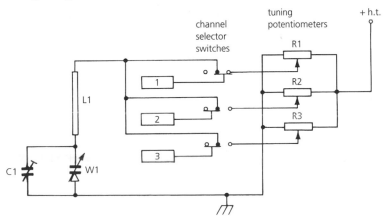

**Fig. 3.8** *Varicap UHF tuner*

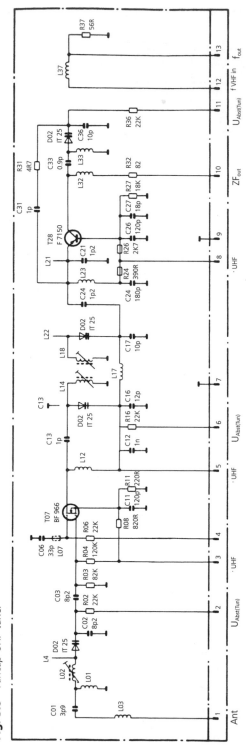

provided by preset tuning control resistors R1, R2 or R3 depending on the channel selected. A variation of approximately 30 V would tune the receiver through all channels in bands IV and V. To ensure frequency stability, a stabilised d.c. supply is used as well as thermistors in some cases to compensate for changes in d.c. voltages due to temperature variation. Where more than one varicap diode is used, they must be matched, i.e. they must have identical characteristics.

An example of a varicap tuner used in a colour TV receiver is shown in Fig. 3.8. The tuning voltage from the channel selector is fed to the cathode of four matched tuning varactors at pins 2, 6 and 11 ($U_{Abst\ Tun}$) via their respective 22k resistors; R02, R16 and R36. The aerial is connected to pin 1 (ANT) with the i.f. output appearing at pin 10 ($ZF_{out}$). The i.f. is also available at pin 13, which is used to feed an a.f.c. system. Field-effect transistor T07 is the r.f. amplifier; common-base T28 is the mixer–oscillator. L14/L18 and associated circuitry provide bandpass coupling.

# The phase-locked loop

Modern TV tuners invariably use a phase-locked loop (PLL) to ensure stability of the i.f. output, avoiding the use of an a.f.c. loop. Available in integrated circuit packages, the phase-locked loop is today widely used in a variety of electronic applications, including chrominance decoding. As illustrated in Fig. 3.9, the phase-locked loop consists of a phase discriminator or detector, low-pass loop filter and a voltage-controlled oscillator (VCO). Without an input signal to the phase discriminator, the VCO is free-running at its own natural frequency $f_2$. When a signal arrives, the phase discriminator compares the input frequency $f_1$ with that of the VCO. A difference results in a d.c. output which after filtering is fed back into the VCO to change its frequency. The process continues until the two frequencies are equal and the PLL is said to be locked.

**Fig. 3.9** *Phase-locked loop (PLL)*

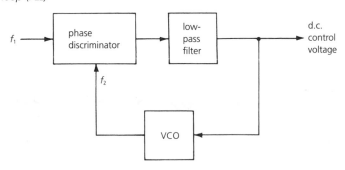

**Fig. Q3.1**

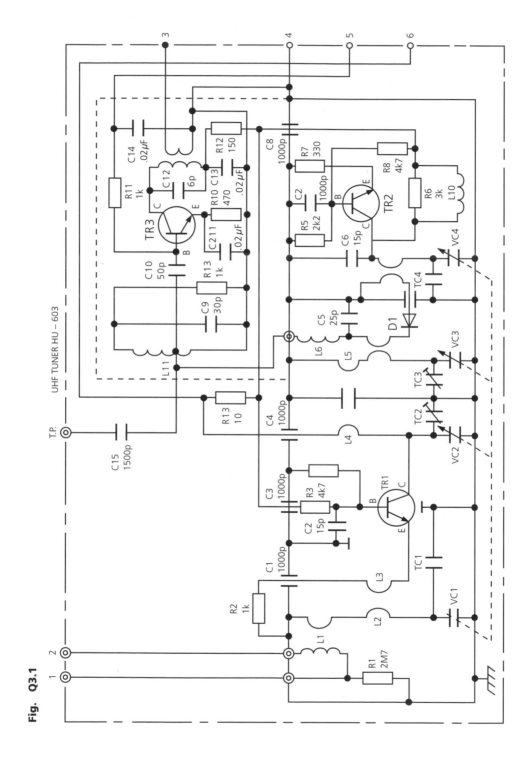

UHF TUNER HU – 603

1.  Explain the reason for using common-base amplifiers in the tuner section of a television receiver.

2.  State the purpose of **each** of the following connections in Fig. Q3.1:
    (a)  1 and 2
    (b)  5
    (c)  6

3.  State the circuit function of **each** of the following components in Fig. Q3.1:
    (a)  TR2
    (b)  D1
    (c)  TR3
    (d)  R2

# 4 The i.f. stage

Recall that the intermediate frequency is derived at the mixer–oscillator stage of the tuner. The local oscillator is made to oscillate at a frequency which is 39.5 MHz greater than the selected carrier frequency. The intermediate frequency is then obtained by selecting the difference between the carrier and the oscillator frequencies. For example, if the TV receiver is tuned to a channel frequency of 511.25 MHz (BBC1 transmission from Crystal Palace) then the oscillator must be tuned to frequency $f_0 = 511.25 + 39.5 = 550.75$ MHz. Similarly for other carrier frequencies. The relative position of the various frequencies of a modulated u.h.f. carrier is displayed in a frequency spectrum.

The frequency spectrum for a modulated 511.25 MHz carrier is shown in Fig. 4.1(a). The 8 MHz bandwidth extends from $f_{min}$ to $f_{max}$ where

**Fig. 4.1** *(a) Frequency spectrum for a modulated 511.25 MHz carrier; (b) intermediate frequency spectrum*

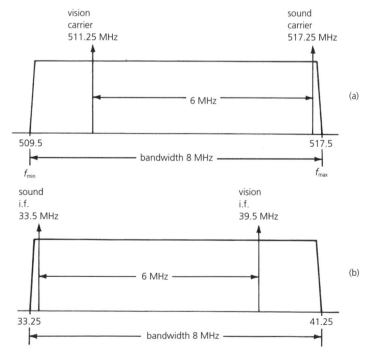

$$f_{min} = 511.25 - 1.75 = 509.5 \text{ MHz}$$
$$f_{max} = 511.25 + 6.25 = 517.5 \text{ MHz}$$

The sound carrier is 6 MHz above the vision carrier, so it has a frequency of 511.25 + 6.00 = 517.25 MHz.

## The i.f. spectrum

After the mixer–oscillator stage, the vision carrier is replaced by an i.f. of 39.5 MHz, giving the intermediate frequency spectrum shown in Fig. 4.1(b) in which every frequency is the difference between $f_0$, the local oscillator frequency and the original frequency in Fig. 4.1(a). Thus the sound carrier is translated into a sound i.f. of

$$f_0 - \text{sound carrier} = 550.75 - 517.25 = 33.5 \text{ MHz}$$

The sound i.f. is now 6 MHz below the vision i.f. Similarly all other frequencies will reverse their position when they are converted to their equivalent on the i.f. spectrum in Fig. 4.1(b).

## The i.f. response curve

Apart from providing sufficient i.f. amplification to drive the detector, the i.f. stage is required to shape the frequency response of the received signal to that shown in Fig. 4.2. The i.f. response curve has four purposes:

**Fig. 4.2** *Intermediate frequency response curve*

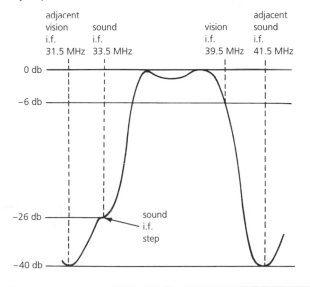

1. To reject the vision i.f. of the adjacent higher channel. The adjacent vision i.f. falls 8 MHz below the vision i.f. at 39.5 − 8 = 31.5 MHz.
2. To reject the sound i.f. of the adjacent lower channel. The adjacent sound i.f. falls 8 MHz above the sound intercarrier at 33.5 + 8 = 41.5 MHz.
3. To provide 26 dB attenuation at 33.5 MHz. This is necessary to prevent any interference caused by a beat between the sound and vision i.f.s. A small step or ledge is provided as shown to accommodate the f.m. deviation of the sound intercarrier. The f.m. step prevents amplitude modulation of the sound carrier; otherwise it would be detected by the vision demodulator, causing a pattern to appear on the screen and a buzz on the sound, a symptom known as *sound on vision*.
4. To provide a steady fall in amplitude from 38 MHz to 41 MHz at the vision i.f. end. This is necessary because the vestigial sideband transmission gives increased emphasis to these frequencies.

These four purposes become very critical in colour TV reception. The 4.43 MHz chrominance subcarrier falls at the higher end of the video spectrum, and when this is converted to an intermediate frequency it becomes 39.5 − 4.43 = 35.07 MHz, only 1.57 MHz away from the 33.5 MHz sound i.f. It follows that, in order to retain the full chrominance information and its correct relationship to the luminance information, the response curve must not be allowed to fall too early at this end, thus restricting the chrominance information, but it must provide sufficient rejection of the sound i.f. Failure to do this produces cross-modulation between the 4.43 MHz chrominance subcarrier and the 6 MHz sound intercarrier. This cross-modulation appears as a 1.57 MHz (6.00 − 4.43) pattern on the screen, known as *herring-bone pattern*.

## The i.f. amplifier

I.f. amplifiers normally employ high-frequency transistors connected in the common-emitter configuration (Fig. 4.3), in which inductor L3 is tuned by its own self-capacitance. A common-emitter amplifier has a low input impedance which shunts the input signal

**Fig. 4.3**  *Intermediate frequency amplifier*

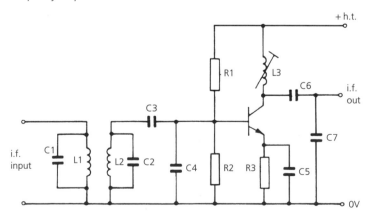

from the preceding stage, an effect known as the *damping* effect. To minimise this, tapped-capacitor or tapped-inductor coupling is used. In Fig. 4.3 the capacitor chain C3/C4 is a *tapped-capacitor* coupling network; it reduces the damping effect on tuned circuit L2/C2. If Rp is the input resistance of the transistor, then the effective shunting or damping resistance Rs appearing across the tuned circuit is

$$Rs = \left( \frac{C3 + C4}{C3} \right)^2 Rp$$

Chain C6/C7 is another tapped-capacitor coupling at the output.

For a tapped-inductor coupling (Fig. 4.4), the relationship between Rs and Rp is

$$Rs = \left( \frac{L1 + L2}{L2} \right)^2 Rp$$

**Fig. 4.4** *Tapped inductor coupling*

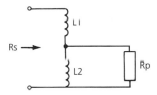

## Staggered tuning

Vision i.f. amplifiers are required to have high gain over a wide bandwidth. This cannot be satisfied by the use of simple tuned circuits which have the response curve shown in Fig. 4.5. Since the product of gain and bandwidth is constant, it follows that any attempt to increase the bandwidth will reduce the gain and vice versa. A high-Q tuned circuit

**Fig. 4.5** *Frequency response of a tuned circuit*

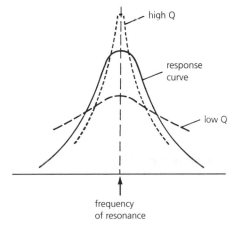

**Fig. 4.6** *Staggered tuning*

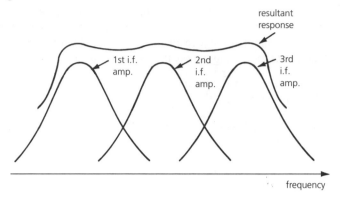

provides higher gain coupled with narrower bandwidth, whereas a low-Q circuit gives increased bandwidth with lower gain. To obtain adequate bandwidth with sufficient amplification, *staggered tuning* may be used.

Instead of using identical tuned circuits for each i.f. amplifier, the tuning frequencies are staggered by tuning each stage to a different frequency. The effect of this on a three-stage i.f. strip is shown in Fig. 4.6. The response curves overlap and the resultant curve combines high gain and wide bandwidth. By spacing the tuning frequencies and using different Q-factors, various shapes may be produced.

## Overcoupling

An alternative to staggered tuning is the use of inductance or capacitive overcoupling. Figure 4.7 shows two types of capacitor overcoupling. Top capacitor coupling is shown in Fig. 4.7(a); capacitor C3 is used to increase the coupling between the primary and secondary of the transformer. This method is also known as series capacitor coupling. In Fig. 4.7(b) capacitor C2 provides the increased coupling between the two windings of the transformer. This method is known as shunt capacitor coupling. As the coupling is increased, the response curve 'opens up' (Fig. 4.8). Undercoupling produces a broad response with a single peak whereas overcoupling produces a wider response with two peaks. Overcoupled tuned circuits are also used as bandpass filters or transformers.

**Fig. 4.7** *Overcoupling: (a) top capacitor coupling; (b) shunt capacitor coupling*

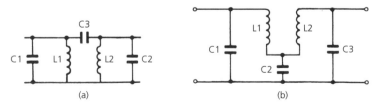

**Fig. 4.8** *Response curve produced by overcoupling*

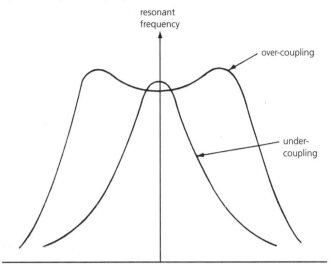

## Rejectors and acceptors

Modern TV receivers use fixed tuned wideband i.f. amplifiers in the form of discrete components or incorporated within individual integrated circuits. The shaping of the response curve is then carried out by a selectivity network consisting of rejector/acceptor circuits that form a complicated filter unit.

A *rejector* is a parallel-tuned circuit whose impedance is a maximum at the frequency of resonance. An *acceptor* is a series-tuned circuit with a minimum impedance at the resonant frequency.

In Fig. 4.9(a) the parallel-tuned circuit L1/C1 inserted between two amplifier stages presents a high impedance at its resonant frequency, thus 'rejecting' that frequency. For frequencies other than the resonant frequency, the tuned circuit represents a low impedance, allowing these signals to flow unaffected. The same effect may be produced by the series-tuned circuit in Fig. 4.9(b). At resonance the tuned circuit presents a very

**Fig. 4.9** *Rejector circuit*

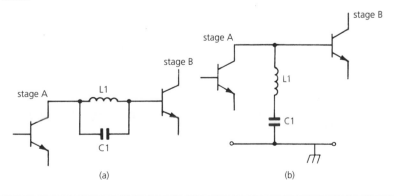

**Fig. 4.10** *Series/shunt frequency trap circuits*

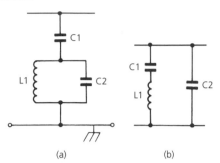

(a)                    (b)

**Fig. 4.11** *Bridge-T rejector circuits*

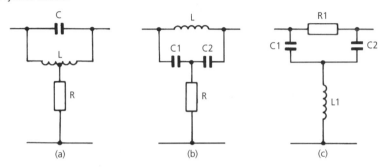

(a)                    (b)                    (c)

low impedance, shorting out these frequencies to the chassis. Other frequencies remain unaffected. These circuits are generally known as *traps*.

To improve the sharpness of a trap circuit, series shunt-tuned or shunt series-tuned circuits are used (Fig. 4.10). These circuits provide frequency emphasis as well as rejection to accommodate the necessary sharpness of the response curve. In Fig. 4.10(a) L1/C2 are chosen to have an inductive reactance at the rejected frequency. This inductive reactance resonates with C1 to form an acceptor, trapping signals at that frequency. The resonant frequency of parallel-tuned circuit L1C2 itself is emphasised since it offers very high impedance to signals at that frequency. A similar effect is obtained from the circuit in Fig. 4.10(b) with L1C1 forming an acceptor and L1C1C2 forming a rejector.

Other rejector circuits using Bridge-T combinations are shown in Fig. 4.11. In each case the circuit appears as a short circuit for all frequencies except the resonant frequency; at resonance it appears as an open circuit. A high degree of attenuation may thus be obtained, coupled with a very sharp response.

Negative feedback may be employed in rejector circuits (Fig. 4.12). At the resonant frequency, L1C1 presents a very high impedance; this introduces a large amount of negative feedback and results in a very low gain. At all other frequencies the gain is normal. In the circuit shown, R1 is the normal emitter resistor and C2 is its decoupling capacitor. L1 is tapped to prevent damping of the tuned circuit by the low impedance of the transistor.

**Fig. 4.12** *Negative feedback rejection*

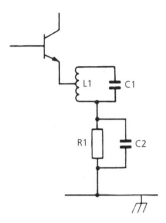

**Fig. 4.13** *Mutual inductance rejector circuit*

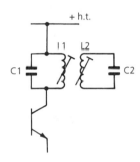

**Fig. 4.14** *Rejector circuit: L1C1 is tuned to the rejection frequency*

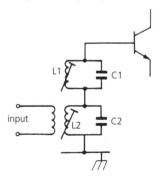

Rejection by mutual inductance is shown in Fig. 4.13. L2C2 offers a very low impedance path to signals at its resonant frequency which, due to the mutual inductance between L1 and L2, absorbs a large amount of energy away from the collector tuned circuit L1C1. A very small signal thus develops across L1C1 at the resonant frequency of L2C2. Normal output is obtained at the resonant frequency of L1C1.

Another rejector circuit is shown in Fig. 4.14, in which L1C1 is tuned to the frequency to be rejected.

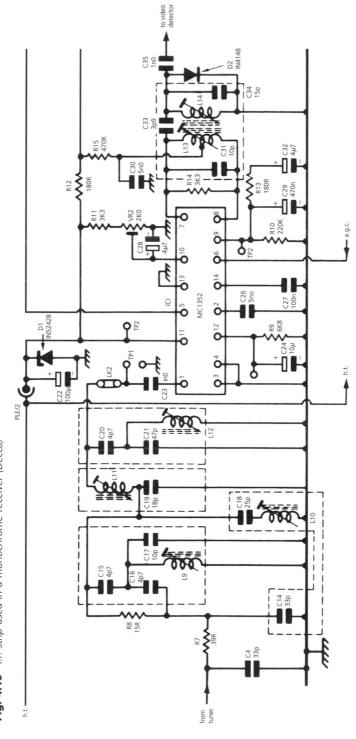

**Fig. 4.15** *I.f. strip used in a monochrome receiver (Decca)*

A complete i.f. strip used in a monochrome receiver is shown in Fig. 4.15. The amplitude modulated signal is fed into a selective or shaping network before going into a wideband amplifier stage employing integrated circuit MC1352. C4/R7 provides matching between the tuner and the i.f. strip. Network R8/C15/C16/C17/L9 forms a Bridge-T rejector network tuned to the adjacent sound i.f. 41.5 MHz. Capacitor C17 is included to improve the sharpness of the trap, C18/C19/L10 is the adjacent vision trap with L10 tuned to 31.5 MHz and L11 is an interstage coupling coil. Bridge-T rejector C20/C21/L12 provides the necessary 26 dB attenuation for the sound i.f. with L12 tuned to 33.5 MHz. Integrated circuit MC1352 is a wideband i.f. amplifier consisting of a number of amplifying stages which require no tuning. The chip also contains a gain-controlled amplifier used for automatic gain control (pin 6). The output of the chip (pins 7 and 8) is fed to the detector via top capacitor coupling unit L13/C33/L14; diode D2 is a clipping diode which removes the positive half of the modulated i.f.

## Surface acoustic wave filters

Increasingly acceptor/rejector networks are being replaced by surface acoustic wave (SAW) filters. SAW filters are very reliable, do not require any tuning, are easily serviceable and comparatively cheap. Figure 4.16 shows a section of an i.f. strip in which CF1 is

**Fig. 4.16**  *Use of a SAW filter in an i.f. strip (Bush)*

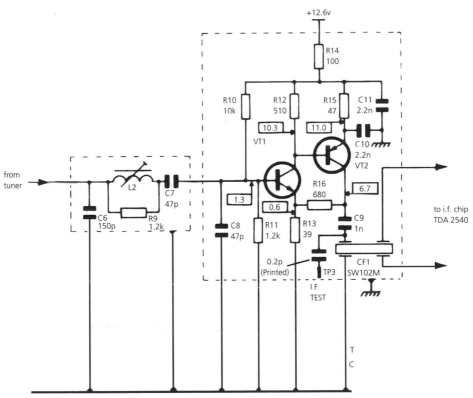

the SAW filter which provides all the necessary i.f. selectivity and rejection. Additional gain to compensate for a loss of approximately 20 dB in the SAW filter is provided by the wideband amplifying stage VT1/VT2. Resistors R13 and R16 provide negative feedback and L2/C7 is a bandpass filter.

## Automatic gain control

The purpose of automatic gain control (a.g.c.) is to vary the gain of the i.f. stage, and in most cases the gain of the r.f. amplifier, to compensate for changes in the strength of the signal received at the aerial. Figure 4.17 shows a block diagram for an a.g.c. system. The strength of the received signal is monitored at the video amplifier stage and fed into an a.g.c. network to produce a d.c. control potential. This voltage is then used to change the gain of the first i.f. stage. The a.g.c. control voltage may also be applied to the tuner, in which case a delay unit is employed to ensure the gain of the r.f. amplifier is reduced only after sufficient reduction in the gain of the i.f. stage. With weak signals the r.f. amplifier will thus function at maximum gain with good signal-to-noise ratio. The precise point at which the tuner a.g.c. begins to operate is determined by a crossover network controlled by a preset potentiometer.

**Fig. 4.17** *Automatic gain control*

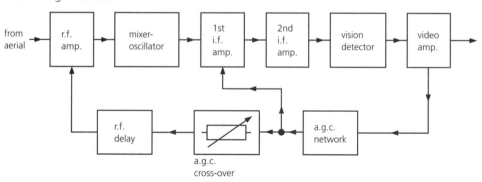

Automatic gain control performs three basic functions in a TV receiver. First, it enables the switching over from a strong channel to a weaker one, or vice versa, without having to adjust the receiver. Second, it avoids overloading the r.f./i.f. amplifying stages; otherwise there would be severe distortion. Third, it attempts to reduce the flutter caused by reflections of transmitted signals from moving objects such as airplanes. The effectiveness of flutter reduction depends on the time constant of the circuit. Ideally, a.g.c. systems should have a short time constant to enable them to follow fast changes in transmitted signal strength. However, this is not always possible since the time constant has to be long enough to decouple, i.e. remove, the video and sync. frequencies from the d.c. control voltage.

There are two types of a.g.c., reverse and forward. *Reverse a.g.c.* uses the fact that the gain of a common-emitter amplifier may be reduced by reducing its current. *Forward a.g.c.* uses the fact that the gain of an amplifier may also be reduced by reducing the voltage between the collector and the emitter which results from an increase in the current through the transistor. Forward a.g.c., in which the gain decreases with increasing transistor current, is universally used in TV receivers since it has a more linear characteristic than reverse a.g.c. Furthermore, since weak signals are amplified at low transistor current, it has a better signal-to-noise ratio.

An example of forward a.g.c. is shown in Fig. 4.18, in which R3 is a d.c. load and C1 is its decoupling capacitor. As the transistor current increases, a d.c. voltage develops across R3; this reduces the collector voltage and with it the gain of the amplifier. The decoupling capacitor ensures that the whole signal develops across the tuned circuit L1/C2, and none across R3. The current through the transistor itself is determined by the a.g.c. control potential. An increase in the control voltage increases the current and therefore reduces the gain; a decrease in the control voltage increases the gain.

**Fig. 4.18** *Principles of forward a.g.c.*

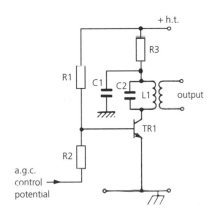

The gain of a tuned amplifier may also be reduced by damping its tuned circuit. In this case a diode is made to conduct when the received signal exceeds a certain level; and when the diode conducts, it places a damping resistor across the tuned circuit.

Both forward-bias and diode damping may be used simultaneously (Fig. 4.19). With a strong signal the a.g.c. potential is high; this causes the current through TR2 to increase, reducing its gain. At the same time, the increase in current causes the TR2 collector voltage to decrease, which forward-biases the damping diode D1. When D1 conducts it places resistor R2 across tuned load L2/C2 to further reduce the gain of the i.f. amplifier. Resistor R5 is the d.c. load for TR2; C5 is its decoupling capacitor. Although R5 is necessary for forward a.g.c., decoupled d.c. load resistors are often incorporated in i.f. amplifiers in order to reduce the d.c. power dissipation of the transistor, e.g. resistor R1 for TR1. D.c. loads such as R1 do not take part in changing the gain of the transistor; for that to happen the d.c. conditions, i.e. the collector current and voltage, have to change first.

**Fig. 4.19** *Forward a.g.c. circuit*

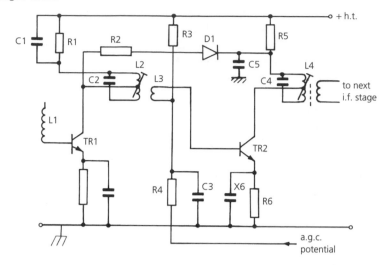

**Fig. 4.20**

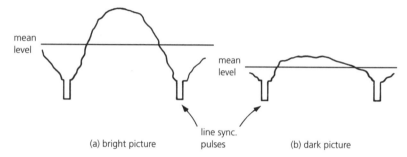

(a) bright picture   line sync. pulses   (b) dark picture

## Peak level a.g.c.

Radio receivers employ what is known as mean level a.g.c., in which the mean level of the received signal is used as a measure of the signal strength. This is unsatisfactory for video reception since the mean level of a video signal does not provide a precise measure of the strength of the signal. As can be seen from Fig. 4.20, the mean value of a video signal reflects the brightness, i.e. luminance content, of the signal. For this reason, peak level (or sync. pulse tip) a.g.c. is employed in which only the peak of the signal, i.e. the tip of the sync. pulses, is monitored. Since the tip of the sync. pulses is always taken to the same level, irrespective of the picture brightness, any variation in amplitude reflects the signal strength and nothing else.

Peak level a.g.c. may be obtained by the use of a simple diode (or transistor) clipping circuit, to allow through only the tips of the sync. pulses, followed by a low-pass filter (Fig. 4.21). TR1 is the luminance amplifier acting as an emitter follower feeding the a.g.c. amplifier TR2. Bias chain R1/R2/R3 is arranged to ensure that TR2 is at cut-off with its base potential higher than that of the emitter. Since TR2 conducts when its base voltage drops below its emitter potential, only the most negative parts of the input

**Fig. 4.21** *Peak level a.g.c.*

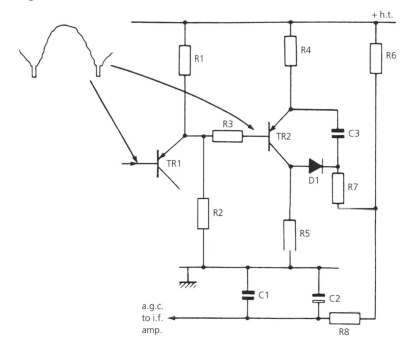

signal, namely the tips of the line and field sync. pulses, will forward-bias TR2 and make it conduct. The pulses appearing at the output of TR2 are smoothed by low-pass filter R7/R8/C2/C1. The charge across C2 is proportional to the magnitude of the sync. pulses, so it provides the a.g.c. control voltage. Diode D1 ensures that C2 does not discharge through R5 during the picture period between one sync. pulse and another. Peak level a.g.c. suffers from the fact that a long time constant is necessary to smooth out the low-frequency 50 Hz field pulses.

## Gated a.g.c.

Peak level a.g.c. suffers from three disadvantages. First, the long time constant of the filter renders the system less sensitive to flutter in signal strength. Second, it may be rendered ineffective if overloading takes place in any stage before the video amplifier. This is because overloading clips the amplitude of the sync. pulses, producing incorrect a.g.c. potential. Third, peak level a.g.c. is susceptible to random noise. The effect of random noise may be removed by using a gate which opens only for the duration of the sync. pulses, hence the name *gated a.g.c.* Further improvements may be introduced by gating line sync. pulses only; this removes the low-frequency field pulses, making it possible to use a shorter time constant to improve the sensitivity to flutter and other fast changes in signal strength.

**Fig. 4.22** *Gated a.g.c. circuit in a monochrome receiver (Bush)*

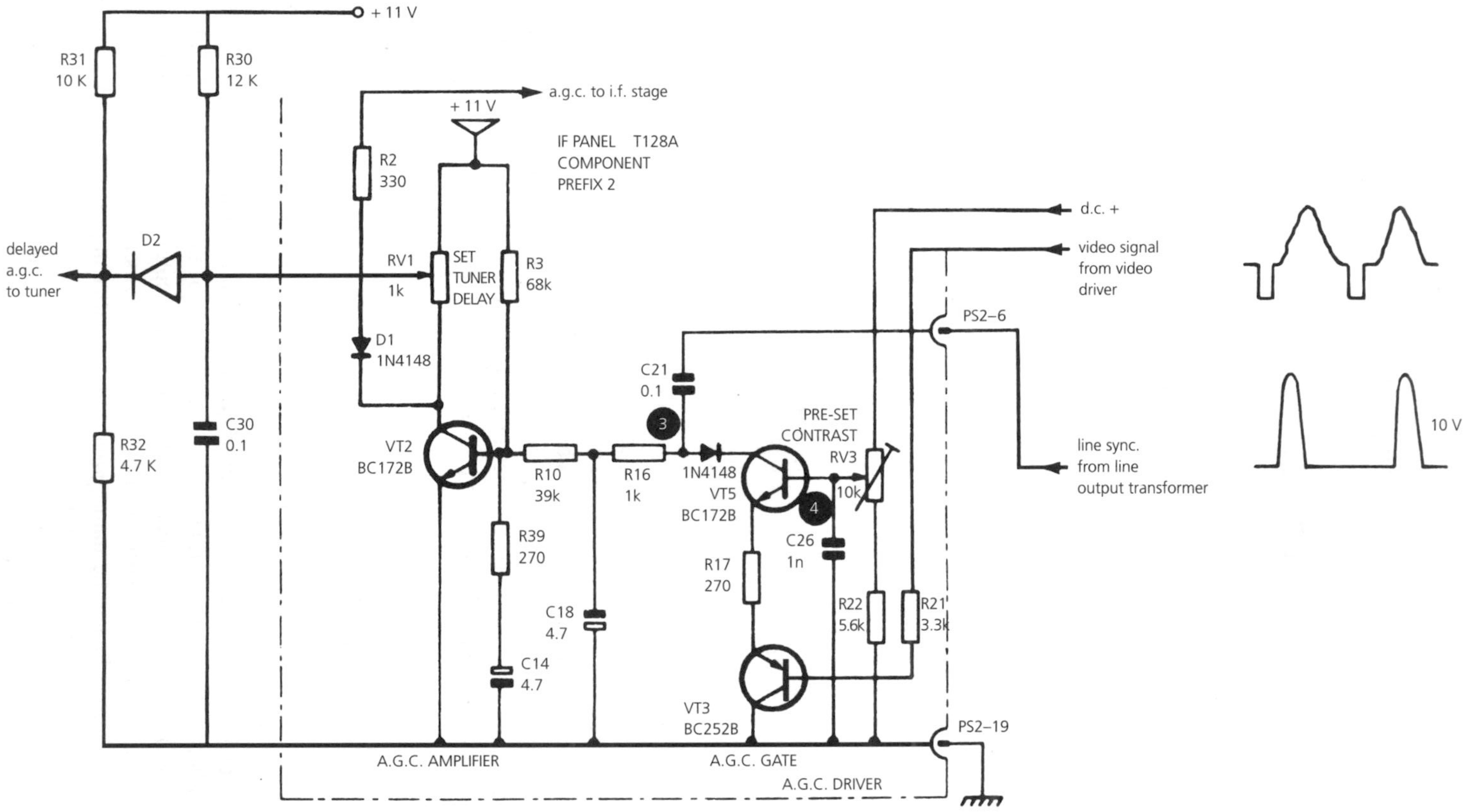

A gated a.g.c. circuit used in a monochrome receiver is shown in Fig. 4.22. Transistor VT3 is the a.g.c. gate and VT5 is the a.g.c. driver. VT5 conducts only when a positive-going line pulse coming from the line output transformer is present at its collector. At the same time, the negative-going line pulse from the video driver forward-biases VT3; this makes it conduct and enables an output to appear at the collector of VT5. In this way only the level of the line sync. pulse is monitored. Network R16/C18/R10 is the smoothing filter. The charge on C14 provided by series chain R3/R39 is varied by the level of the incoming sync. pulses. This charge determines the base bias of a.g.c. amplifier VT2, which provides the a.g.c. control voltage for the i.f. stage. Delayed a.g.c. is provided via diode D2, which is turned on when the i.f. stage has reached maximum gain. The crossover point is set by RV1.

## QUESTIONS

1. For a British UHF TV receiver, state
   (a) the vision i.f.
   (b) the sound i.f.
   (c) the adjacent vision i.f.
   (d) the adjacent sound i.f.

2. In a television receiver, state the purpose of
   (a) a.g.c.
   (b) a.f.c.
   (c) the mixer–oscillator

3. Explain the principles behind the following techniques. For each case, state one advantage and one disadvantage.
   (a) peak level a.g.c.
   (b) gated a.g.c.

4. (a) What is the purpose of a sound trap?
   (b) What is its frequency?

# 5 The video demodulator

## The diode detector

The basic circuit for a conventional diode detector is shown in Fig. 5.1. It consists of a single rectifier diode D1 shunted by a reservoir capacitor C1 and load resistor R1. The diode conducts for the positive half-cycles of the input only. A large input signal is necessary in order to avoid the non-linear part of the diode's transfer characteristics (Fig. 5.2). If the input signal is not large enough, distortion is introduced as well as a number of beat frequencies. The values of C1 and R1 are determined by several considerations, including the required time constant, the forward resistance and the reverse capacitance of the diode.

**Fig. 5.1** *Simple diode detector*

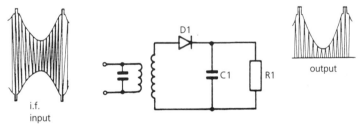

## Time constant

For adequate reproduction of the video information, the time constant of the detector C1R1 must be shorter than the period of one cycle of the highest video frequency, 5.5 MHz, and longer than the duration of one cycle of the 39.5 MHz i.f.

Given that the periodic time of the highest video frequency = 1/5.5 = 0.18 $\mu$s and that of the i.f. = 1/39.5 = 0.025 $\mu$s, the time constant has to be somewhere in between. For this reason 0.1 $\mu$s is normally aimed for. A time constant longer than 0.1 $\mu$s would reduce the high-frequency response of the detector. The output which is the charge across C1 is then no longer able to follow the fast changes in brightness which are represented

**Fig. 5.2** *Detector diode transfer characteristics*

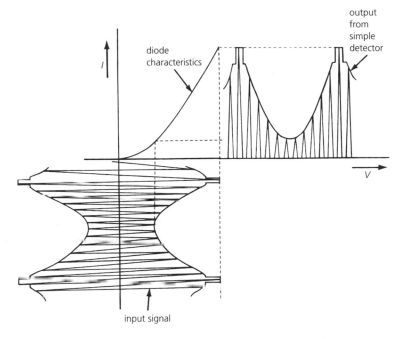

**Fig. 5.3** *Long time constant*

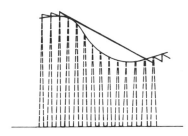

**Fig. 5.4** *Short time constant*

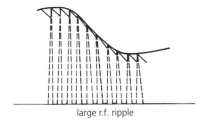

by high video frequencies (Fig. 5.3). On the screen this appears as a smear after a sharp black-to-white edge caused by the inability of the capacitor to discharge quickly. A very short time constant, on the other hand, would retain a high proportion of the i.f. in the form of a ripple (Fig. 5.4).

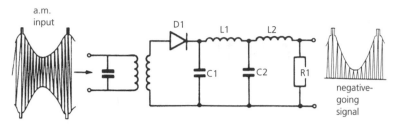

**Fig. 5.5** *Diode detector circuit*

## The load resistance

The load resistor R1 is required to be as large as possible in order to reduce the relative voltage drop across diode D1 when it is forward-biased. Given a predetermined time constant of 0.1 $\mu$s, the precise value of the resistor is determined by the minimum value of the capacitance that can be used. The minimum value of the capacitance itself is limited by the reverse capacitance of the diode, which allows the i.f. to leak through the output when the diode is not conducting. Capacitor C1 must be large enough to minimise this i.f. leak. Taking all these requirements into consideration, a practical value for the detector load resistor is 5k.

Detector diodes are chosen to have a low forward voltage drop, hence a low forward resistance and a low reverse capacitance. Germanium diodes are therefore commonly used. The diode is normally provided with a small forward-bias to reduce the effect of the non-linearity in its transfer characteristics, which is present mainly at low input levels.

As it stands, the output from the basic detector contains a high level of i.f. ripple. This ripple is removed by the introduction of low-pass filter L1/C2 (Fig. 5.5). L1 may be tuned with its own self-capacitance to 39.5 MHz to provide further rejection of the i.f. ripple. A second inductor L2 is normally added for increased ripple filtering and any other undesirable harmonics. The diode may be reversed to produce a positive-going signal. The precise polarity in any one case is determined by the number of phase reversals between the detector and the cathode-ray tube. The purpose is to ensure that the cathode of the c.r.t. is ultimately fed with a negative-going signal so that peak white corresponds to minimum cathode potential, hence maximum electron emission.

A typical video detector circuit used in a monochrome receiver is shown in Fig. 5.6. VT5 is the last i.f. amplifier, L12/L13 is tuned to 33.5 MHz to provide the necessary attenuation to the sound intercarrier before feeding the signal to the detector. Diode D4 is the video diode detector with L5/C31/L14/C33/L15/C34 forming the i.f. and harmonics filter network and R31 is the load resistor. VT7 is an emitter follower employed to prevent the loading of the detector by the subsequent low input impedance video amplifier. The emitter follower also serves as a driver for the video amplifier. It is a common technique to use this stage to feed the sync. separator (and sometimes the sound channel) by including resistor R35 in the collector of VT7. Thus VT7 acts as an emitter follower as far as the video amplifier is concerned but as a common-emitter amplifier to feed the sync. separator. This is also a convenient method for obtaining a positive-going

**Fig. 5.6** *Video detector circuit used in monochrome receiver (Bush)*

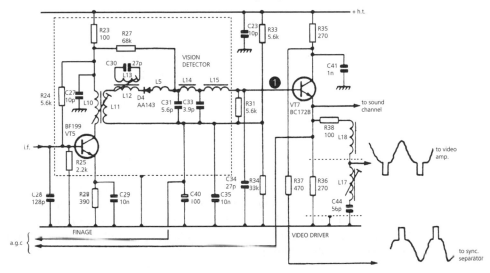

signal (i.e. negative sync. pulses) to feed the video amplifier and a negative-going signal (positive sync. pulses) to feed the sync. separator. The detector diode is given a small forward-bias by resistor R27 going to its anode and potential divider R33/R34 going to the cathode. C35 and C40 are a.g.c. decoupling capacitors.

## Synchronous demodulators

In the amplitude modulated waveform the information is contained in the change of amplitude of the peak of the carrier waveform. By joining the tips of the carrier, an envelope is obtained which represents the original modulating information. The purpose of an a.m. demodulator is therefore simply to retrieve that envelope while removing the carrier wave. This may be carried out by the simple rectifier diode detector. The rectifier is in essence a switch which closes during one half-cycle of the carrier wave and opens during the other half. From this point of view, any switching device may be used, provided it allows the carrier through for one half-cycle only. Since we are only interested in the amplitude of the peak of the carrier, the switch need only be open for the duration immediately before and immediately after the positive (or negative) peak of the carrier. In truth, the modulated carrier is sampled once every cycle of the carrier, a sampling rate equal to the carrier frequency itself.

Figure 5.7 shows a simple block diagram for a synchronous demodulator. The sampling pulses are obtained by the use of a limiter which removes the envelope and leaves a clipped carrier only. The switching or sampling pulses, which have the same frequency and the same phase as the carrier, are used to control a sampling gate that switches on at the peaks of the modulated carrier. The peak levels are then used to charge a capacitor which, given the correct time constant, will reproduce the original modulating signal. An important characteristic of the synchronous demodulator is that it will

**Fig. 5.7** *Synchronous detector*

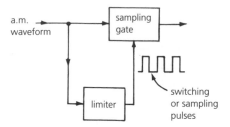

only demodulate those a.m. waveforms which have a carrier that is equal in frequency to and is in phase with the sampling pulses.

The output of the synchronous demodulator may be improved by doubling the sampling rate. Two switching square waves in antiphase to each other are used to operate two separate gates. The two gates are also fed with out-of-phase i.f. signals. The effect is to produce a signal which appears to have passed through a full-wave rectifier. The output contains a carrier component which is twice the frequency of the original carrier, making it easy to filter out. Synchronous demodulators are too complex and expensive for construction from discrete components but lend themselves easily to design on a silicon chip as part of an i.f. or video integrated circuit.

Compared with the conventional diode detector, synchronous demodulators have the following advantages:

1.  More linear characteristics.
2.  Low input levels of the order of 50 mV are adequate.
3.  They only detect information modulated on a carrier which has the same frequency and phase as the sampling pulses.
4.  A simpler RC filter may be used since the ripple has twice the frequency of the ripple produced by the diode detector.

Item 3 means that any stray modulation caused by noise is removed. Also removed are beat frequencies between the sound i.f. on one hand and the adjacent vision carrier and adjacent sound intercarrier on the other. Thus, unlike the diode detector which requires a large attenuation (30–40 dB) of the sound i.f., a lower attenuation (20 dB) is adequate for the synchronous demodulator.

## Practical integrated circuits

Modern i.c.s combine i.f. amplification, vision detection and video driver stages as well as a.g.c. and a.f.c. in a single chip. Figure 5.8 shows a functional diagram of the TBA 440 vision/i.f. chip used in some colour receivers. The i.f. signal is applied to pin 16 via coupling capacitor C314 and fed into a differential amplifier whose gain is controlled by a gated a.g.c. system within the chip. The positive-going video output at pin 11 is taken back to the a.g.c. gate (pin 10) via video level R322, which sets the level of the sync. tips. Gating pulses from the line output stage are applied to pin 7 to

**Fig. 5.8** *TBA 440N vision i.f. chip (ITT)*

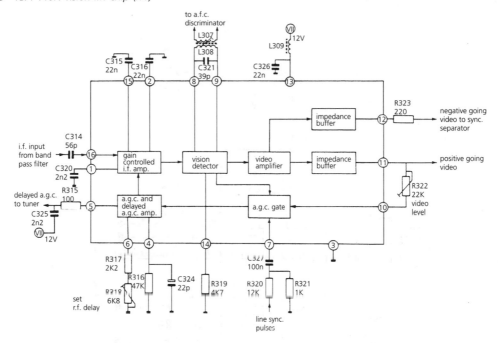

trigger the a.g.c. gate whose output is amplified and used to control the gain of the i.f. amplifier and to provide delayed a.g.c. at pin 5. The i.c. contains a synchronous vision detector and a video amplifier. The reference frequency for the synchronous detector is developed across external tuned circuit L308/C321 at pins 8 and 9. The tuned circuit also provides the reference frequency for the a.f.c. discriminator via transformer coupling L308/L307.

A complete vision i.f. circuit using the TDA 4443 chip is shown in Fig. 5.9. The internal arrangements of the chip are shown in symbolic form, now common to all manufacturers (see the appendix). The i.f. signal from the tuner feeds emitter follower T127 and amplifier T133, which drives SAW filter F129. The SAW filter has two balanced outputs: one for the vision i.f. and the other for the sound i.f. The vision i.f. signal is applied to i.c. I136, which performs the functions of a.g.c. controlled amplifier, synchronous detector and a.g.c. amplifier. The detector is tuned to 39.5 MHz by external coil L141 on pins 8 and 9. The tuner a.g.c. is supplied from pin 5 with the crossover point being adjusted by resistor P136.

## QUESTIONS

1. Explain the difference between a diode modulator and a synchronous modulator.

2. State three advantages of a synchronous demodulator over a diode demodulator.

3. What are the factors that determine the time constant of an a.m. demodulator?

**Fig. 5.9** *Vision i.f. circuit using TDA 4443 chip (Ferguson ICC5 colour chassis)*

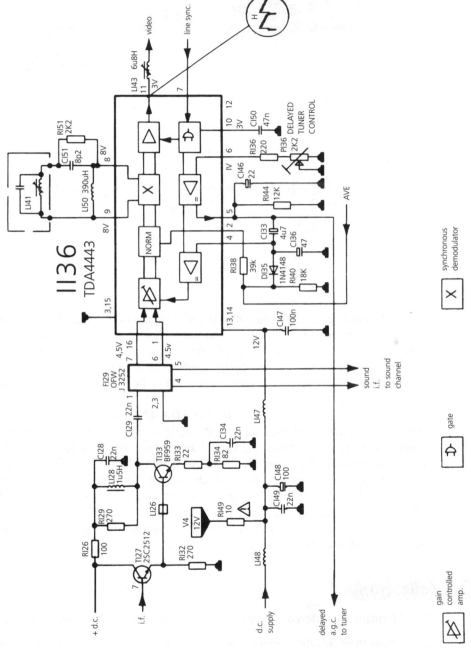

# 6 The monochrome video amplifier

The monochrome video amplifier is a wideband amplifier with a bandwidth extending from d.c. (0 Hz) to 5.5 MHz. It has a moderate gain of between 10 and 25 and an output of 50–100 mV to feed the cathode of a cathode-ray tube. Its frequency response must be such that it preserves the relative phases of all frequencies throughout its bandwidth. In other words, all frequencies must change their phase by the same amount between the input and the output.

## Frequency response

A typical response curve for a common-emitter amplifier is shown in Fig. 6.1. The fall at the low-frequency end is avoided by direct coupling. The fall at the high-frequency end is more difficult to overcome. The drop in the gain at the upper end is caused by stray capacitors $C_{bc}$ and $C_{ce}$ (Fig. 6.2). The input capacitance $C_i$ of the following stage, the tube itself, has a similar effect. These capacitors may be grouped together and represented by a single capacitor $C_t$ falling across the output (Fig. 6.3). The effect of $C_t$ on

**Fig. 6.1** *Frequency response of a common-emitter amplifier*

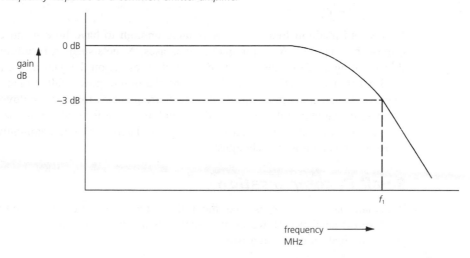

Fig. 6.2

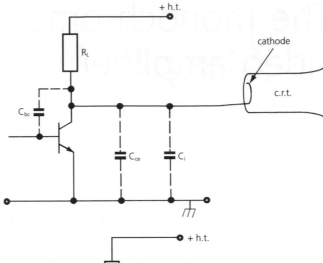

Fig. 6.3

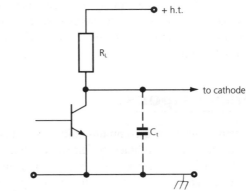

the gain of the amplifier depends on its reactance $X_t$, which varies with the frequency $f$ of the input signal:

$$X_t = \frac{1}{2\pi f C_t}$$

At low and medium frequencies, $X_t$ is large enough to have little or no effect on the gain of the amplifier. As the frequency increases, $X_t$ gets smaller, resulting in a noticeable shunting effect on the output and with it an effect on the gain of the amplifier. At the 3 dB frequency $f_1$ (Fig. 6.1), $X_t = R_L$ and the output power falls to half its midband level. Although transistors with high cut-off frequencies may be employed, the video bandwidth cannot be fully accommodated without some form of compensatory network to overcome the fall at the high-frequency end. There are three commonly used high-frequency compensation techniques.

## Small $R_L$ compensation

If a small load resistor $R_L$ is used, the 3 dB frequency $f_1$ increases, extending the bandwidth of the amplifier. However, this method is of limited use since it also increases the power dissipation of the transistor:

---

power dissipation = d.c. input power − a.c. (or signal) output power

Given that

d.c. input = $I_c$ × h.t.

where $I_c$ is the collector current and h.t. is the d.c. supply, a change in $R_L$ has no effect on $I_c$, keeping the d.c. power unchanged. However, a reduction in the load resistor will reduce the a.c. signal power output with a consequent increase in the power dissipation of the transistor.

## Peaking coil compensation

This technique involves inserting a small coil $L_p$ in series with the load resistor, as shown in Fig. 6.4(a). The coil, known as a *shunt peaking* coil, is effectively connected across the shunting capacitance $C_t$, as shown in Fig. 6.4(b). At low and medium frequencies the coil reactance is too small to have any effect on the output. However, at higher frequencies the coil begins to resonate with $C_t$, lifting the response curve in the manner shown in Fig. 6.5. A very sharp peak is avoided by the suitable choice of the inductor as well as by the damping effect of the load resistor $R_L$ across the tuned circuit.

**Fig. 6.4**  *Peaking coil compensation: (a) circuit; (b) effective tuned circuit*

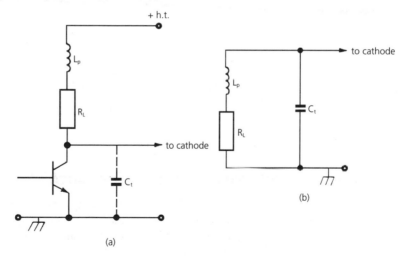

A similar effect may be obtained by the use of a *series peaking* coil $L_s$ (Fig. 6.6). In this case the peaking coil separates the total shunting capacitor roughly into two parts: C1 representing the transistor stray capacitors and C2 representing the c.r.t. input capacitance. C1/$L_s$/C2 functions as a low-pass filter which may be designed to lift the high-frequency end in a similar way to the shunt peaking coil.

In many applications both shunt and series peaking are employed to provide a smooth high-frequency response.

**Fig. 6.5** *Effect of frequency competition on response curve*

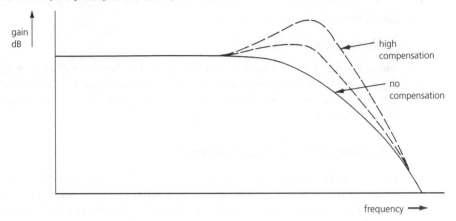

**Fig. 6.6** *Series peaking coil*

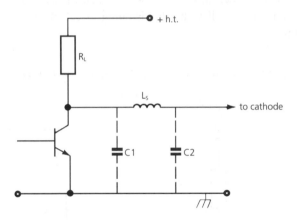

## Frequency-selective negative feedback

The basic principle behind this technique is to arrange for the negative feedback produced by the emitter resistor to be removed gradually as the frequency approaches the high end of the response curve. It involves selecting an emitter decoupling capacitor which does not provide adequate decoupling at low and midband frequencies. Partial decoupling is very simple and extremely popular. However, it does involve a reduction in the overall gain of the amplifier.

In Fig. 6.7(a) the decoupling capacitor C1 is chosen to have a small value (820 pF instead of the normal 100 $\mu$F). At high frequencies the reactance of C1 is very low compared with emitter resistor R1. Negative feedback is therefore removed and the gain is high. At low frequencies, C1's reactance increases, reducing its decoupling effect. Negative feedback is introduced which lowers the gain. The response curve may be further smoothed by the addition of a resistor R2 in series with C1, as shown in Fig. 6.7(b).

**Fig. 6.7** *Frequency-selective decoupling*

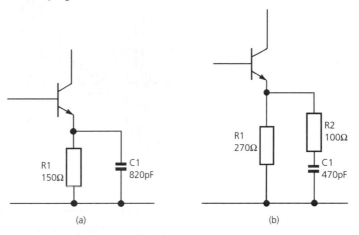

(a)                    (b)

## Tube drive

The cathode-ray tube may be driven by *cathode modulation*, a video signal is sent to the cathode, or by *grid modulation*, a video signal is sent to the grid. Grid modulation is less sensitive and thus requires a larger video signal than cathode modulation. Cathode modulation is universally used in modern TV receivers; it requires a negative-going video signal in which peak white represents a minimum cathode potential.

## Direct video coupling

In order to produce a negative-going signal at the output of the amplifier, it has to be fed with a positive-going signal. The positive-going input is produced by detecting the negative half of the modulated i.f. (Fig. 6.8). As can be seen, the detected signal is wholly negative with peak white at a maximum or zero volts. If this signal is directly coupled to the video amplifier, a large positive bias is necessary to ensure the transistor conducts while the amplitude is above the black level with a maximum current at peak white. This is illustrated in Fig. 6.9(a), in which $I_{b1}$ is the necessary bias current. This arrangement is very wasteful in h.t. supply current and expensive in terms of power dissipation and the working life of the transistor since, even with no signal present, the transistor continues to take maximum current. Direct coupling suffers from another disadvantage: low signal amplitudes caused by reducing the contrast result in the transistor conducting as shown in Fig. 6.9(b). Although peak white is still represented by maximum current, black level current is increased. This second disadvantage may be overcome by using a contrast control which varies the bias of the transistor in such a way as to keep the black level current constant. A typical video amplifier is shown in Fig. 6.10, where RV1 is the contrast control which defines the bias of the transistor. Partial emitter decoupling R4/C1 is employed; together with shunt peaking coil L1, it provides the necessary high-frequency compensation. Video amplifiers have to employ an auxiliary h.t. supply of the order of 60–100 V to provide the necessary high signal amplitude at the output.

**Fig. 6.8**

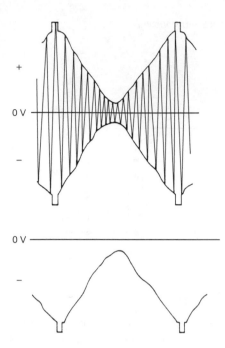

**Fig. 6.9**

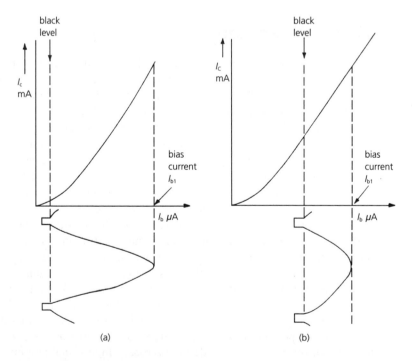

(a)

(b)

**Fig. 6.10** *Directly coupled video amplifier*

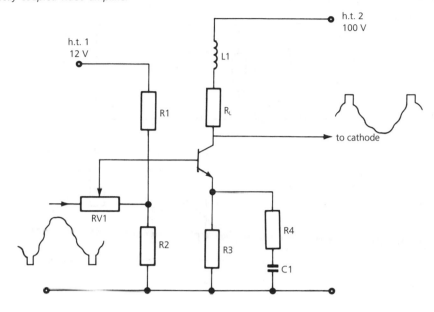

## A.c. coupling

A.c. coupling overcomes the disadvantages associated with direct coupling. It removes the d.c. component of the video signal, allowing the transistor to be biased in class A, i.e. in the middle of its characteristics. The time constant introduced by the coupling capacitor may seem to be a disadvantage in terms of the low-frequency response of the amplifier. However, the time constant may be used to reduce the effect of flutter on the signal. The d.c. component of the video signal represents the average brightness of the picture and must therefore be restored before the signal is fed into the tube. D.c. restoration is carried out by the use of a black level clamping circuit.

A circuit employing a.c. coupling is shown in Fig. 6.11, in which TR1 is the emitter follower stage between the detector and the video amplifier. C1 couples the video signal from the vision detector to the base of the video amplifier TR1. D1 is the clamping diode and the d.c. clamping level is determined by the setting of R2. The operation of the clamper is based on the fact that, during the negative half-cycles of the input signal, D1 conducts and coupling capacitor C1 charges up in the polarity shown. During the positive half-cycles D1 is reverse-biased, preventing C1 from discharging through R2. It can only discharge through R3. Provided time constant C1R3 is large, successive positive half-cycles build up a charge across C1, giving the signal a positive d.c. level. Resistor chain R1/R2 provides a small forward-bias for D1; this lifts the video signal above 0 V by an amount determined by the brightness preset resistor R2.

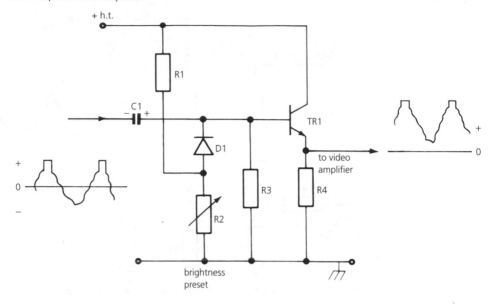

**Fig. 6.11** *A.c. coupled video amplifier*

## Driven black level clamping

The simple clamping circuit described above suffers from the fact that the d.c. level at its output is determined by the mean value of the input, itself dependent on the amplitude of the video signal. Consequently, the black level can be affected by variations in signal strength (changing channels for instance) as well as by the contents of the video signal. To avoid this, black level clamping is carried out during the line blanking period. The clamping circuits are driven by a line pulse derived from the line timebase. In this way the d.c. level is set at the start of each active picture line and, provided the time constant is comparatively long, it remains constant throughout, regardless of the contents of the signal. Driven black level clamping is essential for colour receivers, where three independent signals, red, green and blue, are used to drive the three cathodes of a colour tube.

## Contrast control

The contrast of a picture on the c.r.t. screen may be varied by varying the a.g.c. voltage as explained in Chapter 4. This method involves changing the gain of the i.f./video stages, hence it is only included as a preset contrast control. The variable contrast control available to the user is obtained by varying the amplitude of the signal fed into the tube. This may be carried out by inserting a variable resistor RV1 in series with the input to the video amplifier (Fig. 6.10). A second technique is to vary the amount of emitter decoupling (Fig. 6.12); full decoupling is obtained when the slider is moved to the top end of variable resistor R1 to provide maximum gain and maximum

Fig. 6.12

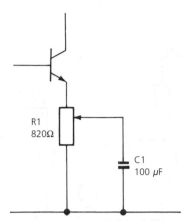

output, hence maximum contrast. As the slider moves towards the chassis, the decoupling effect of C1 diminishes to introduce negative feedback, low gain, low output and low contrast.

## Protection against flashover

The build-up of high voltages inside the tube sometimes results in flashover between the electrodes. This is most likely to happen in colour TV display tubes where voltages of up to 25 kV are used. The large transient current resulting from such a flashover may damage the circuit components outside the c.r.t. To protect the external circuitry, capacitors or sparking gaps are connected between high-voltage electrodes such as the first anode or the focus electrodes and earth, bypassing any transient current.

## Flyback blanking

Ideally the line and field flyback lines should occur below the black level of the video signal, so they are not visible on the screen. However, under certain conditions such as excessive brightness, the flyback trace may become visible and has to be suppressed. Flyback suppression is obtained by turning off the c.r.t. for the duration of the line and field flyback. This is called *flyback blanking*. The simplest blanking technique is to feed a negative-going line and field sync. pulse from the line and field timebases into the first grid of the c.r.t. to turn off the tube. A second method is to feed the sync. pulses into the video amplifier to turn off the transistor and ensure a black level potential at the cathode of the c.r.t. for the duration of the flyback (Fig. 6.13). Negative-going line and field sync. pulses are fed via R1/C1 and R2/C2 respectively to the base of blanking amplifier TR1. Transistor TR1 is biased at saturation with its collector at very low potential (about 1 V), lower than the potential of TR2 emitter. Diode D2 is therefore reverse-biased, acting as an open circuit. When a sync. pulse arrives, TR1 base voltage is brought down, turning the transistor off. The collector of TR1 goes up to HT and

**Fig. 6.13** *Flyback blanking circuit*

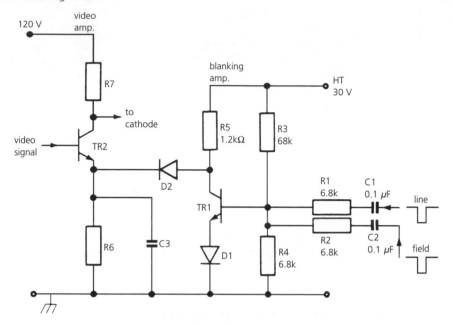

D2 conducts to feed a positive-going pulse into the emitter of video amplifier TR2. For the duration of the pulse, the potential of TR2 emitter remains higher than its base, keeping TR2 off with its collector (and c.r.t. cathode) at 120 V h.t. potential. The tube is therefore blanked.

## Beam current limiting

Changes in the tube beam current between the cathode and the final anode caused by changes in brightness can cause variations in the e.h.t. voltage supply to the final anode. The e.h.t. is regulated up to a maximum beam current; when this is exceeded it causes changes in brightness and variations in picture size. To prevent this, a beam limiter is used. For a monochrome receiver, the limiter is normally a simple diode connected between the video amplifier and the c.r.t. (Fig. 6.14). Under normal conditions D1 conducts, providing direct coupling between video amplifier TR1 and the cathode of the c.r.t. Beam current $I_B$ flows into the cathode of the c.r.t. via the parallel paths TR1/D1 and R2 as shown. The voltage developed across R2 is small enough to keep D1 forward-biased. When the beam current exceeds a certain level, say 100 $\mu$A, the voltage across R2 is high enough to cause D1 to become reverse-biased, thus removing the d.c. coupling. The beam now flows solely through R2, causing its voltage drop to rise. This rise in the potential of the cathode reduces the effective cathode–anode voltage and with it the beam current. The video signal is now a.c. coupled via capacitor C1, which removes the black level set by TR1 collector voltage.

**Fig. 6.14** *Limiting the beam current: D1 is the limiter diode*

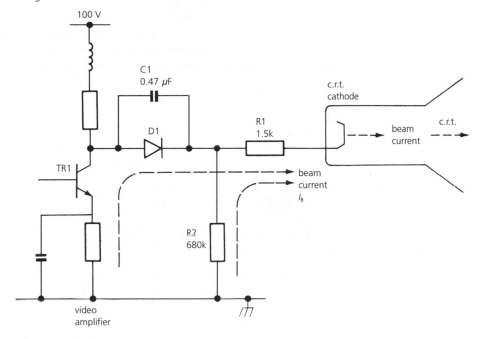

## Typical video stages

A typical video stage for a monochrome receiver is shown in Fig. 6.15. Transistors VT5 and VT7 are the emitter follower video driver and the video output amplifier respectively. The video signal taken from the emitter of VT5 is fed directly to the base of VT7. The same signal is also fed to the sync. separator. The 6 MHz sound intercarrier is taken off at the collector of VT5 using the parallel-tuned transformer L10/C43 with R60 as the damping resistor. Resistor R24 and the d.c. voltage from the diode detector (not shown) provide the bias voltage for VT5. The 6 MHz rejection is provided by L9/C46, and bandwidth extension is provided by selective decoupling capacitors C47 and C48. Contrast control is provided by the partial decoupling of the emitter of VT7. Flyback blanking is obtained by feeding the sync. pulses into the emitter of VT7 to turn off the amplifier during flyback. The line sync. pulses are also fed into the a.g.c. gate transistor VT6 via diode W3. The a.g.c. level is varied by R32 to provide a preset contrast control. Sparking gap SP1 protects the video amplifier against flashover in the tube. The output from VT7 is fed to the cathode of the c.r.t. via time constant R45/C52, which limits the response of the tube to flutter.

Another video output stage is shown in Fig. 6.16, in which TR351 is the video driver, TR353 is the video output, TR352 is the blanking amplifier, C204 is the coupling capacitor, D351 is the video clamp diode and P351 is the black level control. Blanking amplifier TR352 is switched on by positive-going sync. pulses into its base, which then shorts out the base of the video amplifier, turns TR353 off and blanks the tube.

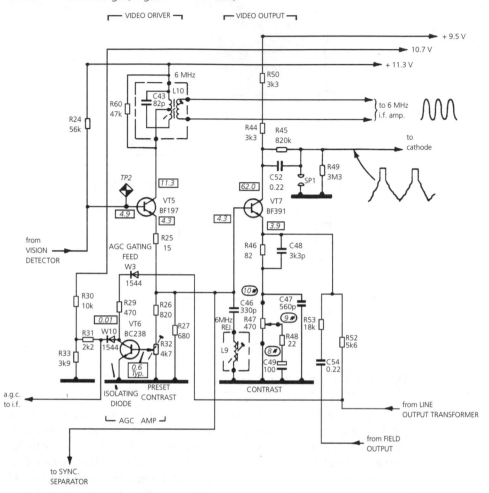

**Fig. 6.15** *Monochrome video stage (Ferguson 1690 chassis)*

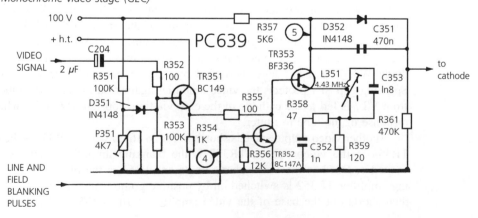

**Fig. 6.16** *Monochrome video stage (GEC)*

Frequency-selective feedback is provided by C352/R358/R359 to improve the frequency response. L351/C353 is a colour subcarrier trap tuned to 4.43 MHz. Its purpose is to remove the subcarrier used in colour transmissions. D352 and its associated C351 and R357 form a beam limiter circuit.

### QUESTIONS

1.  With reference to a video amplifier, explain the purpose of
    (a) frequency compensation
    (b) flyback blanking

2.  State one advantage and one disadvantage of a.c. coupling over d.c. coupling in a video amplifier.

3.  Explain the purpose of beam current limiting and show how it may be achieved.

4.  (a) What is the purpose of black level clamping?
    (b) Draw a simple black level clamping circuit.

# 7 Synchronising separators

The purpose of the synchronising separator is to slice the sync. pulses off the composite video waveform, separate them into line and field, and feed each one to the appropriate timebase. The process must be immune to changes in the amplitude and picture composition of the video signal.

Recall that the sync. pulses are arranged to fall beyond the black level and to occupy 30% of the total amplitude of the composite video. A clipping network is therefore all that is required in order to separate the sync. pulses away from the video information.

A simple sync. separator is shown in Fig. 7.1. Capacitor C1 ensures that TR1 is biased in class C, i.e. beyond cut-off. Without an input, TR1 base is at zero potential. When an input signal is applied to coupling capacitor C1, the positive part of the signal makes the base–emitter (b–e) junction conduct. The resulting b–e current charges C1 in such a way as to make the base go negative. Because of the relatively long time constant C1R1, the capacitor retains the charge, thus providing a reverse-bias (class C) for TR1 that depends on the amplitude of the input signal. With class C biasing, the transistor will only conduct and produce an output at its collector when the signal input is high enough to overcome the reverse-bias. This condition is satisfied by the positive tips of the sync. pulses. The video information falling below this level will therefore be removed.

**Fig. 7.1** *Simple sync. separator*

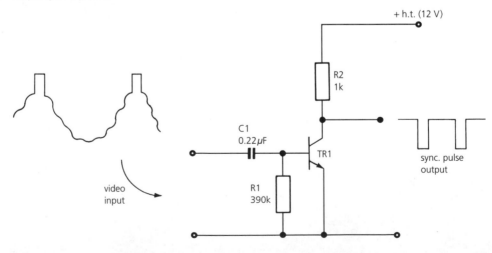

**Fig. 7.2** *Field sync. pulses: (a) odd field; (b) even field*

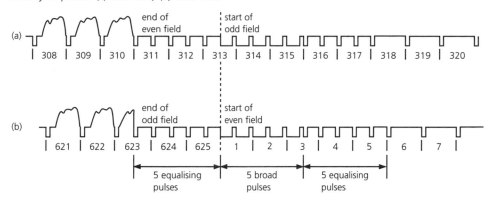

Once the sync. pulses have been separated from the composite video, the receiver must be able to distinguish between the two types of sync. pulses. This is done by send ing the field sync. pulses in the form of a pulse-width modulated waveform consisting of five consecutive broad pulses at twice line frequency (Fig. 7.2). The field flyback lasts for 25 complete lines, giving a total field blanking time of $25 \times 64 = 1600$ $\mu$s. In order to ensure continuous line synchronisation throughout the field blanking period, it is necessary for line triggering edges to occur where a line sync. pulse normally appears. The twice-line frequency broad pulses ensure this takes place. The extra edges during the field flyback are disregarded by the line oscillator. Before and after the five broad field pulses, five equalising pulses are inserted to ensure good interlacing.

## The integrator method

A common method of separating the line sync. pulses from the field sync. pulses is to use an integrator (Fig. 7.3). The time constant CR is chosen to be longer than the duration of a line sync. pulse but slightly shorter than the duration of a complete line, something in the region of 50 $\mu$s. The effect of such an integrator on the sync. pulses is illustrated in Fig. 7.4. The capacitor charges up during the presence of a pulse and discharges when the pulse is absent. The charge developed across C during a narrow

**Fig. 7.3** *Integrator method*

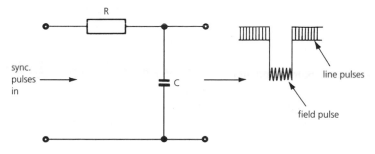

**Fig. 7.4** *Effect of integrator on sync. pulses*

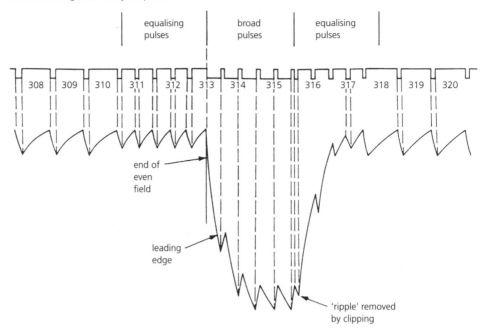

line pulse is small and the capacitor is fully discharged by the time the next line pulse arrives. For the field pulses, the capacitor charging time is long but its discharging time is short. A charge thus accumulates across the capacitor as one broad pulse follows another, lifting the field pulses above the line sync. pulses as shown. At the end of the five broad pulses, the capacitor discharges gradually during the equalising pulses. This produces a rising leading edge, which is subsequently clipped to remove the 'ripple' at its tip, leaving behind the field sync. pulse.

## Equalising pulses

Interlacing involves the scanning of two separate fields: an odd field and an even field. The two fields are not identical. The flyback at the end of an even field starts at the end of a line, whereas the flyback at the end of an odd field starts in the middle of a line. The difference in the starting points of the two flybacks produces different rise times for the odd and even field sync. pulses at the output of the integrator; this leads to bad interlacing. Identical rise times are obtained by the introduction of equalising pulses before and after the broad pulses. The equalising pulses ensure that the pulse waveforms immediately before and immediately after the broad pulses are identical for both the odd and even fields. The effect of the integrator on both odd and even fields is shown in Fig. 7.5. Occurring at twice the line frequency, the equalising pulses coincide with an end of line for the even field and the middle of a line for an add field, producing identical rise times and identical flybacks.

**Fig. 7.5** *Effect of integrator on even and odd fields*

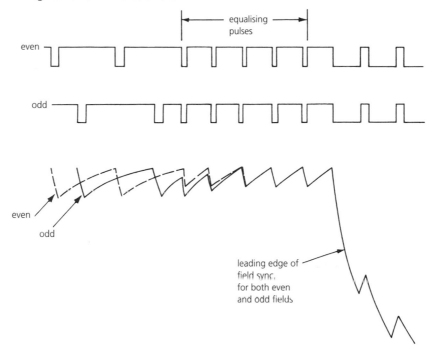

The field sync. pulse produced by the integrator does not have a very sharp edge. The sharpness of the edge may be improved by amplification and clipping. However, the slow response is an advantage in that it reduces the effect of stray or random transient pulses; transients tend to disturb the field synchronisation, producing jitter in the picture on the screen.

## Differentiator method

A very sharp pulse may be produced by using a differentiator. Given a time constant of the same order as the period of the input waveform, the output of a differentiator has the same shape as the input but with a different d.c. level. Figure 7.6(a) shows the output waveform when the input to the differentiator has a high mark-to-space ratio, i.e. a negative-going narrow pulse similar to a line sync. pulse. Figure 7.6(b) shows the effect on the output when the input mark-to-space ratio is reduced to a value much smaller than one, i.e. a positive-going narrow pulse similar to a broad field pulse. In the first case the output pulse is negative, whereas in the second it is positive. In this way the field broad pulses may be lifted above the line sync. pulses (Fig. 7.7). The output is clipped to remove variation in amplitude and to provide a sharp triggering edge. Once again the equalising pulses ensure identical field pulses for both even and odd fields.

**Fig. 7.6** *Differentiator method*

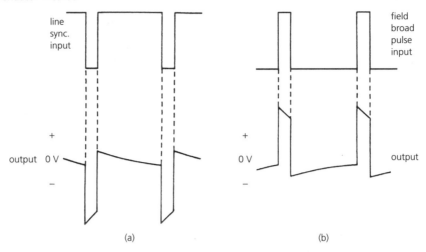

(a)                                                                                                    (b)

**Fig. 7.7** *Effect of differentiator on sync. pulses*

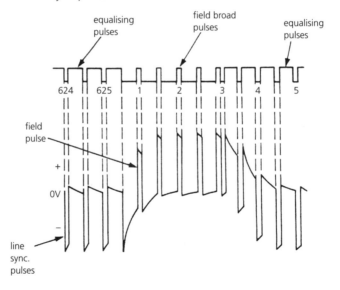

## A typical sync. separator circuit

A typical sync. separator is shown in Fig. 7.8, in which TR5 is the sync. separator transistor biased in class C by coupling capacitor C47. The reverse bias of −0.68 V is determined by resistor R35. Negative-going line and field sync. pulses are produced at the collector of TR5. The line pulses are differentiated by C81 before going to the line timebase. The field pulses are integrated by R39/C49 and clipped by D4 before going to the field timebase. TR5 has a high cut-off frequency in order to maintain the sharp triggering edges of the output.

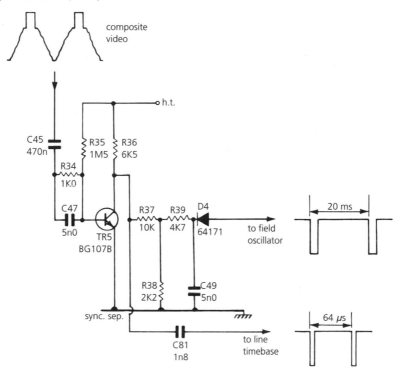

**Fig. 7.8** *Sync. separator circuit (Decca)*

## Noise

Random pulses due to noise and other interference are sometimes present on the composite video signal. These pulses are similar to the sync. pulses themselves and may trigger the timebase at the wrong time. In the case of the field timebase, the integrator which has a slow response, removes most of the noise. However, a random pulse occurring near the end of a field causes what is known as frame or *field slip*. To avoid this, a noise gate (also known as a noise canceller) may be included to obtain a noise-free output from the sync. separator.

In the case of the line timebase, noise causes what is known as *line tearing* as some lines are displaced with respect to the others. On vertical objects, the effect of line tearing is illustrated by ragged edges. Line tearing may be avoided by the use of a flywheel synchronising circuit. The principle of flywheel synchronisation is similar to that of the mechanical flywheel which, due to its large momentum, maintains an average speed unaffected by random changes. The flywheel sync. circuit maintains an average frequency of the sync. pulses by monitoring and taking the average frequency of a number of incoming line pulses so that a random pulse will have very little effect on the frequency.

A block diagram for a flywheel synchroniser is shown in Fig. 7.9. It consists of a flywheel discriminator followed by a reactance stage which controls the frequency of the line oscillator. The flywheel discriminator itself consists of a phase comparator or

**Fig. 7.9** *Flywheel synchroniser*

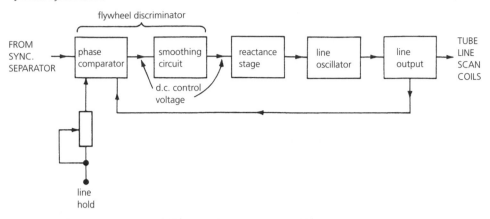

discriminator and a smoothing circuit. A control voltage proportional to the timing, i.e. phase, between the line oscillator and the incoming sync. pulse is obtained from the phase comparator. The voltage is then smoothed by the use of a low-pass filter. For good noise immunity the flywheel discriminator should have slow response, which means a long time constant for the smoothing circuit so that the average frequency is taken over a large number of line sync. pulses. However, the time constant also determines the *pull-in range* of the flywheel discriminator. The pull-in range is the range of oscillator frequency drift over which the discriminator will pull the oscillator into lock without having to adjust the manual line hold control. A short time constant improves the sensitivity, hence it widens the pull-in range of the discriminator. A compromise has to be struck so the oscillator has stability within the pull-in range of the discriminator.

## The flywheel discriminator

A simplified circuit of a flywheel discriminator is shown in Fig. 7.10, in which C4/R4 is a smoothing circuit. Line sync. pulses from the line output transformer are integrated by R1/C1 and the resulting reference sawtooth waveform is fed via C2 into the phase comparator D1/D2. At the same time, negative-going sync. pulses from the sync. separator are fed into the comparator at the cathodes of D1 and D2, pulsing the two diodes into conduction and clamping the reference sawtooth waveform for the duration of the line pulse.

If the line oscillator is running at the correct speed, both the sync. pulse and the sawtooth reference signal will coincide in such a way as to clamp the sawtooth waveform at zero volts (point A in Fig. 7.11). D1 and D2 conduct equally with the two currents $I_1$ and $I_2$ cancelling each other and producing a zero d.c. output.

If the oscillator is running fast, the sawtooth waveform has a higher frequency, hence a shorter periodic time. In this case the sync. pulse will coincide with the negative voltage part of the sawtooth waveform, causing D2 to conduct. A negative d.c. voltage is

**Fig. 7.10** *Flywheel discriminator*

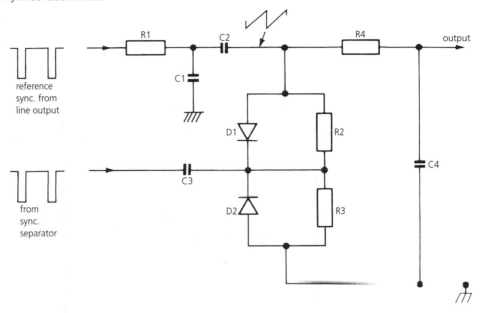

**Fig. 7.11** *Flywheel discriminator with line oscillator running at correct speed*

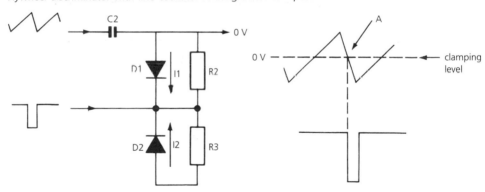

thus obtained at the output. Conversely, if the oscillator were running slow, the sync. pulse would coincide with the positive part of the sawtooth waveform, causing D1 to conduct and producing a positive d.c. output.

A flywheel discriminator circuit employing a phase splitter is shown in Fig. 7.12. The line sync. pulses from sync. separator VT9 are fed into phase splitter VT12 via differentiator C106/R118. The two antiphase sync. pulses, M1 and M2, are then fed into the flywheel discriminator, pulsing W25 and W26 into conduction and clamping the sawtooth reference waveform at the junction of the two diodes in the same manner as described earlier. C109 is part of the smoothing network which feeds the d.c. control voltage to the reactance stage.

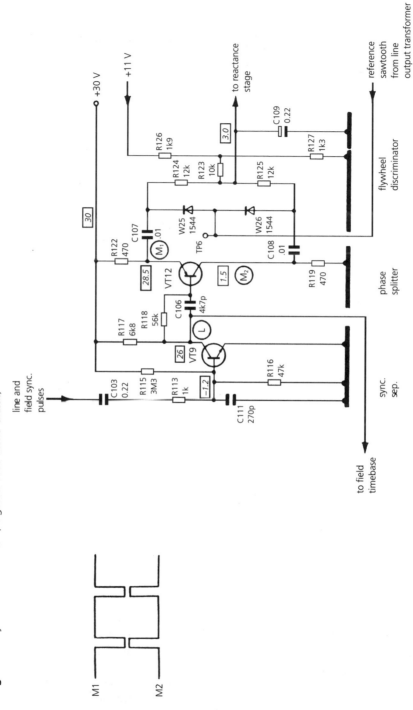

**Fig. 7.12** *Flywheel discriminator (Ferguson 1600 mono chassis)*

**Fig. 7.13** *Reactance stage*

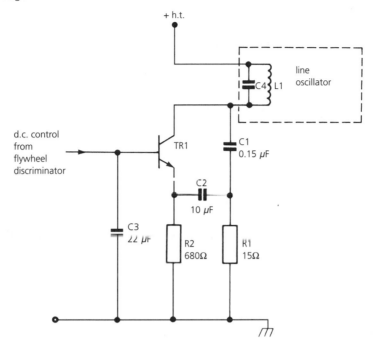

## The reactance stage

The purpose of the reactance stage is to convert a variation in the d.c. control voltage into a variation of reactance. A varactor is commonly employed, although bipolar and unipolar transistors may also be used.

A reactance stage using a transistor in the common-base configuration is shown in Fig. 7.13, in which L1/C4 forms part of the line oscillator tuned circuit. Phase shift network C1/R1 provides signal feedback via coupling capacitor C2. By making the reactance of C1 very much greater than R1, the portion of the output signal that is fed back to the emitter lags the output by almost 90°, representing a capacitive reactance. The value of this reactance, which effectively falls across L1/C1, is determined by the operating point of the transistor, which may be varied by the d.c. control voltage going into the base.

# Stability of line flyback pulses

The stability of the line flyback pulses is important in order to ensure a stable picture. For colour receivers it is critical, since the line sync. pulses are used to perform a number of gating functions in the colour decoder.

Although the line oscillator is locked by the flywheel sychronising circuit, problems of the stability of the line flyback are by no means completely resolved. This is because the line output stage, which utilises the line flyback to generate the e.h.t., has to supply a continuously varying tube beam current. The beam current varies as the brightness along the scan line changes, which in turn varies the demand made on the line output stage, causing the timing of the line flyback to change. The stability of the line flyback may be greatly improved by the use of integrated circuits that employ a more accurate flywheel synchronising system.

## Sync. processing chips

Figure 7.14 shows the basic arrangement of an i.c. sync. processing system in which phase-locked loops act as flywheel synchronisers. The stability of the line flyback is secured by the use of two phase-locked loops, PLL1 and PLL2. Phase detector 1 compares the phase of the square wave output of the line oscillator with the line sync. from the sync. separator, so it ensures the line oscillator is running at the correct frequency and phase. The second phase detector compares the phase of the line oscillator with the line flyback pulse from the line output stage. Any phase error is then corrected by the phase shifter network. The phase shifter is in essence a pulse-width modulator which changes the width and hence the phase of the line oscillator square wave output.

It is usual to include a third phase-locked loop, PLL3, in the line synchronising circuit to control the sensitivity of PLL1 (Fig. 7.15). Upon switching on, changing channels or

**Fig. 7.14** *Basic arrangement for a sync. processing chip*

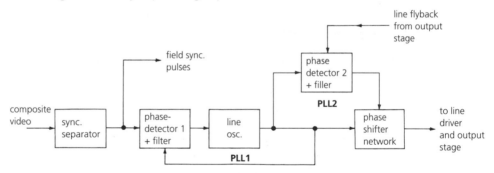

**Fig. 7.15** *Improved arrangement for sync. processor chip*

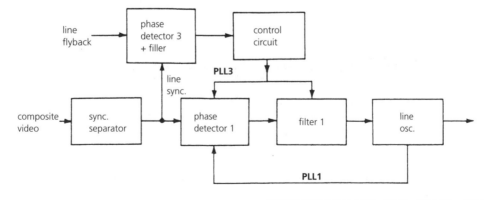

at any time due to weak signal, etc., when a momentary loss of line oscillator lock occurs, it is desirable for PLL1 to have a fast response, i.e. low phase detector sensitivity and a short filter time constant so that good pull-in and quick lock are obtained. Once the oscillator has been brought into lock, a slow response is desirable, i.e. high sensitivity and a long time constant to improve the accuracy of the oscillator and improve the discrimination of the system against noise and interference. To do this, phase detector 3 compares the phase of the line flyback from the line output stage with the sync. pulse from the sync. separator. An output is produced when the two pulses are in phase, in which case the control circuit changes the sensitivity of PLL1 and the time constant of filter 1.

Some integrated circuits employ two phase-locked loops in place of PLL1, a 'slow' phase detector and a 'fast' phase detector (Fig. 7.16). The gating circuit then selects one of the two phase detectors as appropriate under the control of the coincidence detector. The coincidence detector compares the output of the line oscillator with the sync. pulses from the sync. separator. A large phase difference, e.g. during channel change, will cause it to instruct the gating circuit to select 'fast' phase detector 1. The coincidence detector will bring the 'slow' detector 2 into operation during normal reception

**Fig. 7.16** *Basic components of a practical sync. processor (TDA 4578 part)*

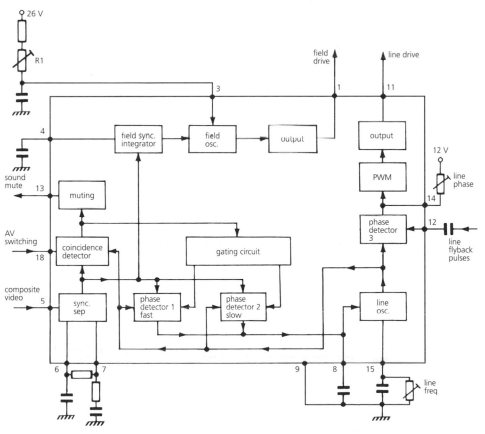

conditions. Similarly, a fast response time is desirable with an off-tape signal. In this case the fast phase detector 1 is brought into operation by the AV switching at pin 18. In the absence of a video signal altogether, the coincidence detector brings the muting circuit into operation, which may be used to provide interchannel sound muting. More than one coincidence detector stage may be used to provide for more than two possible levels of sensitivity of the line sync. system. Optimum sensitivity may then be selected depending on the strength of the received signal. The field sync. pulses from the separator are integrated and used to trigger the field oscillator, whose frequency is set by R1 (pin 3). The sawtooth waveform thus produced then appears at pin 1 for driving the field output stage.

## Sandcastle pulse

Modern integrated circuits incorporate the line and field pulses in a single multi-level pulse known as a sandcastle pulse. A typical three-level sandcastle pulse is shown in Fig. 7.17. The highest level, 7.5 V, provides the narrow burst gating pulse whose average duration is 4 $\mu$s. It is generated by level detection of the line sawtooth signal. The intermediate level, 4.5 V, is derived from the line flyback and has a duration of 12 $\mu$s. At the lowest level, 2.5 V, we have the field blanking pulse with a duration of 21 lines. A level detector or slicer may be used to extract the required pulse from the sandcastle combination as and when required.

**Fig. 7.17** *Sandcastle pulse*

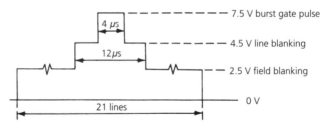

## The use of a common oscillator

Since the field frequency has a fixed ratio to the line frequency, and since the phase of the field deflection waveform is linked to that of the line deflection waveform, it is possible to derive the field trigger pulse from the line pulse. This is done within the i.c. by using a controlled divider network. A further improvement may be made by the PLL-controlled common oscillator for both the line and field frequencies.

The basic arrangement of a sync. processor chip using a common oscillator is shown in Fig. 7.18. The core of the i.c. is a PLL-controlled VCO which oscillates at 500 kHz. The output of the oscillator is divided by 32 (line divider) to give a line scan frequency of 15 625 Hz. Line synchronisation is provided by a multi-stage controlled phase detector which compares the phases of the line pulse from the divider and the

**Fig. 7.18** *Sync. processor chip using a common oscillator*

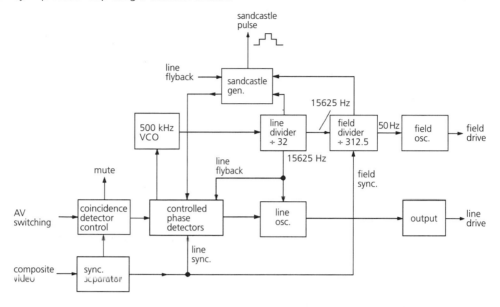

flyback from the line output stage with the sync. pulse to control the phase and frequency of the VCO and trigger the line oscillator. The field frequency is obtained by dividing the line frequency by 312.5 (field divider). The divider is controlled by the sync. pulses from the sync. separator to set the phase of the field signal correctly. The line, field and flyback pulses are built up into a sandcastle pulse by the sandcastle generator as shown.

## QUESTIONS

1. Explain the function of the sync. separator in a TV receiver.

2. (a) Explain why the line and field sync. pulses require separation from each other at the output of the sync. separator.
   (b) Show how this separation may be achieved.

3. In the 625-line British TV system, state
   (a) the number of lines per field
   (b) the number of active lines per complete picture

4. (a) Sketch the sandcastle pulse showing width and voltage levels.
   (b) State the main function of the sandcastle pulse.

# 8 Field timebase

The field timebase in a TV receiver consists of a sawtooth generator followed by a driver and an output stage which feeds the field scan coils. The sawtooth generator contains a free-running pulse oscillator which is triggered by the field sync. pulse derived from the sync. separator. The oscillator includes a facility for manual control of its frequency for 'field hold' as well as one or more controls to ensure vertical linearity.

## Sawtooth generation

Sawtooth generation is obtained by charging and discharging a capacitor as shown in Fig. 8.1(a), in which TR1 forms part of the field oscillator circuit. When TR1 is switched off, capacitor C1 charges up slowly (scan) through R1 and towards h.t. Well before the capacitor voltage reaches h.t., a positive-going field sync. pulse arrives at the base, turning the transistor on and discharging the capacitor very quickly to zero (flyback). The capacitor remains discharged for the duration of the input pulse. When the pulse comes to an end, TR1 is switched off and the capacitor begins to charge up, and so on. The result is the sawtooth waveform shown in Fig. 8.1(b). The frequency of the waveform is the same as that of the input pulse, and its amplitude is a function of time constant C1R1. R1 is made variable to provide amplitude control. By using a small portion of the charging curve, a linear scan may be obtained.

**Fig. 8.1**  *Sawtooth generation: (a) simple transistor circuit; (b) waveforms*

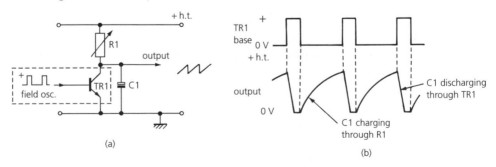

# Blocking oscillator

A typical blocking oscillator circuit is shown in Fig. 8.2. The action of the circuit hinges around the fact that, due to transformer coupling, a voltage is induced into the base only when the collector current is varying up or down. In one case the feedback is positive and in the other it is negative. When the circuit is first switched on, the transistor conducts and the collector current increases, thereby producing a feedback voltage at the base in such a way as to switch the transistor further on. When saturation is reached, the collector current ceases to increase and an opposite voltage is induced at the base, which this time turns the transistor off. The transistor is held in the off state by the negative charge on capacitor C1 until it is sufficiently discharged through resistor R1, when the transistor switches on again and so on.

The output from a blocking oscillator is a narrow pulse waveform (Fig. 8.3). The width, or mark, of the pulse is determined by the parameters of the transformer, whereas the space is determined by time constant C1R1. It follows that the frequency of the oscillator may be varied by changing the value of R1.

Depending on the parameters of the transformer, a large overshoot in the collector voltage may be obtained as the transistor switches off. This overshoot voltage may be of such a magnitude as to exceed the maximum rateable collector voltage, resulting in damage to the transistor. In order to protect the transistor, a diode D1 is connected across the primary winding of the transformer. The diode, which is normally reverse-biased, will conduct only when the collector voltage exceeds the d.c. supply $V_{cc}$.

**Fig. 8.2**  *Blocking oscillator*

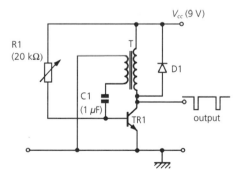

**Fig. 8.3**  *Output of a blocking oscillator*

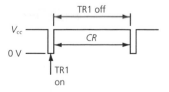

# Multivibrator

The astable multivibrator (Fig. 8.4) may also be used to produce a field timebase waveform. The multivibrator uses two separate time constants, C1VR1 for the mark and C2R2 for the space. The pulse waveform at the collector of TR2 is converted into a sawtooth by capacitor C3.

Time constant C1VR1 determines the period that TR2 remains off, during which time C3 charges up to provide the scan part of the sawtooth. The setting of VR1 thus controls the frequency of the output. The amplitude, on the other hand, may be varied by VR2 to control the picture height. Time constant C2R2 determines the duration for which C3 remains discharged, i.e. the flyback.

A simpler astable multivibrator that is successfully used in TV receivers is the emitter-coupled type shown in Fig. 8.5, in which C1 forms part of both time constants. The negative-going sync. pulse switches TR1 off and TR2 on. C1 with its left-hand plate at +15 V begins to charge up very quickly through TR2, forcing its right-hand plate (TR2 base) towards zero. As TR2 base approaches zero, current through TR2 begins to decrease and its collector voltage begins to increase, which when fed to TR1 base via

**Fig. 8.4** *Astable multivibrator*

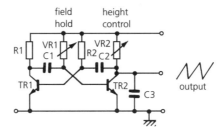

**Fig. 8.5** *Astable multivibrator*

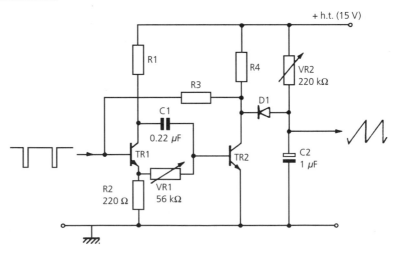

R3 tends to switch the transistor on, further reducing TR2 current, and so on until TR1 saturates and TR2 turns off. The sharp drop in the potential of TR1 collector (from 15 V to about 2 V, a drop of −13 V) is transferred to the base of TR2 via C1. Capacitor C1 begins to discharge through VR1/R2 towards TR1 emitter voltage of +2 V. As it crosses the zero line, TR2 switches on, and so on. It is arranged that just before TR2 switches on 'naturally', the sync. pulse arrives to trigger the multivibrator at precisely the right time. In other words, the natural frequency of the oscillator is adjusted by VR1 to be slightly lower that the field frequency, so the arrival of the sync. pulse ensures field lock.

C2/VR2 is a ramp generator network which provides the sawtooth waveform with VR2 controlling its amplitude, i.e. the picture height. D1 is an isolating diode that conducts only when TR2 saturates. This is necessary to ensure that C2 does not charge up through R4 when TR2 is turned off.

## Thyristor oscillator

The silicon controlled rectifier (s.c.r.) is a four-layer pnpn switching device which turns on when the anode voltage exceeds the breakdown voltage of the device (Fig. 8.6). Switching may also be triggered by the application of a positive pulse to the gate. When the s.c.r. is conducting, the voltage across it falls to a small value known as the *holding voltage*. Once the s.c.r. is switched on, it will only turn off when the anode potential falls below the holding voltage.

A more versatile switching device is the silicon controlled switch (s.c.s.) shown in Fig. 8.7. In this case a second gate G2, known as the *anode gate*, is brought out as a control terminal. This additional gate may also be used to trigger the device on. This time, however, a negative pulse is necessary. Both the s.c.r. and the s.c.s. are commonly known as thyristors.

**Fig. 8.6**  *The silicon controlled rectifier (s.c.r.) and its symbol*

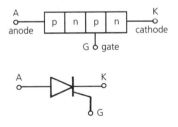

**Fig. 8.7**  *The silicon controlled switch (s.c.s.) and its symbol*

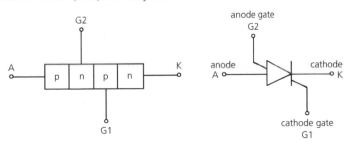

**Fig. 8.8** *Self-oscillating thyristor*

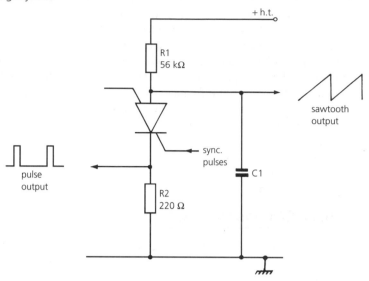

A self-oscillating thyristor using an s.c.s. is shown in Fig. 8.8. When the supply voltage (h.t.) is switched on, capacitor C1 charges up through R1 at a rate determined by time constant C1R1. When the voltage across C1 exceeds the breakdown voltage of the thyristor, the thyristor turns on and discharges the capacitor very quickly through R2. The thyristor remains switched on until its anode potential falls below its holding potential, i.e. when the capacitor is almost fully discharged. When that happens, the thyristor switches off and the capacitor begins to charge up again, and so on. A sawtooth waveform is obtained across C1 while a pulse output is obtained across R2. Synchronisation is obtained by feeding the field sync. pulses into one of the control gates, a positive-going pulse to the cathode gate or a negative-going pulse to the anode gate. The natural frequency of the oscillator is determined by time constant C1R1.

## Field scanning waveform

A linear deflection is obtained when a sawtooth current waveform is fed into the scan coils. However, because of the inductance presented by the scan coils, the voltage necessary to obtain a linear current through the coils may not have a linear shape; it will depend on the ratio of the coil resistance to the coil inductance, $r/XL$. At the low field frequency of 50 Hz the reactance of the field coils is small compared with their resistance, making the coils mainly resistive. In this case the voltage waveform is almost the same as the current.

Consider the inductor shown in Fig. 8.9 in which $r$ represents the effective resistance of the inductor and $L$ its inductance. The applied voltage $v_a$ is equal to the vector sum of the voltage across the resistance $v_r$ and the voltage across the inductance $v_L$. If the applied voltage were a sine wave, then all the other waveforms would be sinusoidal

**Fig. 8.9**

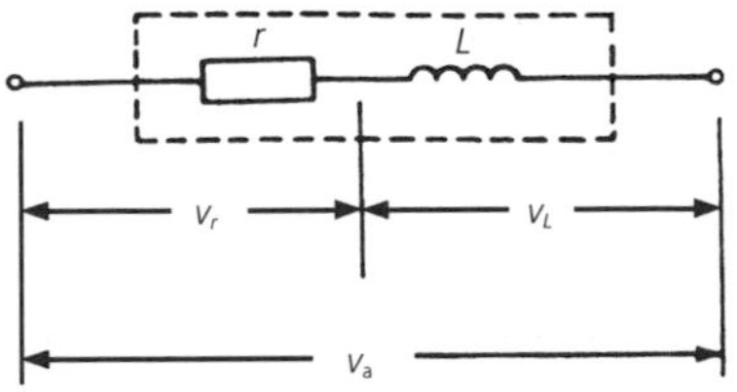

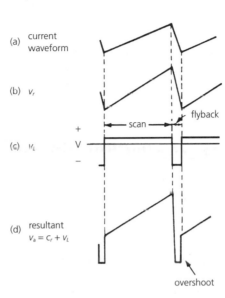

**Fig. 8.10**

(a) current waveform

(b) $v_r$

(c) $v_L$

(d) resultant $v_a = c_r + v_L$

scan

flyback

overshoot

as well. The relationship between these waveforms could then be represented by a simple vector diagram to show their relative amplitude and phase. For an applied waveform other than a sine wave, i.e. a waveform that contains one or more harmonics, each harmonic suffers a different amount of attenuation and phase shift. Current and voltage waveforms will therefore differ not only in their amplitude and phase but in their shape as well. It is not possible in these cases to draw one simple phasor diagram since each phasor represents a sine wave at one particular frequency. A large number of phasor diagrams are therefore necessary, one for each harmonic.

A simple way of deducing the required shape of the applied voltage is to add the waveform across the resistance to that across the inductance, $v_r + v_L$ (Fig. 8.10). Figure 8.10(a) shows the sawtooth current waveform that is necessary for a linear scan. As far as resistance $r$ is concerned, the voltage across it, $v_r$, is in phase with the current, as shown in Fig. 8.10(b). For inductance $L$, the voltage across it

$$v_L = L \, di/dt$$

In other words, the greater the rate of change of current $di/dt$, the greater the voltage across the inductance. For the scan part of the current waveform, the rate of change of current $di/dt = c$, where $c$ is a constant determined by the slope of the scan. Since $v_L = L \, di/dt$, then for the scan part only

$$v_L = L \times c = k$$

where $k$ is another constant. It follows therefore that, in order to obtain a linearly rising current, the voltage applied across a pure inductor $L$ must be of a constant value or d.c. (Fig. 8.10(c)).

At the and of the scan, the current reverses and $di/dt$ sharply increases, producing the large negative-going edge shown in Fig. 8.10(c). For the duration of the flyback, $v_L$ remains constant but this time at a higher value than obtained during the scan; this is because the rate of change of the current is larger during the flyback than during the scan. At the end of the flyback, the current reverses again to produce the positive-going edge shown in Fig. 8.10(c), and so on.

If the two waveforms $v_r$ and $v_L$ are now added, we get the resultant or total waveform $v_a$ shown in Fig. 8.10(d). Although the inductance has an insignificant effect on the scan part of the waveform, it is responsible for the overshoot produced during the flyback.

## Linearity or shaping networks

The field scan waveform requires further modification to ensure a linear display on the tube face. Waveform correction (or distortion) is necessary where transformer coupling is used between the output stage and the scan coils. In this case the sawtooth waveform has to be corrected to overcome the distorting effect of the magnetising current taken by the transformer. Correction is also required because of the flat surface of the tube face. This correction, known as the S-correction, is more noticeable on the line scan and will be discussed in detail in the next chapter. It is normal to include an effective shaping or correction network involving a combination of differentiators and integrators with more than one linearity control as part of the field timebase.

Consider an integrator fed with a sawtooth waveform (Fig. 8.11). The sawtooth consists of a fundamental and a number of harmonics. The integrator, which is a low-pass filter, attenuates the high-frequency harmonics, slowing down the rate of change of the waveform and rounding the edges. The result is the parabola shown in Fig. 8.11(b). An integrator with a longer time constant (i.e. a filter with a lower cut-off frequency) would remove more high-frequency harmonics, resulting in the parabola shown in Fig. 8.11(c).

The effect of a differentiator on a sawtooth waveform is shown in Fig. 8.12. The differentiator is a high-pass filter; it will attenuate the low-frequency components, resulting in sharper corners and faster rising edges, as shown in Fig. 8.12(b). A differentiator with a shorter time constant (i.e. a filter with a higher cut-off frequency) would produce the sharper waveform shown in Fig. 8.12(c).

It is possible therefore to distort the sawtooth in such a way as to increase or decrease the scanning rate for all or part of the scan. Further changes in the shape of the scan may be produced by adding a sawtooth waveform to a parabola to obtain a tilted parabola (Fig. 8.13). This addition is normally carried out by current or voltage feedback. Current feedback involves the monitoring of the scan waveform using a small resistor (1–5 $\Omega$) connected in series with the field scan coils. The voltage which is developed across the resistor is fed back to the input of the field drive amplifier via a correction network.

**Fig. 8.11** *Effect of an integrator on a sawtooth waveform*

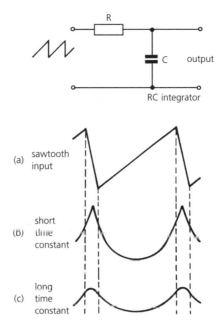

**Fig. 8.12** *Effect of a differentiator on a sawtooth waveform*

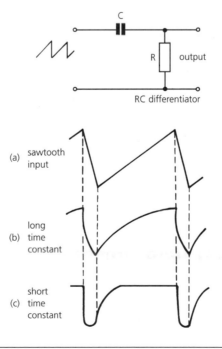

**Fig. 8.13** *Tilted parabola*

**Fig. 8.14** *Class B field output stage (Ferguson 8000 colour chassis)*

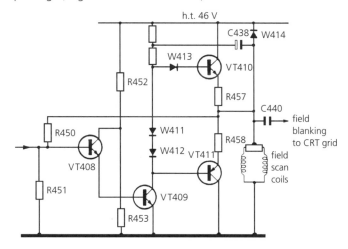

## Class B field output

A typical class B field output stage is shown in Fig. 8.14, in which VT408/VT409 is the driver combination and VT410/VT411 is the complementary output pair. At the start of the scan, VT408 base voltage is low, turning off VT408 and VT409. VT409 collector is at almost h.t. potential; it turns VT410 hard on, driving current into the field scan coils. VT411 is off. At the midpoint of the scan, the centre of the picture, VT408/409 begins to conduct, turning VT410 off and VT411 on; this provides the scan current for the second half of the picture. W411/412 provides a small forward-bias for the output transistors to prevent crossover distortion. Resistor R450 provides d.c. feedback for bias stability.

## Practical field timebase

A practical field timebase with a class B output stage is shown in Fig. 8.15, in which VT23/VT24 is the output pair, VT18/VT19 is the field oscillator and C102/R117/

**Fig. 8.15** *Complete field timebase (Ferguson 1690 mono chassis)*

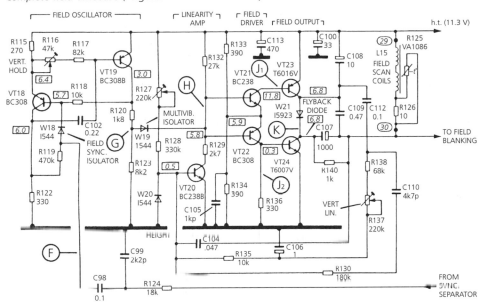

R116/R115 is the time constant network. Positive-going field sync. pulses from the sync. separator are fed via W18 into the base of VT18 which, being a pnp transistor, turns off to start the flyback. Diode W18 isolates the field oscillator from the sync. separator, conducting only when a pulse is present. The ramp is produced by C104, charging up through R127/R128 towards the h.t. supply of 11.3 V. The rising voltage across C104 turns on the high-gain amplifier VT20, which drives the push-pull driver stage VT21/VT22. When VT20 is conducting, VT22 and hence VT24 turn on, whereas VT21 and VT23 turn off. At the end of the scan, VT24 is saturated. At this point, a positive-going pulse from the multivibrator is applied to the bases of VT21 and VT22, via isolating diode W19. Transistors VT22 and VT24 are turned off, whereas VT21 and VT23 are turned on; VT23 collector potential rises to d.c. supply, forward-biasing W21, which short circuits C109. The rapid change in the current through the scan coils produces a back e.m.f., making VT24 collector rise at a rate determined by the *L/r* ratio of the scan coils. When VT24 collector potential exceeds the supply voltage, diode W21 turns off. Capacitor C109 is then placed across the scan coils, which start to oscillate or ring. The flyback rises sharply to the first positive peak voltage of the ringing sine wave then decreases, attempting to go to the negative peak voltage. Well before it reaches negative peak, W21 conducts; this terminates the ringing as the energy stored in the scan coils is fed into the supply rails via W21 and VT23. The process continues until the scan begins again, and so on. Linearity correction is obtained by a feedback network containing linearity amplifier VT20. The field sawtooth waveform is modified with a parabola produced by integrator R138/R137/C106. Further shape correction is obtained by differentiator C110/R130.

**Fig. 8.16** *Mean (d.c.) level of a pulse-width modulated waveform where the mark-to-space ratio is (a) high and (b) low*

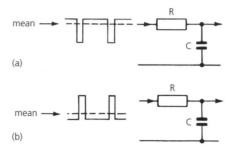

**Fig. 8.17** *PWM waveform for a field scan*

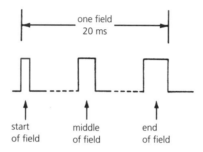

# Switched-mode (class D) field output

The basic principle of the switched-mode (also known as class D) output stage is shown in Fig. 8.16, in which a pulse train is fed into an RC low-pass filter. As can be seen, the pulses are smoothed with the capacitor charging up to the mean value of the input pulse. Since the mean value is determined by the mark-to-space ratio of the input pulse, a varying mark-to-space ratio, i.e. a pulse-width modulated waveform, produces a varying charge across the capacitor. A linearly increasing capacitor voltage, a timebase ramp scan, may thus be produced if the mark-to-space ratio of the input pulse is gradually increased (Fig. 8.17).

A simplified block diagram for a switched-mode output stage is shown in Fig. 8.18. The pulse-width modulated waveform is first fed into the output stage to turn the active device on and off before going into an LC low-pass filter, which produces the sawtooth waveform to drive the deflecting current into the scan coils. The output device, which may be a transistor or a thyristor, thus acts as a switch. The pulse clock rate which drives the pulse-width modulator may be a separately generated waveform or it may be derived from the line scanning pulse. Since the active element is a switch, its power dissipation is extremely low. When the switch is on, its resistance is very small and therefore its power dissipation is very low. When the switch is open, current ceases and power dissipation is nil. The temperature rise is minimised and cost is reduced.

**Fig. 8.18** *Switched-mode (class D) field output stage*

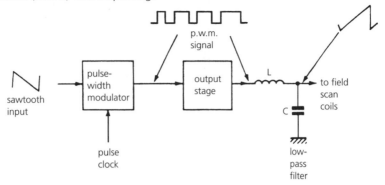

**Fig. 8.19** *Pulse-width modulator (TEA 2029 part)*

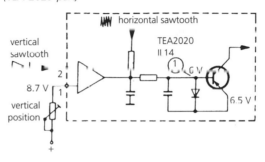

**Fig. 8.20**

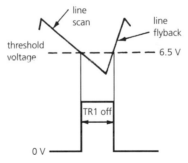

A pulse-width modulator circuit used by sync. processing chip TEA 2029 is shown in Fig. 8.19. The differential amplifier has two inputs, a field sawtooth at pin 2 and a field shift correcting voltage on pin 1. The field sawtooth from the differential amplifier together with a line sawtooth are applied to the modulator transistor TR1. The modulator transistor operates as a comparator with a threshold voltage of 6.5 V, the voltage at its emitter. TR1 will switch off only when the its base voltage falls below 6.5 V, at which point the collector voltage rises to supply (Fig. 8.20). The points on the line sawtooth at which TR1 switches off and on are determined by the ouput voltage of the differential amplifier, which decreases progressively as the scan proceeds from start to finish. This causes the line sawtooth waveform to shift progressively deeper into the switching threshold of TR1 (Fig. 8.21), producing a linearly changing mark-to-space ratio.

Fig. 8.21

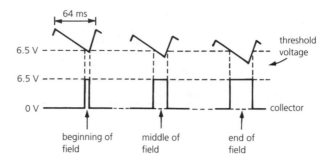

1. Explain why is it necessary to introduce linearity correction to the field waveform and show how this may be achieved.

2. Explain the principle of the pulse-width modulator and show how it is used to generate a sawtooth field waveform.

# 9 Line timebase

Like most small-signal processes, line synchronisation and generation are available as part of an integrated circuit. Where discrete components are employed, two types of oscillators are normally used in the line timebase, the Hartley oscillator and the blocking oscillator. Although the blocking oscillator is more efficient, the Hartley oscillator is more popular because of its good frequency stability; this means that a smaller pull-in range may be used by the flywheel discriminator, thus improving its noise immunity.

A line timebase circuit using a blocking oscillator (VT15) is shown in Fig. 9.1, in which diodes W6 and W7 form the flywheel discriminator with C78/R82/C79/R85/C62 as the smoothing low-pass filter. Feedback transformer L12 is tuned by C83. Note that a reactance stage is not necessary, one of the advantages of the blocking oscillator over the Hartley. VT15 is switched on when C84 discharges through R84/R83, raising the base voltage to a level sufficient to forward-bias the transistor. By varying the d.c. control

**Fig. 9.1** *Line timebase circuit (Ferguson 1690 mono chassis)*

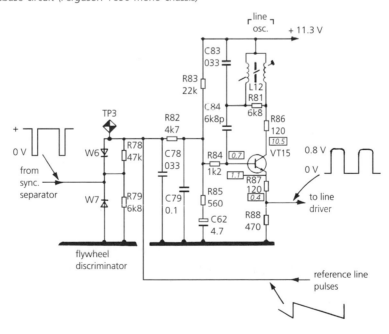

voltage from the flywheel discriminator to VT15 base, the time at which the switching occurs is also varied, thus changing the frequency of operation.

## The Hartley oscillator

The Hartley oscillator is a sine wave oscillator with a tuned circuit as the collector load. Feedback is achieved by tapping the inductor and feeding part of the output back to the base. By using a large reverse bias, it is possible to make the transistor turn off for part of the cycle and saturate for the other part. In this way the transistor is made to act as a switch. A line oscillator circuit employing a Hartley is shown in Fig. 9.2. The tapping in inductor L1 is decoupled to chassis via C1. The signal across the lower part of tuned circuit L1/C1 is fed back to the base of TR1 via C2. A large amount of positive feedback is employed, with C2 providing class C self-biasing for TR1. The level of this reverse bias is determined by time constant C2R1. The tuned circuit oscillates so long as energy in the form of collector current is fed into it at regular intervals. The transistor is biased beyond cut-off, so it will conduct only for the positive peaks of the sine wave fed back into its base. The waveform at the base will have its positive peaks clipped as the b–e junction becomes forward-biased, as shown in Fig. 9.3(a). The pulsating current through the transistor, as it switches on and off, produces a square wave across emitter resistor R3, as shown in Fig. 9.3(b).

**Fig. 9.2** *Hartley line oscillator*

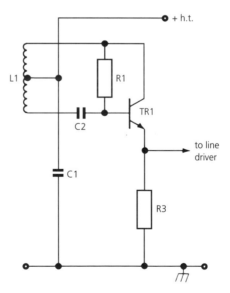

## The line scan waveform

The purpose of the line timebase is to provide the appropriate deflection current through the line scan coils. As in the case of the field scan, the current waveform required to

**Fig. 9.3** *Waveforms for circuits in Fig. 9.2*

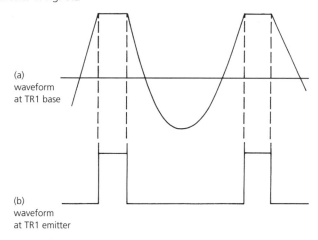

(a)
waveform
at TR1 base

(b)
waveform
at TR1 emitter

**Fig 9.4** *line scan waveforms: coil current and coil voltage*

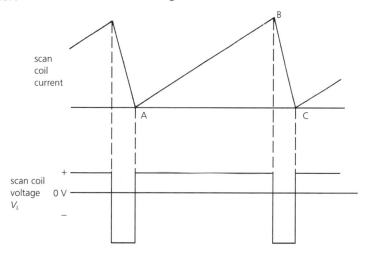

scan
coil
current

scan coil
voltage
$V_L$

produce linear deflection is a sawtooth. However, at the relatively high line frequency, the reactance of the coil $X_L$ is very high compared with its d.c. resistance, so the resistance is insignificant. The line scan coils may then be treated as purely inductive. Given that $V_L = L \, di/dt$ it follows that a linear current waveform in a pure inductor is obtained when a constant or d.c. voltage is applied across it. To obtain a sawtooth current waveform, the voltage waveform must be the pulse shown in Fig. 9.4. For the scan period AB, the current is increasing at a small and constant rate. Consequently, voltage $V_L$ is a small positive value which remains constant for that duration. For the flyback period BC, the current is decreasing at a high and constant rate; $V_L$ is again constant but this time large and negative.

**Fig. 9.5**

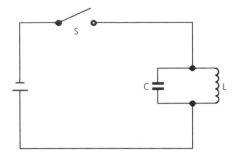

**Fig. 9.6** *Damped oscillation*

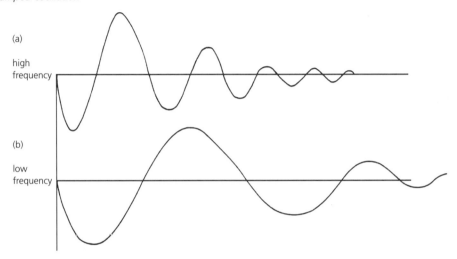

(a)

high
frequency

(b)

low
frequency

Consider the tuned circuit in Fig. 9.5. When the switch is closed, a step waveform is applied across the tuned circuit and energy is fed into it. Oscillation, known as ringing, takes place at a resonant frequency $f_0 = 1/2\pi\sqrt{LC}$. These oscillations take place because electromagnetic energy in the scan coils is continuously transformed into electrostatic energy in the capacitor, and vice versa. Current therefore flows from the coil to the capacitor and back again. Ideally, this ringing should continue indefinitely since there is no power or energy loss in either a pure inductor or a pure capacitor. However, due to losses caused mainly by the very small resistance of the inductor, ringing gradually dies out, producing what is known as damped oscillation (Fig. 9.6).

Similar oscillation or ringing occurs when a sharp change in voltage is applied across an inductor. The tuning frequency in this case is the self-capacitance of the coil as well as any stray capacitance due to other components. In the case of the line scan, ringing occurs at the beginning of each flyback (Fig. 9.7) with a consequent distortion on the left-hand side of the picture on the tube face.

It is possible to remove the effect of ringing by shunting the scan coils with a damping resistor. This, however, will result in a large waste of energy, reducing the power available for beam deflection and reducing the angle of deflection of the tube.

**Fig. 9.7** *Ringing*

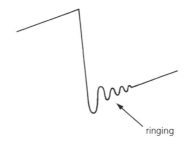

ringing

**Fig. 9.8** *Use of efficiency diode D1*

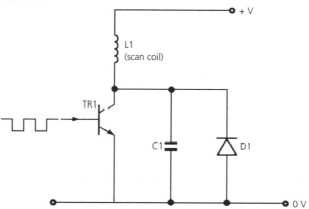

## Efficiency diode

To avoid ringing but without loss of power, an efficiency diode is used. This technique is based on using the energy stored in the scan coils due to flyback to provide the first half of the scan. It involves a switching network which directs the transfer of energy to and from the scan coils to obtain the required waveform. A circuit using a parallel transistor–diode switch is shown in Fig. 9.8, in which L1 is the scan coil, D1 is the efficiency diode and TR1 is the line output transistor. When TR1 is switched on by a positive edge to its base at time $t_1$ (Fig. 9.9), a constant h.t. voltage is applied across scan coil L1. A linearly increasing current is therefore obtained, forming part AB of the scan. The current continues to rise until point B, when at time $t_2$ a negative step to the base switches off TR1. At this point TR1 collector suffers a sudden jump from almost chassis potential to +h.t. This positive voltage ensures that D1 remains non-conducting. C1 is now effectively connected across L1. The large change in current in L1 produces ringing at a frequency determined by C1 and other stray capacitors. Energy due to the sudden change of current through L1 is transferred to C1 to commence ringing oscillation at point B. When C1 is fully charged (point C on the flyback), the charging current drops to zero as the ringing comes to the end of the first quarter-cycle of oscillation. The second quarter-cycle begins as energy from C1 is transferred to L1. The current

**Fig. 9.9** *Waveforms for circuit in Fig. 9.8*

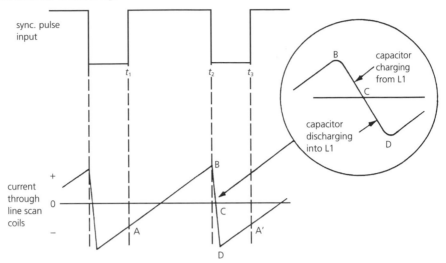

reverses as the capacitor begins to discharge. When the cycle reaches its negative peak at the end of the first half-cycle (point D), the current begins to decrease, attempting to go to zero again. The rate of change of current $di/dt$ suffers a change of direction. Before the negative peak (point D), the rate of change of current is positive. This induces an e.m.f. across L1, which makes TR1 collector (and D1 cathode) positive, ensuring the diode is off. At the negative peak itself, $di/dt = 0$ and the induced e.m.f. is also zero. After the negative peak, the current begins to decrease. The rate of change of current is therefore negative, reversing the induced e.m.f. and making TR1 collector (and D1 cathode) negative. The diode conducts. Its effect is similar to TR1 conducting, placing the h.t. across L1 to start the scan. Current through L1 rises linearly to form about 30% of the scan up to A′, when TR1 is switched on at time $t_3$ by a positive edge to the base, and so on.

## S-correction

In order to compensate for the flat surface of the tube face, correction of the line scan waveform is necessary. This correction, known as symmetrical or S-correction, becomes increasingly important with wide-angle deflection tubes.

As can be seen from Fig. 9.10, an equal angular deflection of the beam scans a smaller distance at the centre compared with the distance it scans at the two ends of the line. Thus, to obtain a linear picture scan using a flat tube face, a non-linear angular deflection is necessary. The purpose of the non-linearity is to slow the rate of change of the angular deflection at both ends of the scan (Fig. 9.11). Since the current through the line scan coils is responsible for the angular deflection, the corrected waveform must be of the same shape.

**Fig. 9.10**  *Effect of flat tube screen*

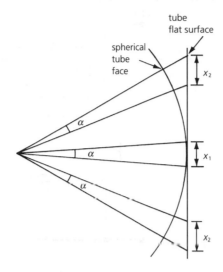

**Fig. 9.11**  *(a) S-correction waveform; (b) half-line frequency sine wave*

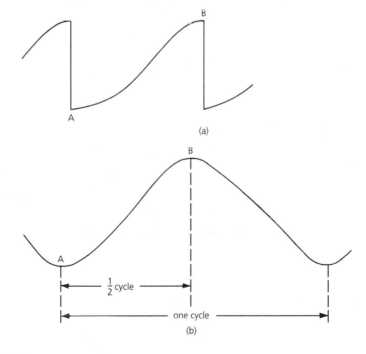

The scan part of the waveform in Fig. 9.11(a) approximates a half-cycle of the sine wave in Fig. 9.11(b) and may be simply obtained by connecting a capacitor in series with the scan coils. The value of the capacitor is chosen so that it resonates with the scan coils at a frequency slightly higher than half the line frequency, approximately 7.8 kHz.

**Fig. 9.12** *Line output stage with S-correction capacitor C2*

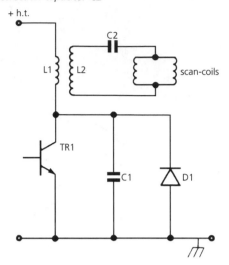

**Fig. 9.13** *(a) Directly coupled line scan coils; (b) current through scan coils*

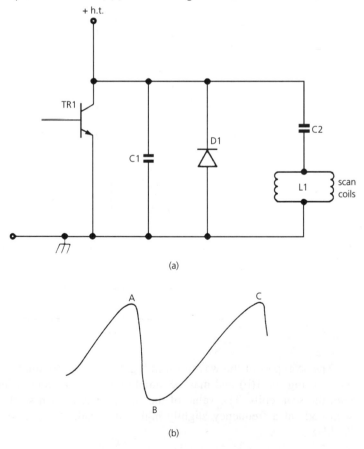

(a)

(b)

**Fig. 9.14** *Line output using b–c junction as efficiency diode*

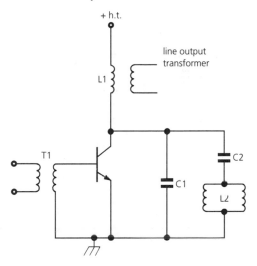

This provides a time duration from A to B of

$$\frac{1}{2}\,\text{period} = \frac{1}{2} \times \frac{1}{7.8} = 64\,\mu s\ \text{(approx.)}$$

In the line output stage of Fig. 9.12, C2 is the S-correction capacitor which has a value of between 1.5 and 3 $\mu$F. The scan coils are transformer-coupled to the line output transistor TR1. D1 is the efficiency diode and C1 is the flyback tuning capacitor.

Modern TV receivers employ direct coupling (Fig. 9.13). At the end of the scan, TR1 is switched off, tuned circuit L1/C1/C2 is pulsed into oscillation to provide the flyback AB (Fig. 9.13(b)). At the negative peak (point B) diode D1 conducts, placing C2 across L1 to commence an oscillation at a frequency of

$$\frac{1}{2\pi\sqrt{LC}} = \text{half-line frequency}$$

The first half-cycle of this oscillation, BC, provides the scan. At approximately one-third of the scan, TR1 is switched on and takes over from the diode, and so on.

Further simplification of the line timebase circuit may be obtained by using the b–c junction of the output transistor TR1 as the efficiency diode (Fig. 9.14). The polarity of the b–c junction is the same as the polarity of an efficiency diode, had it been connected; the n-region collector (cathode) is connected to h.t. whereas the p-region base (anode) is connected to chassis via the secondary winding of T1. At the end of the scan, the b–c junction is forward-biased in the same manner as an efficiency diode.

## Line output transformer

Although the scan coils are not normally transformer-coupled, a transformer known as the line output transformer (LOPT) is employed to provide a number of functions,

**Fig. 9.15** *Line output with boost capacitor C1 and diode D1*

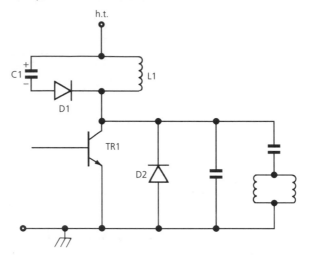

including the extra high tension (e.h.t.), the auxiliary d.c. supplies, the boost voltage, the gating pulses for a.g.c. and the reference pulse for the flywheel discriminator (Fig. 9.14). In choosing the value of the tuning capacitor for the line output stage, the inductance introduced by the line output transformer must be taken into account.

## Boost voltage

One important function of the line output transformer is to provide the high d.c. supply voltage of 30–90 V required by the video, line and field output amplifiers. For receivers operating from the mains supply, this d.c. voltage may be obtained by rectifying the mains voltage. However, this is not possible for battery-operated receivers, so a boost voltage from the LOPT is used.

The boost voltage is obtained by the use of an efficiency diode and a storage capacitor. Consider the circuit in Fig. 9.15, in which L1 is the primary winding of the line output transformer, D1 is the boost diode, C1 is the storage or boost capacitor and TR1 is the line output transistor. D1 is connected in such a way that it only allows charging current to flow through C1 and prevents the discharging current from flowing through L1, thus maintaining the charge across C1. When TR1 is switched on, D1 conducts; this places C1 across L1. Ringing occurs and electromagnetic energy in L1 is transferred to C1, which charges up to h.t. When C1 attempts to discharge through L1, the current reverses and D1 stops conducting. Provided C1 is large, in the region of 200 $\mu$F, it will retain the charge across it. When TR1 is switched off to start the flyback, D1 remains off. At the end of the flyback, TR1 collector goes negative due to the reversal of the rate of change of current in the scan coils. D1 and D2 conduct. Energy in L1 is transferred to C1 to replace any loss in its charge. Excess energy is fed back into the h.t. supply line. C1 thus remains charged up to h.t.

**Fig. 9.16** *Line output with boost capacitor in series with h.t. line*

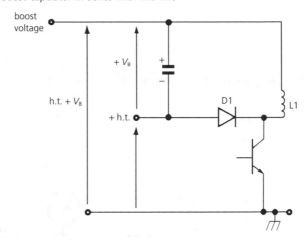

It is possible to connect the boost capacitor in series with the h.t. line to produce a boosted voltage (Fig. 9.16). The voltage $V_B$ across the capacitor, which could exceed the h.t. potential, is added to the h.t. line.

## Practical circuit

A line timbase circuit for a monochrome receiver using discrete components is shown in Fig. 9.17, in which W9/W10 is the flywheel discriminator, VT23 the reactance stage and VT24 is a Hartley oscillator. L14 and C99 form the oscillator tuned circuit. Self-biasing is obtained by C99/R121. The output from the reactance stage VT23 is effectively connected across the left-hand side of the winding of L14 and as such will determine the frequency of oscillation. W11 and C107 are the boost diode and capacitor respectively. The 25 V boost voltage is obtained from an h.t. line of only 11.6 V. VT26 is the line output transistor, C108 is the S-correction capacitor, C109 is the tuning capacitor and L15 is the line linearity control. Capacitor C106 is a flashover capacitor used to protect the output transistor from flashover inside the tube.

The line and field timbase (synchronisation) part of the STV2110 chip is shown in Fig. 9.18. Composite video is fed into an onboard sync. separator (pin 12) which feeds pulses into the field sync. and the flywheel line sync. system. Chip technology uses two phase detectors in the flywheel line sync. system. The first (phase det 1) is used to lock the crystal-controlled 503 kHz main oscillator to the incoming line sync. pulses from the sync. separator. The oscillating frequency is divided to produce the 15.625 kHz line frequency and 50 Hz field frequency. The phase of the line frequency is controlled by the second phase detector (phase det 2), which receives the feedback signals from the line output transformer (pin 13). The line output drive is obtained at pin 15; the field output drive pulse is produced at pin 14.

**Fig. 9.17** *Line timebase circuit (Ferguson 1590 mono chassis)*

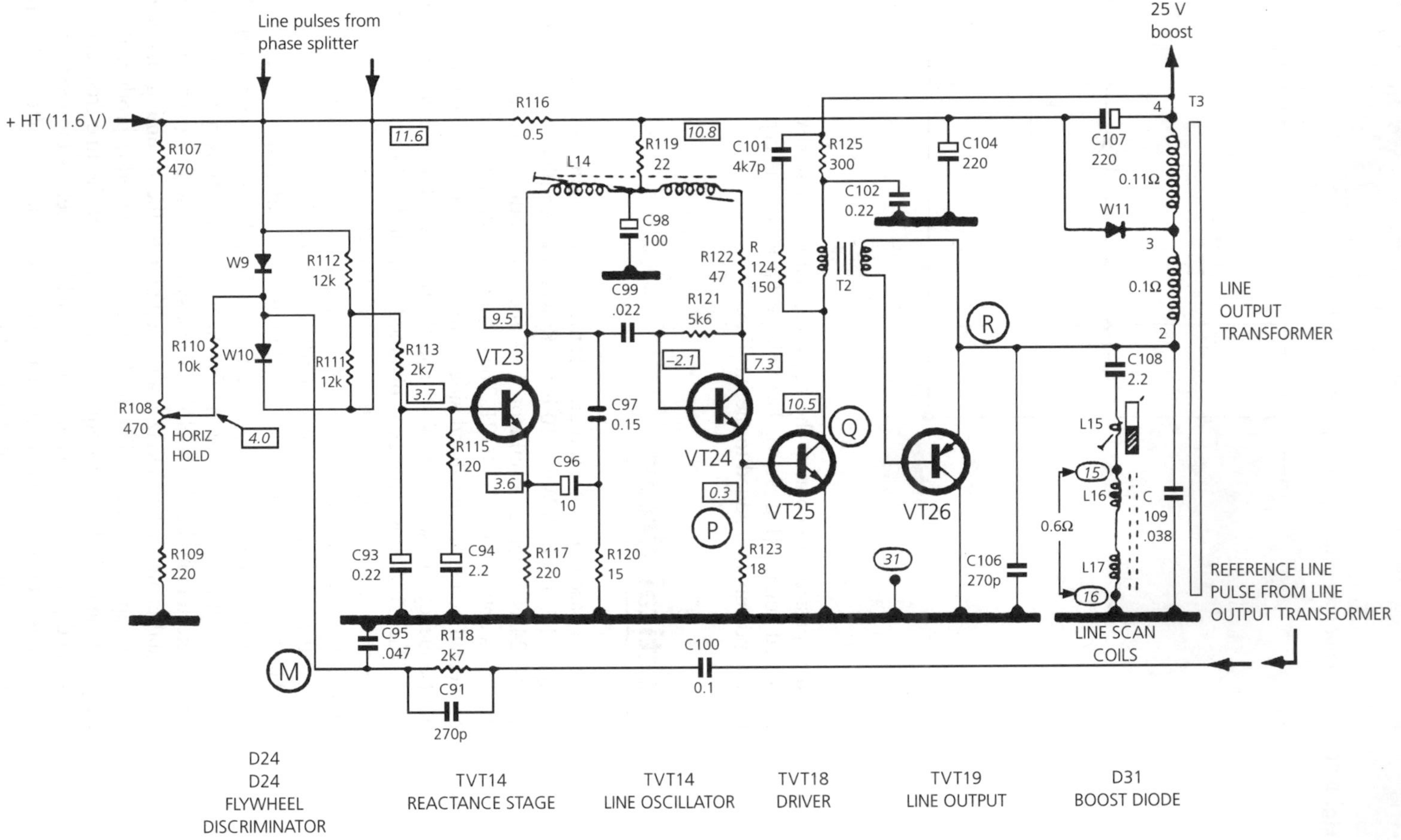

**Fig. 9.18** *Line and field time base*

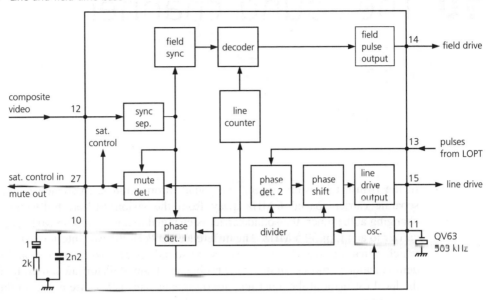

QUESTIONS

1. Sketch the waveform of the line scan signal going into the line scan coils.

2. Explain the purpose of S-correction and show how it may be achieved.

3. Explain the function of the efficiency diode in a line scanning circuit.

4. What is the purpose of the boost diode?

# 10 The sound channel

In the British television system, the sound information is frequency modulated on a separate carrier spaced 6 MHz away from the vision carrier. Following frequency changing at the tuner, the sound carrier is changed to an i.f. of 33.5 MHz and the vision carrier to an i.f. of 39.5 MHz. The difference between the two intermediate frequencies, which remains at 6 MHz, is known as the sound intercarrier frequency and is used for demodulation purposes. In this way, problems of sound distortion caused by the drift in the local oscillator at the tuner unit are overcome. This is because a drift in the oscillator frequencies causes the two i.f.s to change by the same amount, keeping the difference between them constant. Frequency drift does not have the same effect on the vision carrier; this is because the vision carrier is amplitude modulated.

The intercarrier 6 MHz beat frequency is obtained from a device such as a diode which, when used over the non-linear part of its characteristic, produces the sum and difference of two separate input frequencies. The resulting 6 MHz beat frequency retains the frequency modulated information which, when demodulated, reproduces the original sound signal. Ideally the sound intercarrier should have a constant amplitude. However, some amplitude modulation will be present, caused by the vision i.f. which is amplitude modulated. This interference will cause what is known as *vision buzz* on the audio output, unless it is removed by a limiting circuit before demodulation.

A block diagram of a sound channel is shown in Fig. 10.1. The output of the video detector stage is fed into a 6 MHz filter or trap before going into a limiting stage containing one or more clipping (or limiting) amplifiers. After frequency demodulation, the sound signal is amplified by a driver before going into the a.f. output stage and subsequently to the loudspeaker.

**Fig. 10.1** *Sound channel*

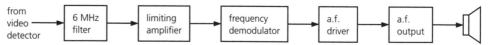

## Frequency demodulators

In frequency modulation the carrier frequency deviates above and below its centre frequency in accordance with the amplitude of the modulating signal. The f.m. detector or demodulator thus has to convert the frequency deviations back into the original signal.

**Fig. 10.2** *(a) Tuned circuit and (b) its phasor diagram*

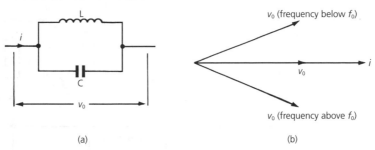

There are two main types of f.m. demodulator that are used in TV receivers: the ratio detector and the quadrature (coincidence) detector. The ratio detector uses discrete components and has the advantage of providing its own rejection of amplitude modulation. Although more complex, the quadrature detector lends itself more easily to i.c. packaging; it is extensively used in modern TV receivers.

The operation of the f.m. detector is based on the fact that the impedance of a tuned circuit is resistive at the resonant frequency but becomes inductive at lower frequencies and capacitive at higher frequencies. Consider the LC circuit in Fig. 10.2(a). At the tuned frequency $f_0$, the circuit is purely resistive with voltage $v_0$ in phase with the current $i$, as shown in Fig. 10.2(b). If the frequency falls below the resonant frequency, voltage $v_0$ leads the current; if the frequency rises above the resonant frequency, the voltage lags the current. The amount of phase shift is determined by the deviation away from the tuned frequency of the circuit. If the input to the circuit is at the sound intercarrier, then provided the circuit is tuned to 6 MHz, the phase shift represents the original modulating sound signal. In the f.m. detector this phase shift is translated into a voltage variation to reproduce the audio signal.

## The quadrature (coincidence) detector

A schematic diagram of a quadrature frequency detector is shown in Fig. 10.3, in which L1/C1 is tuned to 6 MHz. The input to the detector is a clipped sine wave from the preceding limiting stage. The two leads carrying the input signal connected directly to TR1 and TR3 are also applied across the tuned circuit via two capacitors, C2 and C3. With respect to chassis, the two leads carry two clipped sine waves in antiphase. At the centre frequency, L1/C1 is purely resistive with its current $i$ in phase with the voltage across it, $v_0$. However, the input voltage $v_1$ is applied across a circuit consisting of L1/C1 in series with C2 and C3 (Fig. 10.4(a)). Since the reactance of C2 and C3 is large compared with the resistance of the tuned circuit, the current $i$ leads the input voltage $v_1$ by almost 90° (Fig. 10.4(b)). Thus at the centre frequency, $v_1$ and $v_0$ have a phase difference of 90°, usually known as phase quadrature, hence the name of the detector. From Fig. 10.3 it can be seen that current may flow into load resistor $R_L$ provided one of the following two coincidences occur: either TR1 and TR2 are switched on or TR3 and

**Fig. 10.3** *Quadrature (coincidence) detector*

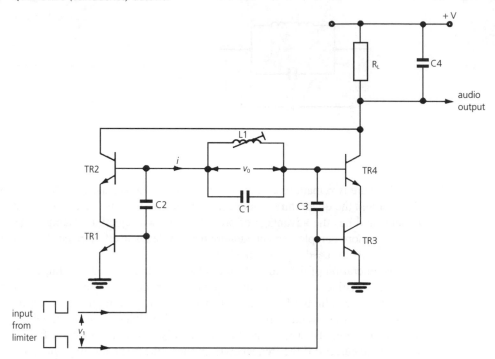

TR4 are switched on. Hence the name 'coincidence' that is sometimes used for this type of demodulator. TR1 and TR3 conduct on alternate half-cycles of input $v_1$. TR2 and TR4, on the other hand, conduct on alternate half-cycles of voltage $v_0$, which is applied to the two transistors. However, since voltage $v_0$ is 90° out of phase with $v_1$, the area of coincidence for TR1/TR2 and TR3/TR4 occurs for one-quarter of a cycle only (90°), as shown in Fig. 10.5. As the input frequency deviates, voltages $v_1$ and $v_0$ are no longer at 90° but either greater or less than 90°. When the phase shift is greater than 90°, the coincidence area increases and more current flows into $R_L$. Conversely, when the phase shift is smaller than 90°, the coincidence area decreases and so does the current through

**Fig. 10.4**

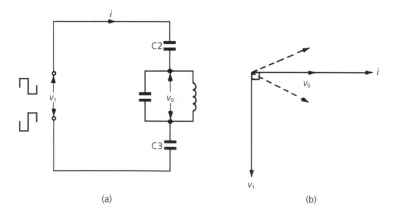

(a)                                        (b)

**Fig. 10.5** *Waveforms for coincidence detector in Fig. 10.3*

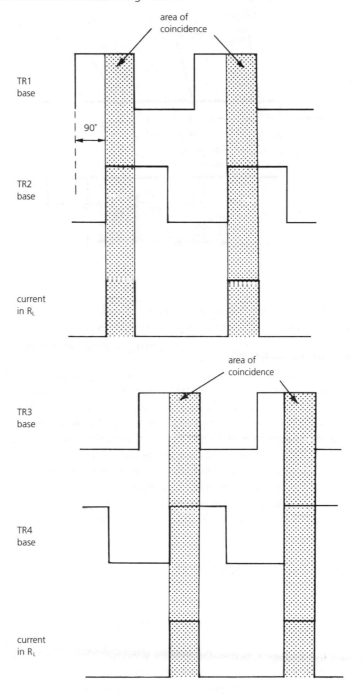

**Fig. 10.6** *Sound processing chip TBA 120 (Philips TX mono chassis)*

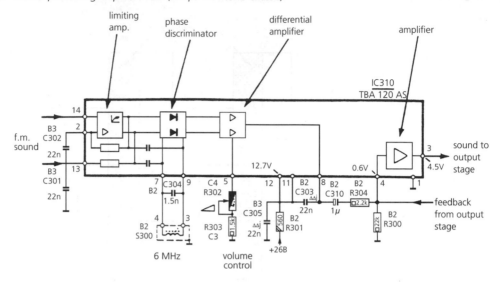

$R_L$. The mean current through load resistor $R_L$ thus varies with the phase shift, which itself changes in accordance with the deviation of the f.m. intercarrier. After smoothing out the ripple by capacitor C4, the output voltage across $R_L$ is an accurate representation of the modulating signal.

The quadrature detector has several advantages. The ripple at the output is twice the frequency of the incoming 6 MHz intercarrier and thus may be easily removed by a capacitor connected across the load resistor. The same capacitor, if correctly chosen, produces the necessary de-emphasis. This type of detector requires a single tuned circuit, making alignment easy to carry out. And finally, since the circuit is basically a switching network, the output depends solely on frequency deviation, not on the amplitude of the input; provided the amplitude is large enough for proper switching, the detector automatically suppresses any a.m. interference.

An i.c. incorporating a quadrature detector is shown in Fig. 10.6; R302 is the volume control and inductor S300/C304 is the detector tuned circuit.

## The ratio detector

The basic circuit for a ratio detector is shown in Fig. 10.7, in which C2 and C3 are for r.f. decoupling and C4 is an a.m. decoupling capacitor. At the centre frequency, the voltages induced in the secondary, $v_a$ and $v_b$, are 90° out of phase with the primary voltage and the voltage $v_t$ across the tertiary winding. The circuit is completely balanced, with diodes D1 and D2 conducting equally to give a zero output. As the frequency deviates away from 6 MHz, the phase difference becomes either greater or less than 90° and the currents through the diodes are no longer equal. An output voltage is produced which is a measure of the deviation.

**Fig. 10.7** *Ratio detector circuit*

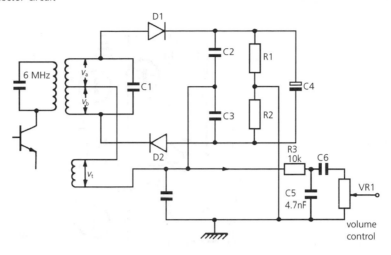

# De-emphasis

It is common for f.m. broadcasting to introduce pre-emphasis at the transmitter; high-frequency audio signals are boosted in comparison with middle and low frequencies. The purpose is to swamp most of the noise, the greater part of which tends to be high frequencies. At the receiving end, the signal must be subjected to de-emphasis by attenuating the treble, or high-frequency, content in a similar but opposite way to the emphasis at the transmitter. Emphasis is carried out by the use of a filter and is expressed in terms of the time constant of the filter. In the UK a pre-emphasis of 50 $\mu$s is used. A de-emphasis filter with a time constant of 50 $\mu$s must therefore be used in the receiver to restore the audio frequencies to their proper relative levels. In Fig. 10.7 R3/C5 is a filter circuit with a time constant of 47 $\mu$s; it provides the necessary de-emphasis.

# A practical audio output stage

The circuit of an audio output amplifier used in a monochrome receiver is shown in Fig. 10.8, in which 3VT1 and 3VT2 form a d.c. coupled driver pair and 3VT3 and 3VT4 are the complementary output transistors. A number of feedback loops are used. The main one, 3R15/3C10, provides d.c. and selective a.c. feedback. Loop 3R12 provides only a.c. feedback to the emitter of 3VT1, with the d.c. feedback path blocked by 3C8.

An audio stage employing i.c. SN76033N is shown in Fig. 10.9. The integrated circuit consists of preamplifier, driver and class B complementary stages. The audio signal from the detector is applied to pin 1 via a tone control network R801/R802/R803/C802/C805/C804. Resistors R804 and R805 provide the d.c. bias for the preamplifier; C803 is the decoupling capacitor. C812 and R810 are used to compensate for the reactive load of the loudspeaker. Integrated circuit amplifiers are capable of very high gains, so their frequency response must be tailored to obtain the desired bandwidth and

**Fig. 10.8** *Sound output stage (Bush)*

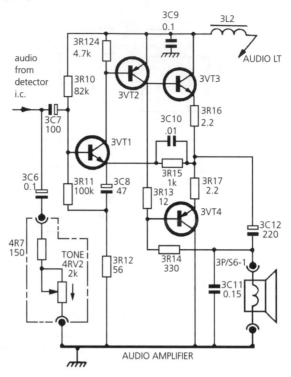

**Fig. 10.9** *Sound output using the SN76033 amplifier chip (Ferguson)*

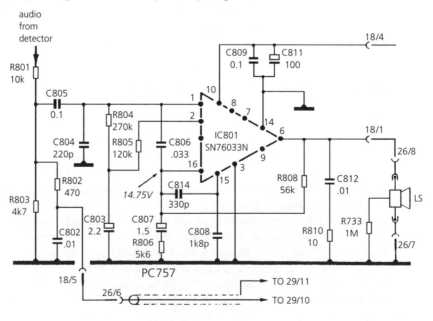

to prevent instability. R806/C807/C806 provide frequency tailoring, and high-frequency compensation is provided by C808. The i.c. does not operate until C807 has charged to approximately half the supply potential. R808 is chosen to provide a suitable time constant so at switch-on the amplifier is muted for a short time to eliminate undesirable noise.

## QUESTIONS

1. What is vision buzz and how is it generated?

2. (a) What is pre-emphasis as applied to f.m. TV sound?
   (b) Show how it may be removed at the receiving stage.

3. List the two main types of f.m. sound detector; give one advantage and one disadvantage of each.

4. What is the bandwidth of the sound component of the TV signal?

# 11 Colour burst processing

Recall that the chrominance information is contained in a 4.43 MHz modulated sub-carrier which forms part of the incoming composite video signal. Colour difference signals B' − Y' and R' − Y' are used to modulate the subcarrier, which is then suppressed to leave two quadrature components, $U$ and $V$. At the receiver, the colour subcarrier is separated from the luminance signal and the two colour difference signals are then recovered by a synchronous demodulator using a phase-locked reference oscillator controlled by the colour burst. The luminance and colour difference signals are then applied to a matrix network, which performs the operation necessary to reproduce the original colour signals, R', G' and B', that can be applied to a colour tube.

Colour decoding consists of four distinct parts: a colour burst processing section, a chrominance signal processing section, a matrix or mixer and a colour drive amplifier.

## Colour burst processing

Colour burst processing separates the burst signal from the chrominance information so it may be used to recreate the subcarrier which has been suppressed at the transmitter. The subcarrier has to be restored in both its frequency and phase to ensure correct colour reproduction. Two subcarriers at 90° to each other have to be produced, one for each colour difference demodulator, B' − Y' and R' − Y'. Furthermore, in the PAL system, the subcarrier for the R' − Y' demodulator has to be phase-reversed on alternate lines. The burst processing section is also used to provide *automatic chrominance control* (*a.c.c.*) as well as the colour killer signal for monochrome-only transmission.

A block diagram for colour burst processing is shown in Fig. 11.1. The colour burst, which consists of about 10 cycles of the original subcarrier, is mounted on the back porch of the line sync. pulse. It is separated from the rest of the composite video by the burst gate amplifier. The burst gate amplifier is turned on by a delayed line flyback pulse. The delay in the flyback pulse is necessary because, having been placed on the back porch, the burst does not coincide with the line sync. pulse; rather, it arrives immediately after it. The delay ensures that the amplifier begins to conduct on the arrival of the burst. The burst gate amplifier allows the burst to go through a phase-locked loop network known as the *automatic phase control* (*a.p.c.*) consisting of the phase discriminator, filter, react-ance stage and the voltage-controlled oscillator. The phase discriminator compares the phase of the burst with the 4.43 MHz output of the voltage-controlled crystal oscillator.

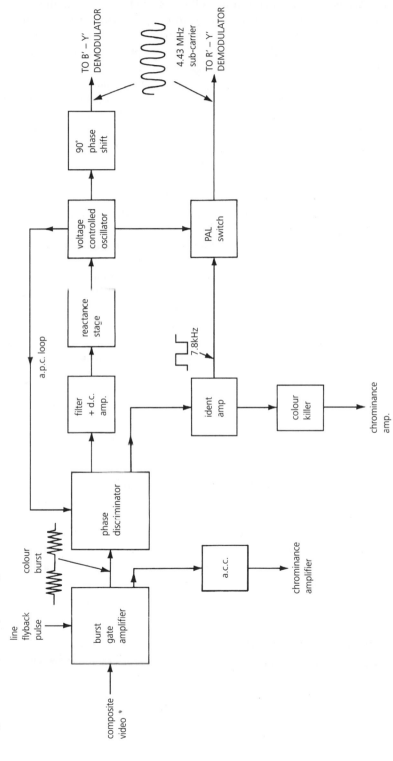

**Fig. 11.1** Colour burst processing

If there is an error, a correction voltage is produced, which after going through a low-pass filter is fed to a reactance stage to bring the frequency and phase of the oscillator into step with the burst. As was explained earlier, the burst signal is not of constant phase but swings ±45° either side of the $-U$ phasor component to convey information of the phase reversal of the $V$ component at the transmitter. For this reason, the phase discriminator must compare the phase of the oscillator with the average phase of the burst signal. Averaging is accomplished by the introduction of a low-pass filter, normally in the form of an a.c. negative feedback at the discriminator stage.

Referring to Fig. 11.1, the subcarrier for the $B' - Y'$ demodulator is obtained by the insertion of a simple 90° phase shift network at the output of the oscillator. The subcarrier for the $R' - Y'$ demodulator must suffer a phase reversal on alternate lines. To achieve this, a square wave at half the line frequency is needed. A component of this frequency is present at the incoming signal because of the ±45° phase change of the colour burst on alternate lines. Since one complete swing of the burst phase takes place every two lines, the frequency of the 'swing' is half the line frequency, i.e. 15 625/2 = 7.8125 kHz, which is normally quoted as 7.8 kHz and referred to as the identification or *ident* signal. After amplification, the ident signal is fed into a PAL switch that reverses the phase of the 4.43 MHz oscillator signal on alternate lines.

Two other functions are derived from the burst processing section: colour killing and automatic chrominance control (a.c.c.). The purpose of the colour killer circuit is to close down the chrominance amplifier path on monochrome-only transmissions to prevent random colour noise appearing on the screen. The presence or otherwise of the colour burst indicates the type of transmission. The ident signal is therefore used to provide a normal bias for the chrominance amplifier, which will be turned off if the signal is absent.

Automatic chrominance control prevents varying propagation conditions from changing the amplitude of the chrominance signal in relation to the luminance. To realise this, the gain of the chrominance amplifier is made variable by a control voltage in a similar way to the control of the gain of the i.f. stage by an a.g.c. signal. The a.c.c. control voltage must be proportional to the amplitude of the chrominance signal. This voltage cannot be derived from the actual chrominance signal during the active picture scan, since this varies in amplitude as the colour information itself changes. It is derived instead by monitoring the amplitude of the colour burst. A fall in the amplitude of the burst signifies an attenuated chrominance. This is corrected when the a.c.c. control voltage into the chrominance amplifier increases its gain, and vice versa.

An alternative method for producing the 7.8 kHz switching signal is to use a bistable multivibrator which is triggered once per line by pulses from the line timebase and synchronised by the ident signal (Fig. 11.2).

## Circuit using discrete components

Figure 11.3 shows the essential elements of a burst processing circuit using discrete components; VT109 is the burst gate amplifier, W106/W107 form the phase discriminator, VT110 is the d.c. amplifier and VT111 is the crystal voltage-controlled oscillator. Burst gate amplifier VT109 is switched on during the burst period by a positive-going delayed

**Fig. 11.2** *Use of a bistable to produce the ident signal*

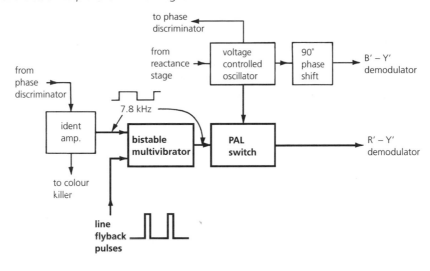

**Fig. 11.3** *Colour burst processing using discrete components (Ferguson 8000 chassis)*

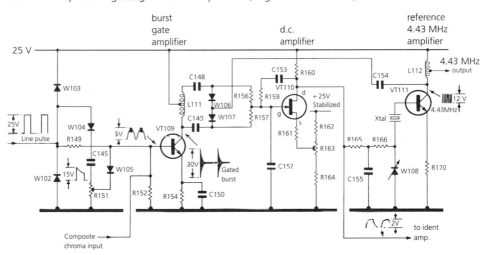

line flyback pulse. The line pulse is clamped by W102 to chassis and by W103 to the 25 V d.c. supply, giving the pulse an amplitude of 25 V. After attenuation by R149 and R151, the line pulse appears at the base of burst gate amplifier VT109, together with the composite chrominance signal including the colour burst. At the same time, the 25 V line pulse forward-biases diode W104, charging C145 via R151. The voltage across R151 rises and then decays as the capacitor begins to charge up. When the voltage across R151 falls to a value lower than the pulse voltage on VT109 base, diode W105 conducts and cuts off VT109. The time constant of C145/R151 is chosen such that VT109 is switched off at the conclusion of the burst. VT109 is held off during the active period of the line scan by the charge on capacitor C150, which holds the emitter positive with respect to the base until the next line pulse arrives. The amplified gated burst is then applied to the phase discriminator. Inductor L111 is centre-tapped, so it will apply the burst in opposite phase across discriminator diodes W106 and W107. The burst is then compared with the signal from the reference oscillator applied to the centre of the diodes. Any difference in phase or frequency between the burst and the oscillator will result in a correction, or error, voltage across C152. The correction voltage is amplified by f.e.t. d.c. amplifier VT110 before going to the reactance stage, varicap diode W108, which controls the frequency and phase of reference oscillator VT111. The averaging of the phase of the swinging burst is carried out by the filtering action of C153 and R159 together with the filtering and storage circuit R165/C155, which prevents sudden changes in oscillator frequency due to the intermittent nature of the burst. VT111 is a conventional crystal-controlled oscillator with the 4.43 MHz subcarrier being selected by self-tuned inductor L112. Resistor R151 sets the timing of the burst gate and R163 provides the initial setting of the oscillator frequency and phase.

## The use of integrated circuits

In modern receivers, integrated circuits are extensively used for all major functions of the decoding channel. They invariably employ different circuit techniques to those used in discrete circuitry. In addition, a variety of functions are included in any single i.c., such as subcarrier regeneration, demodulation and mixing. Figure 11.4 shows the

**Fig. 11.4** *Reference combination TBA 450*

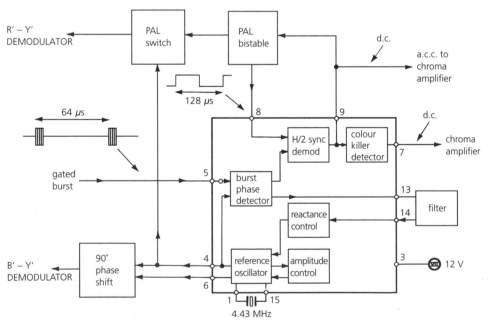

**Fig. 11.5** *Colour burst processing employing chip TBA 395Q (Ferguson 9000 chassis)*

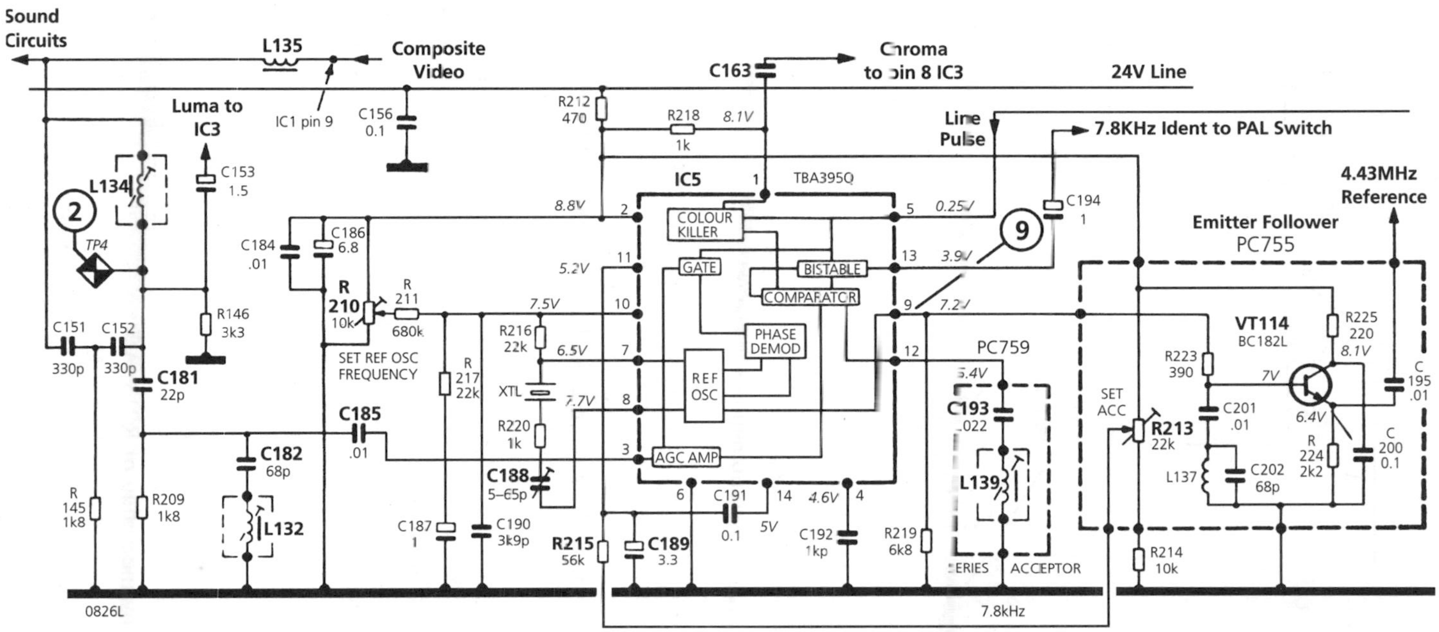

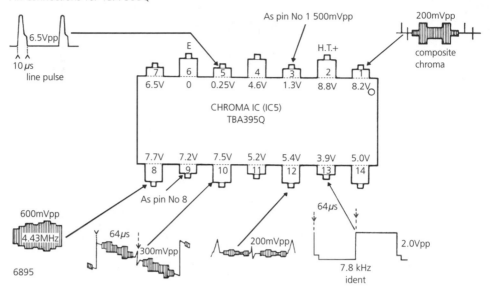

**Fig. 11.6** *Pin connections for TBA 395Q*

TBA 450, an early Mullard i.c. for automatic phase control of the subcarrier; it is known as the reference combination i.c. The reference oscillator has its frequency determined by a 4.43 MHz quartz crystal connected externally between pins 1 and 15. An output of the oscillator is compared with the gated burst signal by the burst phase detector. Any difference in phase or frequency produces an error voltage, which is fed back to the oscillator via a reactance stage and an externally connected filter. The amplitude of the output from the reference oscillator is stabilised by an amplitude control unit incorporated within the i.c. Two antiphase subcarrier signals are available at pins 4 and 6, and they are used to provide the quadrature subcarriers for the R′ − Y′ and B′ − Y′ demodulators. The H/2 or half-line frequency synchronous demodulator compares the phase of the 7.8 kHz ident signal from the burst phase detector with the phase of the switching square wave produced by the bistable fed into pin 8. If the bistable phase is incorrect, the output voltage of the H/2 sync. demodulator (pin 9) rises to force the bistable to miss a step. When synchronism is obtained, the colour killer detector output at pin 7 rises and opens up the chrominance channel.

A circuit employing the TBA 395 is shown in Fig. 11.5. The waveforms present at each pin are shown in Fig. 11.6. The composite video is fed via harmonic suppressor and 6 MHz filter L134/C151/C152 to the chrominance high-pass filter L132/C182/C181. The resulting chroma signal is coupled into pin 3 of the integrated circuit via C185. The a.c.c. (a.g.c.) amplifier maintains the chrominance signal at a constant level, predetermined by the setting of R213. After passing through the colour killer stage, it emerges at pin 1. C193 and L139 connected to pin 12 are tuned to 7.8 kHz to make the comparator sensitive at this frequency. The subcarrier output appears on pin 9 and is applied via emitter follower VT114 to the R′ − Y′ demodulator and via a phase shift to the B′ − Y′ demodulator.

**Fig. 11.7**  *Using an 8.86 MHz reference frequency*

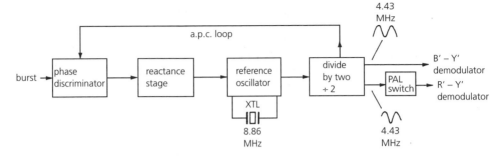

**Fig. 11.8**

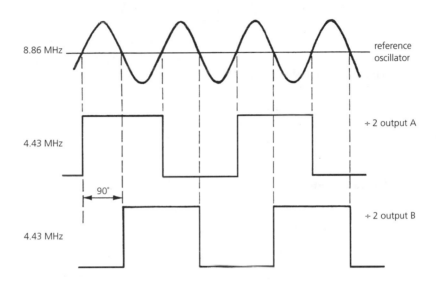

## 8.86 MHz reference oscillator

The 90° phase-shift network used to provide the quadrature subcarrier waveform suffers from phase drift due to temperature change and ageing components. A more precise 90° phase may be provided by the use of a reference frequency twice the subcarrier, i.e. 8.86 MHz. This technique dispenses with the phase-shift network altogether (Fig. 11.7). The 4.43 MHz subcarrier frequency is obtained by dividing the output of the reference oscillator by 2. The output of the ÷2 network is then fed back to the phase discriminator to complete the automatic phase control (a.p.c.) loop. The divider provides two quadrature subcarriers, which may then be fed directly to the B′ − Y′ demodulator and via a PAL switch to the R′ − Y′ demodulator.

The divider may consist of two bistables, one is made to trigger on the positive half-cycle and the other on the negative half-cycle of the oscillator frequency (Fig. 11.8). Each bistable divides the 8.86 MHz by 2 to provide a subcarrier of 4.43 MHz, and at the same time it provides a 90° phase difference between them.

**Fig. 11.9** *Burst-phase equaliser*

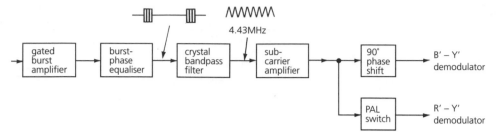

## Passive subcarrier regeneration

It is possible to obtain the subcarrier directly from the burst, thus avoiding the use of a reference oscillator and its associated circuitry. If the burst is to be used directly, it must have a constant phase. The PAL system, however, provides for a 'swinging' burst which suffers a 90° phase shift, line by line. A burst-phase equaliser is therefore necessary (Fig. 11.9). A subcarrier is then extracted by a very narrow, highly selective crystal band-pass filter. The output of the filter is a continuous 4.43 MHz sine wave which, when amplified, is ready to provide the quadrature subcarriers to the two demodulators as shown.

### QUESTIONS

1. What is the frequency of the colour ident signal and what is its relationship to the line frequency?

2. In the PAL receiver, why is it necessary to have a PAL switch?

3. State the function of the colour killer signal and explain how it is produced.

4. Colour burst processing uses an a.p.c. What is it and what is its function?

# 12 Chrominance processing channel

Recall that the chrominance information forms part of the composite video, from which it has to be separated before demodulation. The chrominance information is centred on a 4.43 MHz subcarrier with a bandwidth limited to $\pm 1$ MHz. The first task of the chrominance channel is therefore to separate the chrominance from the composite video. This is carried out by the bandpass chrominance amplifier (Fig. 12.1). Before demodulation can take place, the composite chrominance signal must be separated into its two component parts, $U$ and $V$, each of which must be demodulated separately to recreate the original colour difference signals. The weighted colour difference signals, $U$ and $V$, are separated from each other by a unit consisting of a delay line (usually known as PAL delay line) and an add/subtract network. Two separate signals, $U$ and $V$, are produced, which are fed to their respective demodulators, $B' - Y'$ and $R' - Y'$. Each demodulator is fed with a 4.43 MHz signal at the correct phase from the reference oscillator and burst channel. Two gamma-corrected colour difference signals, $B' - Y'$ and $R' - Y'$, are thus obtained. The third colour difference signal, $G' - Y'$, is obtained from the first two by the $G' - Y'$ network as shown. The three colour difference signals, together with luminance signal $Y'$, are then fed into the RGB matrix. By adding $Y'$ to the three colour difference signals, the original gamma-corrected $R'$, $G'$ and $B'$ colours are reproduced, and after amplification they are fed directly into the appropriate c.r.t. gun. A luminance delay line is inserted into the luminance signal path to ensure that both signals arrive at the matrix at the same time.

The chrominance channel must also provide a facility for some or all of the following functions:

- Colour kill to turn the chrominance amplifier off during monochrome-only transmission.
- Manual saturation (or colour) control to allow the user to change the colour intensity of the display by varying the gain of the chrominance amplifier.
- Automatic chrominance control (a.c.c.) which varies the gain of the chrominance amplifier.
- Burst blanking to turn the amplifier off during the subcarrier burst. Failure to do this will result in a greenish striation appearing on the left side of the screen.
- Intercarrier sound rejection. The 6 MHz sound intercarrier must be removed by one or more 6 MHz traps in the amplifying stage.
- D.c. clamping to reintroduce the d.c. level lost during the processing channel.

**Fig. 12.1** *Chrominance processing*

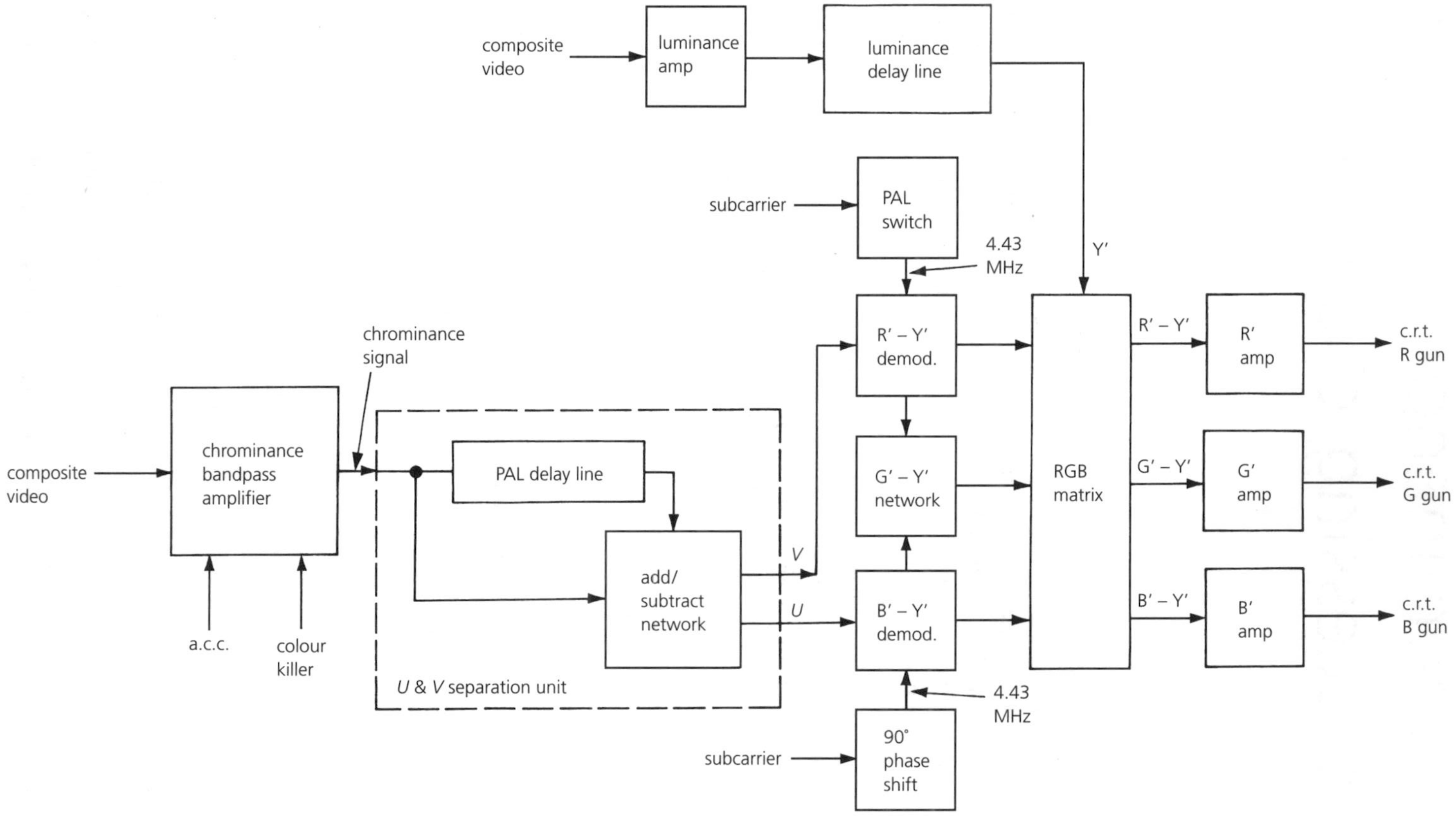

Where a.c. coupling is used throughout the channel, d.c. clamping is carried out at the RGB drive amplifier stage. However, in modern receivers d.c. coupling is employed for the early part of the processing system. In these cases, d.c. clamping is used at an earlier stage. It is necessary to d.c. clamp all three colour signals to ensure a common black level for the red, green and blue guns. Any drift in the d.c. level of one amplifier with respect to any of the other two would lead to overemphasis, producing an unwanted colour tint. For this reason, driven black level d.c. clamping is used in colour receivers.

# Colour-bar test display

Before embarking on a detailed examination of the various parts of the chrominance channel, a study of the colour-bar test display and its associated signals is necessary.

The colour-bar signal provides a rigorous test for colour transmission and reception. The BBC transmits a 95% saturated and 100% amplitude colour-bar signal. A saturation of 95% indicates that each colour is 95% pure hue with 5% dilution of white. A 100% amplitude indicates that at least one colour is at maximum amplitude. This is the colour-bar display that will now be considered.

A standard colour-bar display consists of eight vertical bars of uniform width: three primary colours, three complementary colours, white and black. They are arranged in descending order of luminance from left to right as follows:

*white yellow cyan green magenta red blue black*

A monochrome receiver displays the luminance component of the colour-bar display (Fig. 12.2). This is known as the grey scale, with peak white on the left followed by grey stripes which grow progressively darker until black is reached on the right-hand side. Notice that the luminance steps are not uniform.

The RGB components of the colour-bar display are shown in Fig. 12.3; they represent the waveforms at the red, green and blue guns of the cathode ray tube.

**Fig. 12.2** *Luminance components of colour bars*

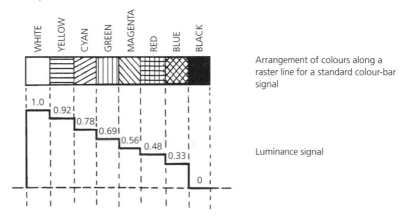

Arrangement of colours along a raster line for a standard colour-bar signal

Luminance signal

**Fig. 12.3** *RGB components of colour bars*

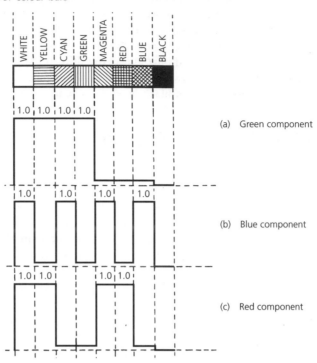

(a) Green component

(b) Blue component

(c) Red component

**Fig. 12.4** *Colour difference components of colour bars*

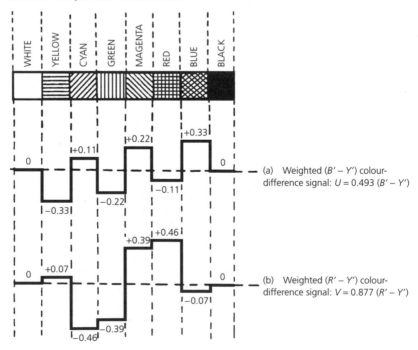

(a) Weighted $(B' - Y')$ colour-difference signal: $U = 0.493 (B' - Y')$

(b) Weighted $(R' - Y')$ colour-difference signal: $V = 0.877 (R' - Y')$

Fig. 12.5

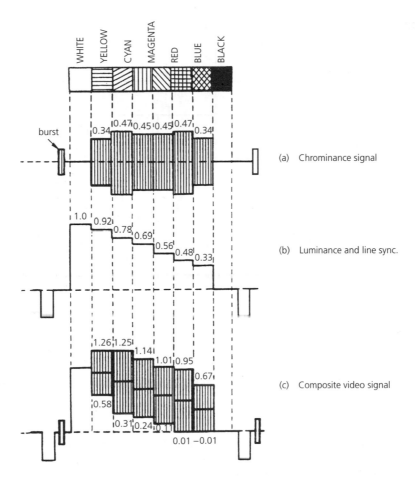

(a) Chrominance signal

(b) Luminance and line sync.

(c) Composite video signal

The actual components of the chrominance signal, i.e. the colour difference signals, may be calculated by summing the amplitudes of the various components at each bar. First the RGB signals are gamma-corrected to produce R′, G′ and B′. Then, by subtracting each one from the luminance Y′, gamma-corrected colour difference signals are obtained. The B′ − Y′ and R′ − Y′ colour differences are then weighted to produce $U$ and $V$ respectively (Fig. 12.4). These components are amplitude modulated using two 4.43 MHz quadrature subcarriers. The carrier is then suppressed and the phasors representing the two components are added to produce the complete colour-bar chrominance signal shown in Fig. 12.5(a). This chrominance signal is then mounted on the luminance signal in Fig. 12.5(b) to produce the complete composite video signal for a colour-bar display, as shown in Fig. 12.5(c).

## Chrominance amplifier

The primary function of the chrominance amplifier is to extract the chrominance information from the composite video and amplify it to a level that can drive the delay line and

**Fig. 12.6** *Two-stage chrominance amplifier (Bush)*

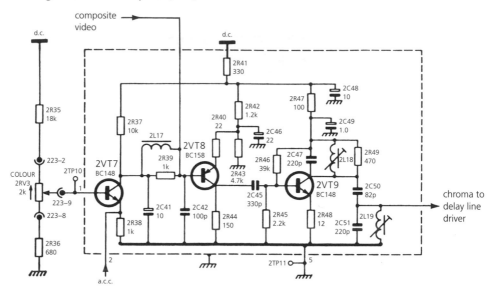

the add/subtract network. It is essentially a bandpass amplifier consisting of two or more stages of amplification. A typical two-stage chrominance amplifier used in early colour receivers is shown in Fig. 12.6, in which the tuned pair 2L18/2C47 and 2L19/2C51 with top capacitance coupling 2C50 provide the required bandpass of ±1 MHz. Two RC-coupled amplifiers are used, 2VT8 and 2VT9. Automatic chrominance control is achieved using the principle of reverse a.g.c., by which the gain of the npn transistor 2VT7 is controlled. The a.c.c. control voltage is applied to the emitter of d.c. amplifier 2VT7. The amplified control voltage at the collector of 2VT7 is then applied to the base of 2VT8 via 2L17 to determine its bias and hence its gain. Similarly, adjustment of the colour saturation control 2RV3 causes the 2VT8 bias voltage and hence the gain to change accordingly.

## *U* and *V* separation

Figure 12.7 shows how the PAL-D system separates the $U$ and $V$ components; the chrominance signal of the previous line is stored or delayed for one line duration (64 $\mu$s). The delayed signal is then added to and subtracted from the signal of the current line to produce separate $U$ and $V$ signals. The precise value of the required delay is 63.943 $\mu$s, slightly less than one line duration; this is because it must correspond to a whole number of subcarriers.

Let us assume that the chrominance signal going into the delay line driver in Fig. 12.7 is unswitched, i.e. $U + V$. The chrominance signal at the output of the delayed line is that of the preceding line, i.e. switched line $U - V$. Addition of the two signals gives

$$(U + V) + (U - V) = 2U$$

whereas subtraction gives

**Fig. 12.7** *U and V separation*

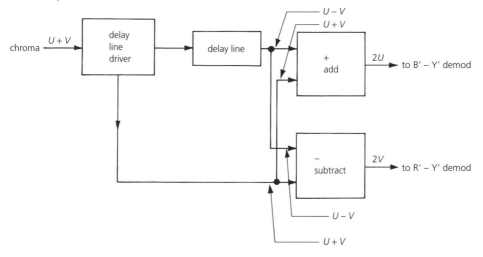

$$(U + V) - (U - V) = 2V$$

Alternatively when the input to the line driver is a switched line $(U - V)$, the stored signal is unswitched $(U + V)$. The result of the adder remains as

$$(U - V) + (U + V) = 2U$$

But the subtraction is reversed:

$$(U - V) - (U + V) = -2V$$

The $U$ and $V$ separator thus retains the phase reversal of the $V$ component and also removes the phase error.

## The ultrasonic delay line

To provide the comparatively long delay required by the PAL-D receiver, it is not practical to use electrical methods. Instead, ultrasonic delay lines are used; a transducer converts the electrical signal into an ultrasonic wave which is made to travel along a solid glass block. Another similar transducer at the other end reverses the process and the chrominance signal is recovered. The speed of propagation of ultrasonic waves is about 2750 m/s, far lower than electric waves in a conductor. Provided a suitable length of the glass block is used (about 17.6 cm), a delay of 64 $\mu$s may be obtained.

An early construction of an ultrasonic delay line is shown in Fig. 12.8, in which a single reflecting surface is used. The length of such a delay line is approximately 8.8 cm (1/2 × 17.6 cm). Modern delay lines employ multiple reflecting surfaces (Fig. 12.9). The required path length is obtained by arranging for the ultrasonic wave to reflect back and forth within the delay line. A delay line with smaller dimensions is therefore possible.

It is essential for the delay time of the line to remain substantially constant at 63.943 $\mu$s. A specially selected glass is thus used in the manufacture of the delay line.

**Fig. 12.8** *Ultrasonic delay line*

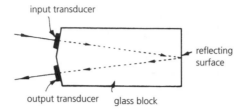

**Fig. 12.9** *Multiple surface ultrasonic delay line*

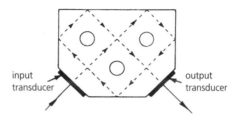

**Fig. 12.10** *Add/subtract network*

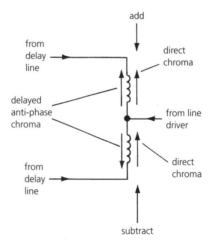

# *U* and *V* separator circuit

A typical add/subtract network is shown in Fig. 12.10. It consists of a centre-tapped transformer fed with a delayed chrominance signal across its two terminals. Two equal and opposite (antiphase) delayed chrominance signals thus develop across each half of the transformer winding. The direct (undelayed) chrominance signal is fed into the centre tap so that it adds to the delayed signal in one half (top) and subtracts from the delayed signal in the other half (bottom). These two signals are then used to feed the $R' - Y'$ and $B' - Y'$ demodulators respectively.

**Fig. 12.11** *Circuit incorporating chrominance chip SN76226 (Ferguson 9000)*

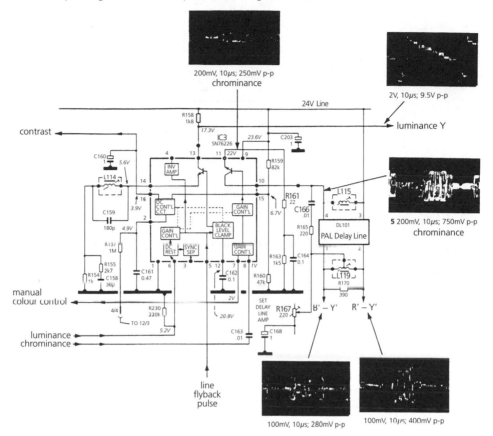

200mV, 10μs; 250mV p-p
chrominance

2V, 10μs; 9.5V p-p

luminance Y

5 200mV, 10μs; 750mV p-p
chrominance

line
flyback
pulse

100mV, 10μs; 280mV p-p

100mV, 10μs; 400mV p-p

# Use of integrated circuits

A circuit using the SN76226 chip incorporating a chrominance amplifier and $U - V$ separator is shown in Fig. 12.11. The chrominance signal applied to pin 8 of the i.c. is amplified by a two-stage gain-controlled amplifier. The gain of the first stage is set by a manual colour or saturation control at pin 7. The gain of the second stage is adjusted by the contrast control applied at pin 16, which also adjusts the gain of the luminance amplifier. The chrominance signal emerges at pin 10, the collector of an internal transistor. The gain of this transistor is determined by its emitter resistor connected externally at pin 11. From pin 10 the chrominance is fed to the PAL delay line. The direct signal is taken from L115 via C166 to the centre of the bifilar L119. Amplitude and phase adjustments are provided by R167, L115 and L119. Black level clamping is driven by line pulses applied to pin 5 of the chip. The output from the black level clamp is fed into the base of an internal transistor to provide the necessary d.c. restoration for the luminance signal. The SN76226 also incorporates a sync. separator with line and field pulses appearing at pin 3.

**Fig. 12.12**  *Colour difference processing*

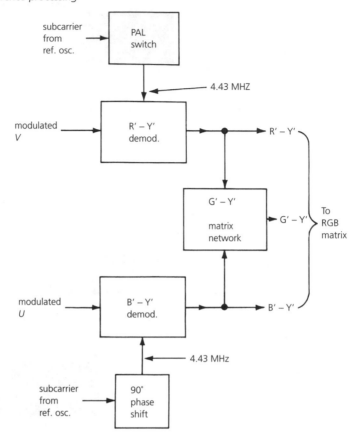

## Colour difference demodulators

The weighted chrominance components $U$ and $V$ from the separator unit are applied to their individual colour difference demodulators to obtain $B' - Y'$ and $R' - Y'$. The third colour difference signal, $G' - Y'$, is obtained by a $G' - Y'$ matrix (Fig. 12.12). For the $B' - Y'$ demodulator, the subcarrier is shifted by 90°; for the $R' - Y'$ demodulator the subcarrier is phase reversed every line through the PAL switch.

Recall that the $U$ and $V$ components of the chrominance signal are the weighted colour difference signals, whereby

$$U = 0.493(R' - Y')$$

and

$$V = 0.877(R' - Y')$$

It follows that, before $B' - Y'$ and $R' - Y'$ are recovered, the weighted components must be deweighted. This is usually carried out at the demodulation stage by the inclusion of colour difference amplifiers which provide the $B' - Y'$ channel with more gain than

**Fig. 12.13**  *Colour difference demodulation*

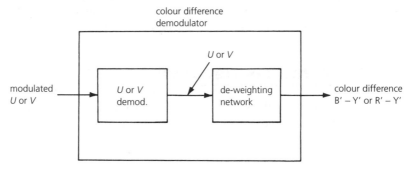

**Fig. 12.14**  *Operation of colour difference demodulator*

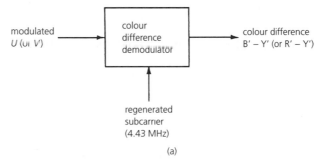

(a)

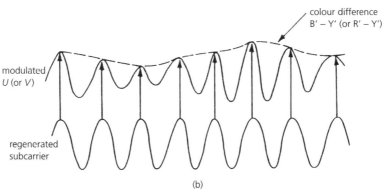

(b)

the R′ − Y′ channel. The whole unit is then known as the colour difference (B′ − Y′ or R′ − Y′) demodulator (Fig. 12.13).

In amplitude modulation the information is contained in the change of the amplitude of the peak of the carrier. When the carrier is suppressed, the modulating information continues to reside in the changing amplitude of the modulated signal. To recover the original information, the amplitude of the modulated signal has to be detected when the carrier is at its peak. In order to do this, a synchronous detector is used. In essence the colour difference synchronous demodulator is a switching or sampling device which detects the level of the incoming modulated *U* (or *V*) signal every time the regenerated subcarrier is at its positive peak (Fig. 12.14).

Unlike video demodulation, where the use of synchronous detectors is a preferred option, the suppressed subcarrier amplitude modulation employed in colour TV transmission makes it essential to use synchronous demodulators.

## G′ – Y′ matrix network

The proportions of R′ – Y′ and B′ – Y′ components which are necessary to obtain the third colour difference component G′ – Y′ may be mathematically determined as

$$G' - Y' = 0.51(R' - Y') - 0.186(B' - Y')$$

However, if G′ – Y′ matrixing takes place before deweighting of the $U$ and $V$ components is carried out, different proportions are necessary:

$$G' - Y' = -0.29(R' - Y') - 0.186(B' - Y')$$

Since the required R′ – Y′ and B′ – Y′ levels are both less than unity, they may be derived by using a simple resistor network such as Fig. 12.15. The ratios of $R_x$, $R_y$ and $R_z$ are chosen to provide the correct proportions of B′ – Y′ and R′ – Y′.

**Fig. 12.15** *Simple resistor matrix*

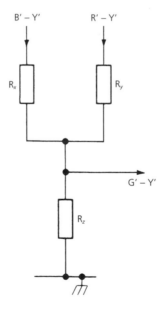

## Luminance delay

The next stage of the chrominance processing channel is the addition of the luminance signal to the colour difference signals to obtain the three gamma-corrected primary colours, R′, G′ and B′. At this point the chrominance and luminance signals must be in step.

For this to happen, the time taken by the two signals to pass through their amplifiers and associated circuitry must be equal. The chrominance signal, which has a relatively narrow 1 MHz bandwidth, suffers a greater delay than the luminance, which has a bandwidth of 5.5 MHz. To compensate for this, an extra delay has to be introduced into the luminance signal path. A luminance delay of 0.5–1.0 $\mu$s is thus inserted before the luminance is fed into the RGB matrix.

A typical delay consists of a single solenoid wound over earthed metal strips. The cut-off frequency is designed to fall well above the highest luminance frequency, and the fall in output at the high-frequency end is compensated by 'free' metal strips positioned along the coil. The delay line must be properly terminated at each end by its characteristic impedance to avoid reflection of the luminance signal from one end to the other.

The absence of a delay in the luminance signal results in a double image on the screen, one image due to the luminance and the other due to the chrominance signal, an effect similar to ghosting. Multiple images will also be observed if the delay line is not properly terminated.

Modern delay line constructions include an internal 4.43 MHz subcarrier rejector circuit, thus dispensing with yet another alignment adjustment at the manufacturing stage.

## RGB matrix

At the tube a picture is produced by modulating the three c.r.t. electron beams using the gamma-corrected R', G' and B' signals. These are derived by the addition of the colour difference signals to the gamma-corrected luminance signals Y':

$$Y' + (R' - Y') = R'$$
$$Y' + (G' - Y') = G'$$
$$Y' + (B' - Y') = B'$$

During addition, the difference in the bandwidth of the luminance signal and the colour difference signals introduces *glitches* or notches at points of fast colour transition. The narrow bandwidth of the colour difference signals restricts the rise time of rapidly changing signals. When they are added to the rapidly changing luminance signal, a glitch is introduced (Fig. 12.16). On a colour-bar display, the effect of the glitch appears as a dark band between the colour bars.

The addition may be carried out by a special matrix before the R', G' and B' colour signals are applied to the cathode of the c.r.t., a technique known as *direct drive*. Alternatively, matrixing may be carried out by the tube itself, a technique known as *colour difference drive*. Colour difference drive involves feeding the colour difference signals to the control grids of the c.r.t. while applying a negative luminance signal −Y' to the cathodes. Applying a negative luminance to the cathode is equivalent to applying a positive luminance to the grid; it results in the mathematical addition of the two signals.

In the colour difference drive, separate colour difference and luminance signals are fed into different electrodes of the c.r.t. This creates problems with timing, accentuated

**Fig. 12.16**

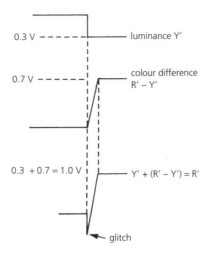

because the two signals have different bandwidths, 5.5 MHz for the luminance and 1 MHz for the colour difference. Further problems arise from the relative sensitivities of the electrodes, which have to be maintained the same for all levels of the input signals. Direct RGB drive overcomes these problems by performing the necessary matrixing close to the point of demodulation. For these reasons, direct RGB drive is used in modern TV receivers.

## Beam limiting

Most TV receivers employ some form of beam current limiting to ensure the electron emission (beam current), hence the brightness, does not exceed a predetermined limit. Beam limiting is optional for monochrome TVs but essential for colour TVs. A very high beam current overloads the line output amplifier and the e.h.t. tripler, causing deterioration in focus; it overdrives the c.r.t., causing limited highlights; and it leads to excessive power dissipation in the mask within the tube, which may cause misconvergence of the three primary colours on the screen.

Beam limiting involves sampling or monitoring the strength of the beam current directly at the cathode, or indirectly by monitoring the d.c. current taken at the line output stage or the e.h.t. winding. The beam current itself is then controlled by reducing the black level (i.e. the brightness) or the amplitude of the luminance signal (i.e. the contrast) or both.

The circuit in Fig. 12.17 shows a typical beam limiter arrangement used in colour TV receivers; the e.h.t. current is monitored by a diode D1 placed between the earthy end of the e.h.t. overwind and chassis. Two types of current flow through D1: forward current $I_D$ of about 600 $\mu$A to chassis, caused by voltage $V_{cc}$, and beam current $I_B$ flowing in the opposite direction. When the beam current goes above 600 $\mu$A, D1 switches off and its anode goes negative; this reduces the base voltage of emitter follower TR1, hence it varies the brightness/contrast controls. For larger tubes, the present beam current limiting is raised to 1 mA. C1 and C2 are decoupling capacitors.

**Fig. 12.17** *Beam limiter*

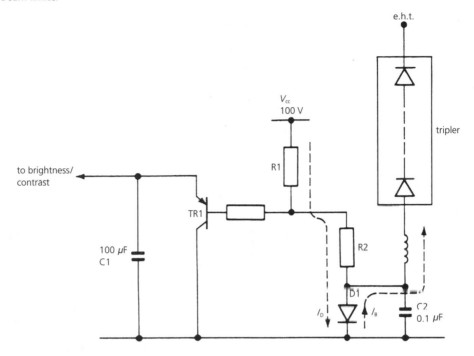

# Video output stage

With RGB drive, the output stage must be able to deliver a peak-to-peak signal of 80–150 V to the c.r.t. cathode. The signal drive to each gun is different because of the different efficiencies of the electrodes; the red gun requires the largest drive. The large signal drive requires a high d.c. supply voltage and often two transistors are connected in series to share this high voltage. To ensure adequate bandwidth, series peaking coils may be employed. Power transistors in class A configuration are used with the necessary heat sinks to deliver the relatively large power necessary to drive the red, green and blue guns of the c.r.t. A small load resistance or an emitter follower buffer are used to ensure low output impedance. This low impedance allows for fast charge and discharge of the c.r.t.'s cathode input impedance, giving good frequency response at the upper end of the bandwidth.

Output stages are normally mounted on the c.r.t. base panel to remove the bandwidth limitations associated with long leads.

Early output stages were also used for matrixing by feeding the luminance to the base and the colour difference signal to the emitter of the transistor. Black level clamping and brightness control of the three guns were also incorporated at the video output stage as well as grey scale adjustment. The purpose of grey scale adjustment, or *tracking*, is to ensure that a purely monochrome picture has no traces of colour tint at any level of brightness from lowlights to highlights across the grey scale display. With the introduction of more advanced integrated circuits, matrixing is now carried out separately and clamping is introduced at the luminance signal channel before matrixing.

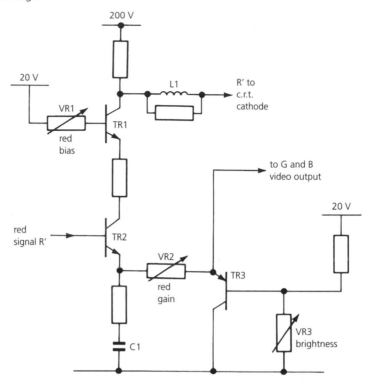

**Fig. 12.18** *Video output stage*

Modern integrated circuits provide complete luminance and colour processing on a single chip, including such functions as blanking, black level clamping, beam limiting, and brightness and contrast control. RGB buffer amplifiers are also included to provide large colour drives (about 3 V) for the output stages. Some i.c.s include automatic grey scale correction.

Figure 12.18 shows the essential elements of a video output stage suitable for driving a precision-in-line (PIL) tube. Only the red stage is shown since the other two stages are identical. TR1 and TR2 are connected in cascode to share the 200 V d.c. supply and to provide the necessary controls of the output signal: the biascontrol VR1, which determines the cut-off point of the gun (lowlights), and the gain control VR2, which controls the highlights. The bias control is necessary since the A1 electrodes in a PIL tube are strapped together and must be adjusted at the video stage. The base voltage of TR1 is set by VR1, which determines the collector potential and hence the potential of the red cathode. TR2 is driven by the red signal R′ from a preceding i.c., which also provides the d.c. base bias voltage. The gain of TR2 is set by VR2, which varies the d.c. current and hence the gain of the amplifier. This control and similarly located preset resistors in the green and blue output stages are used to set the individual drives of the three guns when carrying out the grey scale highlight corrections. TR3 provides the path for the TR2 d.c. current. By changing VR3, the current through TR2 and TR1 is varied and with it the red cathode potential. Since TR3 also forms the

**Fig. 12.19** *Complete RGB output stage (Ferguson 9000 chassis)*

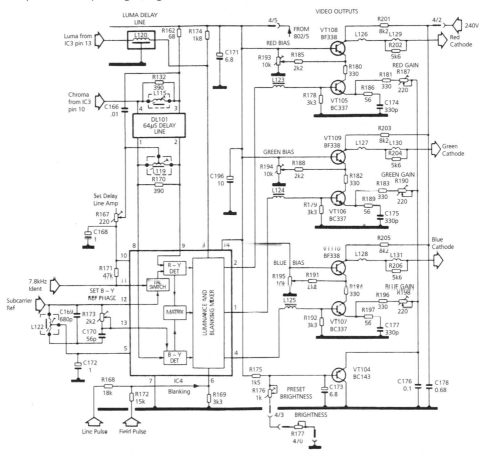

d.c. path for the green and blue cascode arrangements, VR3 causes the d.c. potential of the three cathodes to vary together, thus providing brightness control. High-frequency response is maintained by peaking series coil L1 at the output, VR2 and TR3 provide negative feedback for low and medium frequencies. High frequencies are decoupled via C1 to maintain the gain at the upper end.

A practical circuit employing cascode output stages is shown in Fig. 12.19, in which IC4 (SN76227) provides R′ − Y′ and B′ − Y′ demodulation, G′ − Y′ matrixing, RGB matrixing and blanking. The chrominance signal is fed to the chroma delay line DL101 and subsequently split into B′ − Y′ and R′ − Y′ by bifilar L119; the two difference signals are fed to pins 8 and 9 of IC4 respectively. Amplitude and phase adjustments are provided by R167, L115 and L119. The luminance signal is fed to pin 3 of IC4 via luminance delay line L120. Within IC4 the colour difference signals are demodulated and matrixed to yield the G′ − Y′ component. The three colour difference signals are then mixed with the luminance to provide the R′, G′ and B′ signals at pins 2, 1 and 4 of IC4 respectively. Line and field blanking pulses applied at pin 6 are also added to the colour signals at the mixing stage. The RGB signals are then fed to their respective

**Fig. 12.20** *Circurt Showing RGB output stage*

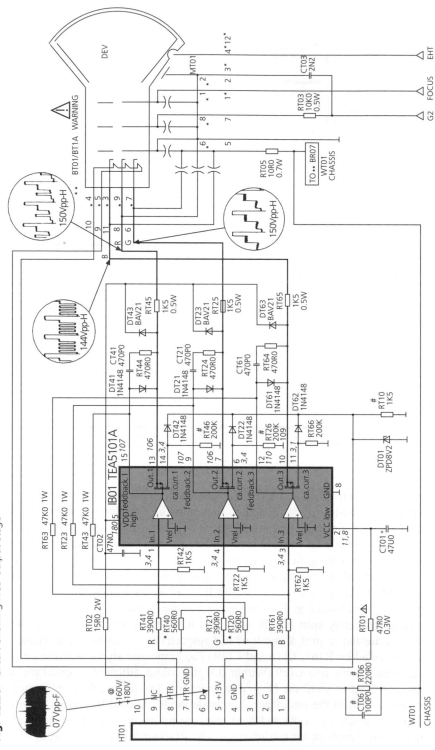

**Fig. 12.21** *Video output stage with emitter follower at the output*

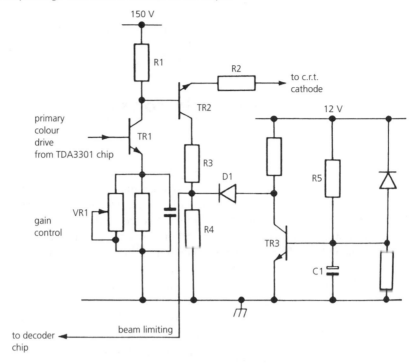

output stages: VT108/VT105, VT109/VT106 and VT110/107. Bias presets R193, R194 and R195 provide lowlight controls; gain presets R187, R190 and R198 provide the highlight controls. Brightness control is achieved by returning the video output stages to chassis via VT104. The bias of VT104 is varied by R176 (preset brightness) and R177 (brightness) connected to the base of the transistor.

An RGB output stage using chip TEA 5101A is shown in Fig. 12.20. The chip is fed with two d.c. supply voltages: VCC low (pin 2) and VDD high (pin 5). R, G and B signals entering the chip at pins 1, 4 and 3, respectively, are amplified and buffered by the appropriate f.e.t. stage. CRT drive signals are obtained at pins 13, 7 and 10.

Improved bandwidth may be obtained by including an emitter follower buffer transistor TR2 at the output of the video output stage (Fig. 12.21). A single class A transistor amplifier TR1 is used in conjunction with the TDA 3301 colour decoder chip; this provides a 3 V peak-to-peak RGB drive for the output transistors, which have maximum dissipation of 400 mW.

The circuitry associated with TR3 (Fig. 12.21) is a *slow start* switching circuit that overrides the system at switch-on. When the receiver is first switched on, TR3 is off and its collector potential is at 12 V. D1 conducts, taking TR2 collector to approximately 12 V, which causes the beam limiting system in the decoder chip to come into operation, cutting the beam current. After a short period of time (determined by time constant C1R5), capacitor C1 charges up through R5, and when it reaches a certain level, TR3 turns on, bringing its collector down to chassis potential. D1 is now reverse-biased, isolating TR3 from the rest of the circuit and releasing the beam current. The potential at the junction of R3/R4 is fed to decoder chip TDA 3301 for the purpose of normal beam limiting. As the beam current increases, it increases the potential at junction R3/R4, which is then used to set the contrast control accordingly. TR3 is also included to avoid a peak white flash caused by the automatic grey scale adjustment produced by the processor chip during the heater warm-up period.

## Automatic grey scale correction

Automatic grey scale correction is carried out by clamping the three beams R, G and B. The clamp levels are set up individually by comparing the beam current level of each gun with an internal reference generated by a timing and logic counter. The error voltages produced are stored for each individual gun. A switched sampling system performs the comparison and storage for each beam current in turn. A scanner then adjusts the luminance signal for each beam before it is fed to the matrix. This system corrects the lowlight drive levels only . The highlights are corrected in the traditional way by adjusting the gain of the red and green output amplifiers. Following normal practice, the gain of the blue channel is fixed at close to maximum.

Automatic grey scale tracking is carried out during the field blanking period. In the HA11498 processing chip, automatic grey scale tracking drive pulses are inserted on blanked lines 17 to 23 (Fig. 12.22). White level pulses (highlights) for green, red and blue are inserted on lines 17, 18 and 19 respectively; dark level pulses (lowlights) are inserted on lines 21, 22 and 23 respectively; line 20 is reserved for a beam cut-off test pulse. In each case the drive current for each gun is measured in sequence and set to the appropriate level.

**Fig. 12.22**  *Automatic grey scale drive pulses*

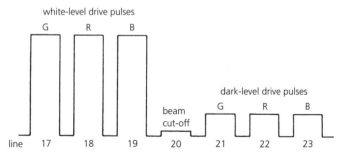

The luminance signal is first amplified within the i.c. before emerging at pin 35 to pass through the luminance delay line package. It re-enters the chip at pin 36. The luminance delay line represents a thick film circuit and includes the subcarrier rejecter. Internal systems operating on the luminance channel perform black level clamping, contrast control and blanking. The contrast and saturation controls are interlinked. The effect of the brightness control at pin 30 is to shift the d.c. level on which the luminance signal sits. A preset brightness control is also provided.

The chrominance circuitry follows conventional practice. The reference oscillator is tuned to 4.43 MHz and the 90° phase shift is carried out within the i.c. After filtering, the colour difference signals are matrixed with the luminance signal. The blanking signals are inserted at this stage. Two timing and logic counters, A and B, are employed to extract blanking, clamping and burst gate pulses from the sandcastle pulse input.

The switched beam current clamping unit provides the control voltage for beam limiting. When the current for any of the three beams reaches the maximum allowable value, the beam limiter reduces the drive via the contrast control.

The chip provides for a changeover to direct RGB video text input. The changeover is controlled by the fast blanking at pin 23, which operates the switching arrangements at the output amplifier stage.

## The TDA 3301 colour processing chip

One of the most comprehensive colour decoder chips is the 40-pin TDA 3301 (Fig. 12.23). It incorporates automatic grey scale correction to ease colour balance adjustment during manufacture and provide compensation as the tube ages. Fast blanking (pin 23) blanks out input from the i.f. and allows external RGB signals from the SCART connector to be processed.

### QUESTIONS

1. What are the $U$ and $V$ signals, and what is their relationship to the colour difference signals?

2. State the two colour difference signals used in a colour processing channel of a TV receiver. Explain how the third colour difference signal is produced.

3. State the colours of the colour bar in the correct order from left to right.

4. Sketch a waveform to show the green component of a colour-bar signal.

5. Show the luminance component of the colour-bar signal.

6. Using a block diagram, explain how the $U$ and $V$ signals are separated.

7. What is meant by grey scale correction and how is it carried out?

**Fig. 12.23** TDA 3301 colour processing chip and associate circuitry

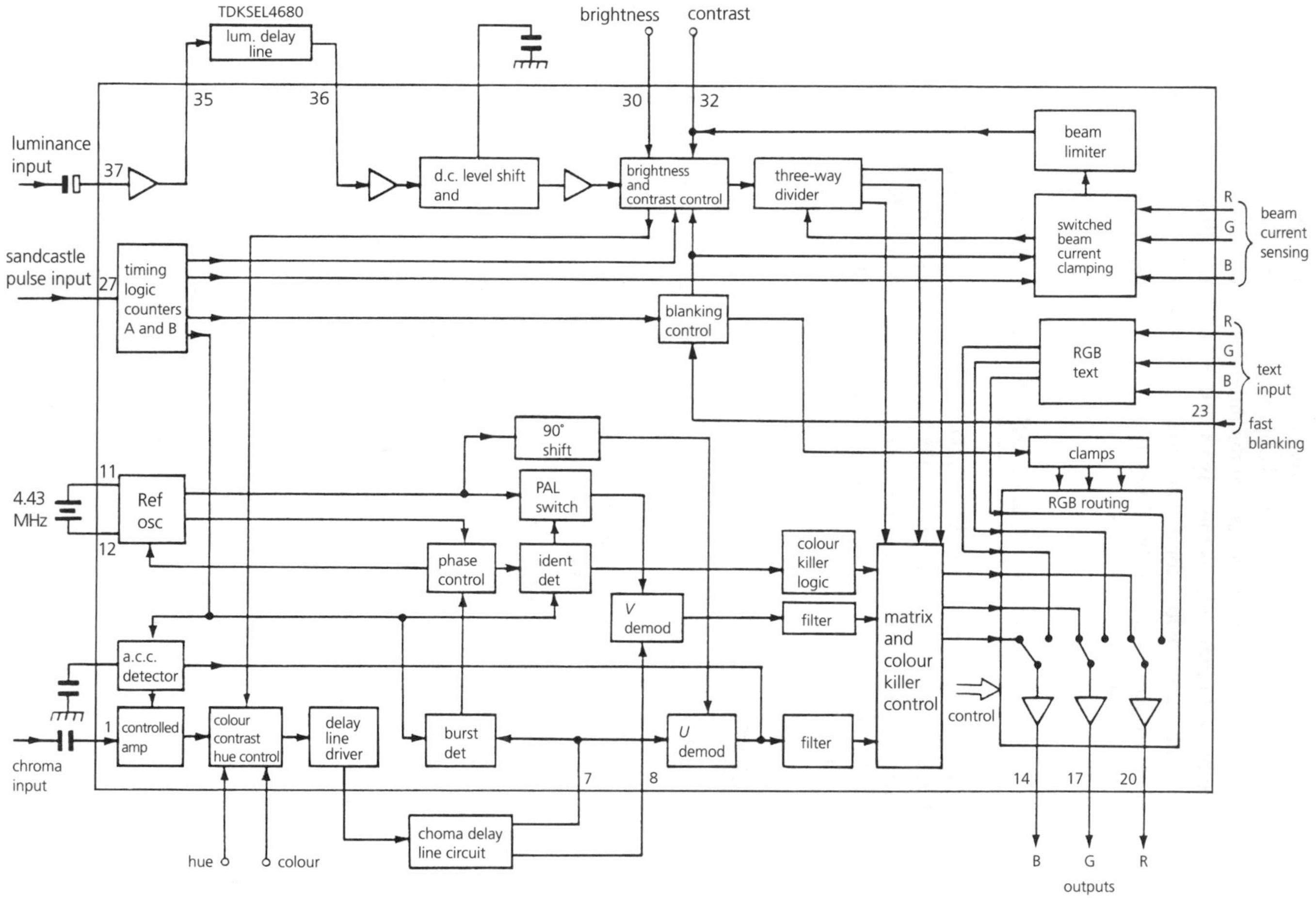

# **13** Power supply circuits

The various sections of the TV receiver demand different power requirements depending on their function and the level of the signal that is being handled. Large amplitude signals from the video and the line output stages require a high or boost voltage of 150–250 V known as high tension (h.t.). On the other hand, a low voltage supply of 10–40 V, sometimes known as low tension (l.t.), is required for those sections such as the tuner, the i.f. strip, video or sound drive and field or line oscillators that handle small signal levels. Furthermore, extra high tension (e.h.t.) in the region of 25 kV is necessary for the final anode of the cathode-ray tube.

Apart from the a.c. supply to the tube's heater and d.c. supplies to the field and line oscillators, regulated power supplies are necessary to ensure steady output levels. The power supply must also ensure safe operation of the receiver under normal conditions as well as under faulty conditions such as excessive current or voltage requirements. A great diversity of approaches and techniques are employed in the design of power supplies for TV receivers. The aim is to improve regulation and efficiency and to minimise power dissipation, thus reducing the cost and weight of the receiver.

## **Unregulated power supply**

A simple unregulated power supply is shown in Fig. 13.1, in which D1/D2 is a mains full-wave rectifier. C1 is a reservoir capacitor and R1/C2 is a smoothing or low-pass filter which removes the 100 Hz ripple appearing at the output. For a more effective smoothing, the series resistor R1 may be replaced by a large inductor. In this simple circuit the d.c. output decreases as the load current increases. For a constant d.c. output, a regulator or a stabiliser must be used. It is essential that the d.c. output is maintained constant against both changes in the mains supply and changes in the load current, *mains regulation* and *load regulation* respectively.

## **Series stabilisation**

A simple series stabiliser or regulator is shown in Fig. 13.2, in which TR1 is the series regulator. Load regulation is obtained by zener diode Z1, which provides the reference voltage for the base of common emitter transistor TR1. Z1 maintains the base of TR1

**Fig. 13.1** *Unregulated power supply*

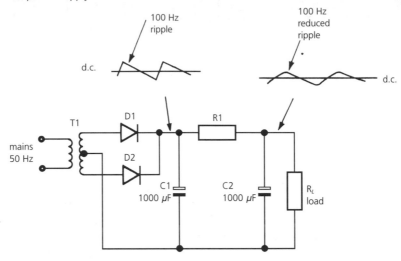

**Fig. 13.2** *A series stabilised power supply*

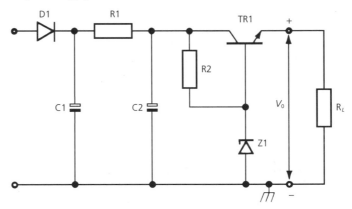

at a constant potential determined by its breakdown voltage. The d.c. output taken at the emitter is thus maintained at 0.6 V below the zener voltage. Changes in the output level produce changes in the b–e bias of the series stabiliser in such a way as to keep the output constant.

For higher load currents, a Darlington pair may be used (Fig. 13.3). Both transistors are connected in the emitter follower configuration with an overall gain equal to the product of the gain of each transistor. A Darlington pair i.c. package is normally used.

The sensitivity of the regulator may be improved by the incorporation of a voltage comparator, also known as an *error detector* (Fig. 13.4). TR1 is the normal series regulator. TR2 compares a portion of the output voltage with the reference voltage of the zener. Changes in the output level are amplified and fed into the base of TR1, which maintains the output constant. A shunt bypass resistor, R2, is sometimes connected across the c–e junction of TR1 to reduce the power dissipation of the series stabiliser

**Fig. 13.3** *Darlington pair stabiliser*

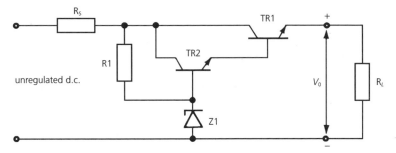

**Fig. 13.4** *Stabilised supply with error comparator TR2*

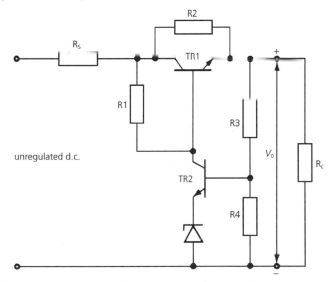

by diverting a portion of the load current away from the transistor. Shunt resistor R2 also provides the initial start-up voltage for the regulator by bleeding the current across TR1 to the output; this provides the necessary potential for TR2 and the zener to begin to conduct and start the regulator functioning.

## Practical regulated power supply

A regulated power supply used in a monochrome receiver is shown in Fig. 13.5, in which VT21 is the series regulator and VT22 is the comparator or regulator control amplifier. The receiver allows battery operation. The power supply provides a d.c. rail at a nominal value of 11.6 V when the receiver is connected to the mains. W7/W8 is a full-wave rectifier; C88 and C89 are diode bypass capacitors; C85, C86 and C87 are smoothing capacitors. W6 is a protection diode which causes the fuse to blow if the battery is connected the wrong way round.

**Fig. 13.5** *Regulated power supply (Ferguson 1590 mono chassis)*

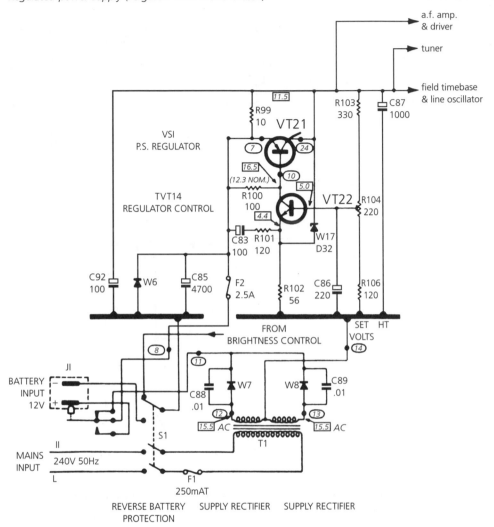

Zener diode W17 is connected to the h.t. rail with R102 providing its breakdown current. This method reduces the voltage rating of comparator VT22 and provides current limitation and short-circuit (s/c) protection. A fall in the h.t. rail due to an increase in load current is compensated by increased conduction of the regulator VT21. The circuit continues to operate in this way until the load current reaches a level where the output voltage falls to or below the breakdown voltage of the zener. At these levels, the zener ceases to conduct, reducing the voltage across R102, hence VT22 emitter, to almost chassis potential. With the base of VT22 at a certain positive voltage, VT22 bottoms, saturating VT21 as well. VT21 current is now at its maximum value, thus limiting the load current. This maximum current continues to flow so long as the output

**Fig. 13.6** *Colour receiver supply employing regulator chip SRT 371*

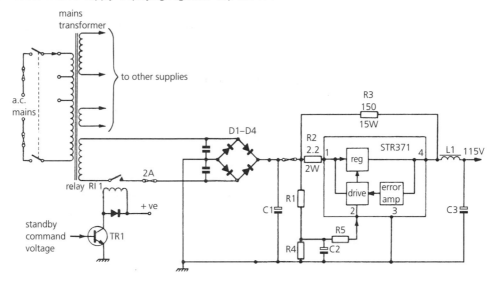

voltage is high enough to provide the necessary forward-bias to the base of VT22. If the output voltage falls below that level, as in the case of a short circuit, VT22 base drops to zero potential, turning it off. This in turn switches off VT21 and protects the regulator.

The ripple which is present at the input of the regulator is fed to the emitter of VT22 via phase shift network C83/R101. The phase-shift network ensures that the ripple at the emitter is equal in amplitude but out of phase with the ripple present there, as a result of the feedback from the output to VT22 base via resistor chain R103/R104/R106. R99 is a regulator shunt resistor which bypasses approximately one-third of the current drawn from the supply, and R104 sets the output level.

Mains switch S1 incorporates a contact which is shorted to the chassis when the receiver is switched off. This contact is connected to the brightness control which, upon switching off, is taken to chassis potential; this ensures the screen is blacked, removing the bright spot which would otherwise be observed in the centre of the screen for a short period after switch-off. This process is known as *spot suppression*.

A series stabilised power supply suitable for a colour TV receiver using regulator chip STR371 is shown in Fig. 13.6. Within the i.c., regulation is achieved by the drive controlling the series regulator. The isolating mains transformer drives a full-wave rectifier D1–D4 which, together with reservoir capacitor C1, provides the regulator chip with an unregulated voltage of about 135 V. This input voltage charges capacitor C2 via R1, which feeds the regulator drive to ensure mains regulation. Load regulation is obtained by the error amplifier, which also feeds into the regulator drive. R3 is a bypass resistor used to reduce the power dissipation of the chip, and L1/C3 is a low-pass filter. Relay RL1 controls the input to the full-wave rectifier to provide standby operation which maintains a d.c. supply to the remote control unit when the receiver is switched off. The standby command voltage from the remote control chip is fed into the base of TR1, which turns the transistor on and operates the relay.

**Fig. 13.7** *The thyristor*

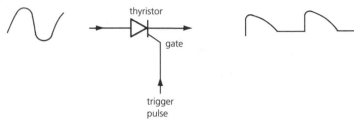

**Fig. 13.8** *Controlled conduction*

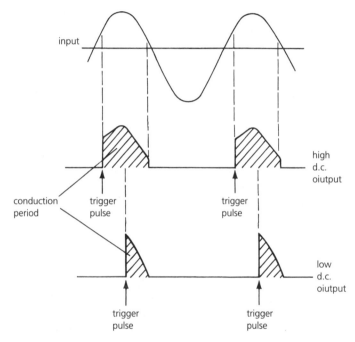

## Thyristor-controlled rectifier

Recall that the thyristor (Fig. 13.7) may be triggered into conduction by a positive voltage applied to its gate, provided that its anode is positive with respect to the cathode. When fed with an a.c. voltage, the thyristor can only conduct during the positive half-cycle. The conduction period is determined by the timing of the trigger pulse to the gate. The output level may thus be controlled by switching the thyristor for longer or shorter periods of time (Fig. 13.8).

## Switched-mode power supplies

The switched-mode power supply (SMPS) is in essence a converter. It converts un-regulated d.c. into a switched or pulsating d.c. and back again into a regulated d.c.

**Fig. 13.9** *Switched-mode power supply (SMPS)*

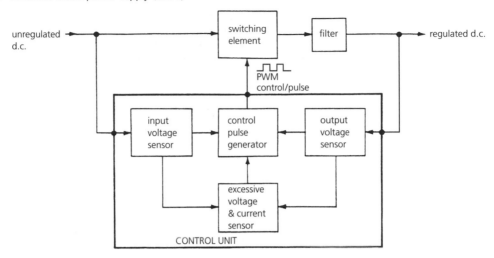

The switching speed determines the a.c. or ripple frequency at the output. A variety are used in TV receivers. They range from the chopper type and the Syclops to the self-oscillating power supply (SOPS).

The basic function blocks of a switched-mode power supply are shown in Fig. 13.9. The switching element may be a transistor or a thyristor; opened and closed at regular intervals by a pulse from the control unit, it is used to charge up a reservoir capacitor. The charge across the capacitor is determined by the period during which the switch is closed. Regulation is obtained by making the time intervals when the switching element is open and closed (i.e. the mark-to-space ratio of the control pulse) depend on the d.c. output. They may also be made to depend on the input voltage; this guards against changes in the mains (or battery) supply voltage. The control unit, which is normally an integrated circuit, provides protection against excessive current and voltage as shown.

The control pulse generator is in essence a pulse-width modulator which may be driven by line sync. pulses or by a free-running oscillator.

The switching action of the SMPS, which may involve large currents, can introduce interference known as *mains pollution* in the form of sharp transients, spikes or glitches superimposed upon the mains waveform. This is overcome by the introduction of high-frequency chokes or decoupling capacitors at the input terminals; they prevent high-frequency pulses from going back into the mains supply.

## Use of the energy reservoir inductor

The switching element which provides control and hence regulation is only one requirement of the switched-mode power supply. The other requirement is the efficient use of energy, and this is achieved by using an inductor as an energy reservoir. The inductor may be connected either in series or in parallel with the load.

**Fig. 13.10** *Series type SMPS*

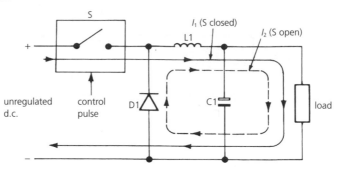

**Fig. 13.11** *Shunt type SMPS*

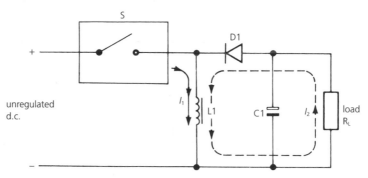

A series connection is shown in Fig. 13.10, in which D1 acts as an efficiency diode. When the switching element S is closed, current $I_1$ flows from the positive side of the unregulated input into the load as shown. The magnetic field set up by the current flowing through L1 causes energy to be stored in the inductor. When the switch is open, the current ceases and the magnetic field collapses. A back e.m.f. is induced across the inductor in such a way as to forward-bias D1, causing a current $I_2$ to flow into the load in the same direction as before. The energy stored in the inductor when the switch was closed is therefore consumed when the switch is open. The ripple at the output has a frequency that is twice the switching speed and is easily removed by smoothing capacitor C1.

A shunt type switched-mode power supply is shown in Fig. 13.11. When the switch S is closed, D1 is reverse-biased and current $I_1$ flows into L1, feeding energy into the inductor. When the switch is open, $I_1$ collapses and the back e.m.f. across the coil drives current $I_2$ into $R_L$, transferring the stored energy from the inductor to the load. One disadvantage of this technique is that, since the current flows into the load during the 'open' period of the switch, the output ripple is larger in amplitude and lower in frequency compared with the series connection. However, the parallel connection technique makes it possible to achieve electrical isolation by simply replacing the inductor with a transformer.

**Fig. 13.12** *Using a chopper transformer*

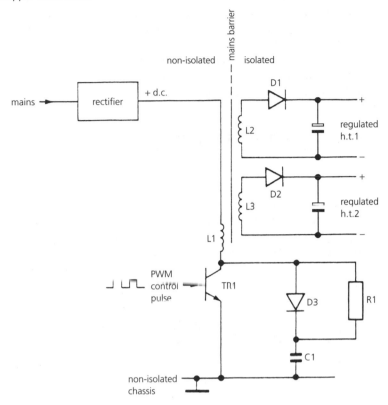

## Chopper transformer

Figure 13.12 shows a shunt type power supply using a chopper transformer with chopper transistor TR1 as the switching element. Energy is stored in the primary winding L1 when TR1 is on. When the chopper transistor is turned off, energy stored in the primary is transferred to the secondary windings to produce two regulated supplies, h.t.1 and h.t.2.

Network D3/R1/C1 provides protection for the chopper transistor against excessive rise of collector voltage caused by the back e.m.f. generated when the chopper transistor switches off. When TR1 is turned off, its collector voltage rises, turning D3 on. This effectively places *snubbing capacitor* C1 across TR1, diverting the major part of the back e.m.f. energy away from the transistor. When TR1 is switched on, C1 is discharged through R1.

The use of a chopper transformer makes it possible to obtain a number of regulated d.c. rails by merely adding windings to the transformer, as shown in Fig. 13.12. There is thus no need for a mains-isolating transformer, which saves on the cost and weight of the receiver.

It is normal to derive the control pulse going into the base of TR1 from the line sync. pulse. This makes the functions of the chopper transistor and the line output transistor very similar. This similarity prompted some manufacturers to combine the functions of both into a single transistor, a technique known as *synchronous converter and line output stage* (*Syclops*). Other techniques involve combining the functions of the chopper transformer and the line output transformer into a single transformer, a technique known as *integrated power supply and line output* (*IPSALO*).

## Start-up and soft start

The switching element and the control unit require a d.c. supply before they can begin to function. This voltage may be obtained from the nominal 12 V regulated supply for the control unit or other regulated or unregulated h.t. rails. It is normal that, once the control chip is brought into operation, the SMPS itself is used to provide the required regulated voltage to the chip. A slow or soft start is desirable for the switching element to prevent it from overworking at switch-on when the output voltage of the SMPS is zero. Soft start also ensures that the autodegaussing is completed before the tube starts to be scanned.

## Series chopper SMPS

The basic arrangement for a series chopper power supply is shown in Fig. 13.13, in which D2 is an efficiency diode and L1 acts as the reservoir inductor. Rectified mains is fed to the collector of the chopper transistor TR1. Control pulses from the oscillator are fed to the e–b junction of TR1 via driver TR2 and transformer T1. The oscillator may be a monostable driven by line sync. pulses. This means that the power supply may only start when the line oscillator is made operational. To overcome this disadvantage, a free-running oscillator may be used, which ensures the system is self-starting. Line pulses may still be used for synchronous operation. The mark-to-space ratio of the control

**Fig. 13.13** *Series chopper SMPS*

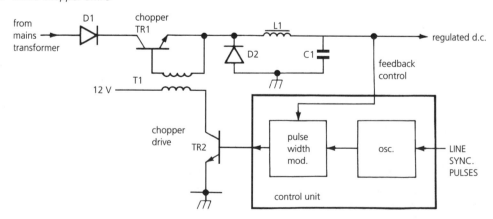

pulse, i.e. the time TR1 is on compared with the time it is off, is determined by the pulse-width modulator. Regulation is achieved by the feedback control voltage, which varies the mark-to-space ratio to compensate for changes in the output voltage. The control chip also includes mains regulation, excessive voltage and current protection, a slow start and a standby facility to provide the d.c. supply for the remote control board when the receiver is switched off.

## Shunt type chopper SMPS

Figure 13.14 shows the essential elements for a shunt type switched-mode power supply. The shunt chopper transistor TR6 is controlled by chip TEA 2018A. Three separate rails are derived from the chopper transformer: 13 V for the line driver and audio stages, 17 V regulated supply for the signal processing sections of the receiver and 95 V for the line and field output stages. The transformer also provides isolation as shown.

Bridge rectifier D3–D6 and reservoir capacitor C69 provide the start-up voltage for the control chip and the d.c. supply for the chopper transistor. At switch-on C71 charges up from the rectified mains via resistor network R60/R89/R91. When the voltage across C71 reaches 3.8 V, the i.e. swings into operation and delivers drive pulses to the base of the chopper transistor. The transformer is energised and a voltage develops across secondary winding 2–3, which is then rectified by D7 to establish the normal chopper-derived 11 V supply for the control chip. The voltage provided by secondary winding

**Fig. 13.14** *Shunt type chopper SMPS*

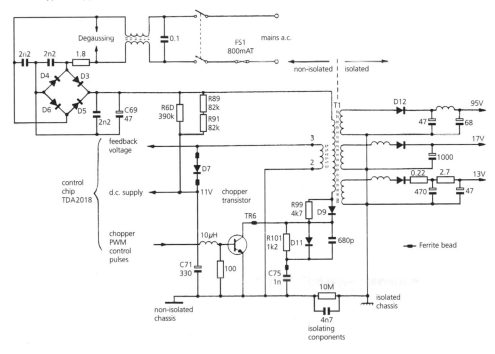

2–3 is also used to provide the feedback voltage to the control chip. The mark-to-space ratio of the chopper drive pulses is then adjusted by the chip to keep the output constant.

Network C75, R101 and D11 provides protection of TR6 against excessive back e.m.f. Further protection of the transistor is provided by limiting network D9 and R99, which prevents excessive ringing in the transformer due to the back e.m.f.

## Thyristor-controlled SMPS

A switched-mode power supply using thyristor SCR1 as the series switching element is shown in Fig. 13.15, in which L65 is the reservoir inductor, W77 is the efficiency diode and C147 is the reservoir capacitor. To ensure that SCR1 is turned off before W77 comes on, the thyristor cathode is connected to a tapping point on the inductor L65 as shown. The rectified mains is used to charge C137 to about 28 V via potential divider R165/R152. As a result, C138 begins to charge up (via R168) with its bottom plate negative. When this voltage is less than the voltage at the gate of SCR3, W73 and consequently SCR3 switch on to drive a trigger pulse to the regulator via transformer T1. At the end of each half-cycle, the cathode of W68 goes to 0 V. With its anode at a positive potential, W68 conducts, switching on VT62 for a very brief period and discharging C138 ready for the next half-cycle. Regulation is obtained by varying the voltage at the gate of driving thyristor SCR3, which varies the firing time of regulator SCR1. Error amplifier VT65 compares a portion of the output voltage fed into its base via R184/5/6 with a reference voltage at its emitter, produced by zener W78. Its collector is then used to control the firing of SCR3. Mains regulation is obtained by feeding the voltage developed across C137 to the base of error amplifier VT65 via resistor R179.

C143 provides a slow start facility. At switch-on C143 starts to charge up via R175 and R177. The values of these two resistors are such that SCR3 cannot conduct at the start. As the capacitor charges up, the voltage to the gate of SCR3 rises, firing the SCR first late in the cycle then moving progressively earlier as the voltage across C143 builds up. This results in a gradual rise in the h.t. voltage. When normal operation of the power supply has been established, the charge across C143 reverse-biases W76, thus isolating the slow start network from the rest of the circuit.

Excess voltage protection is provided by the *crowbar* action of SCR2. Should the voltage exceed the breakdown voltage of zener W85, then SCR2 conducts, taking the cathode of regulator SCR1 to chassis and blowing the fuse. Excess current protection is provided by VT66. Should the current consumption of the receiver rise beyond its normal level, the voltage across smoothing resistor R197 increases, causing VT66 to conduct. This fires SCR2 and blows the fuse. VT66 conducts when its base voltage falls below the voltage set by zener W83 at the emitter. To avoid false alarms due to flashovers of transients, the time constant of R197/C152 is chosen to introduce a delayed response.

## SMPS control chip

A number of control chips are used in TV receivers. The precise interconnections with other sections of the receiver vary between one set and another. A simplified block diagram

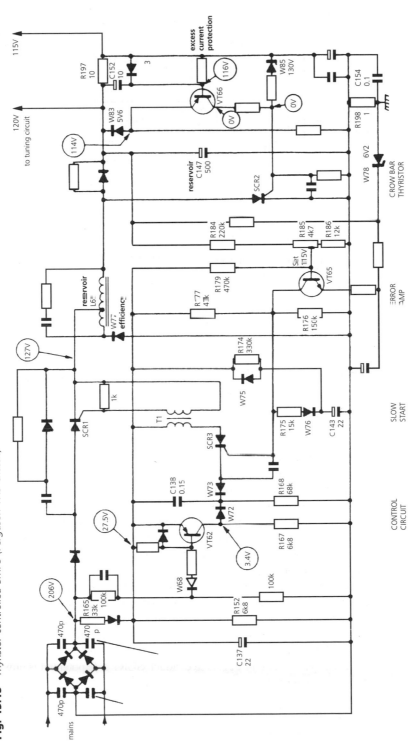

**Fig. 13.15** *Thyristor-controlled SMPS (Ferguson TX9 chassis)*

**Fig. 13.16** *SMPS control chip*

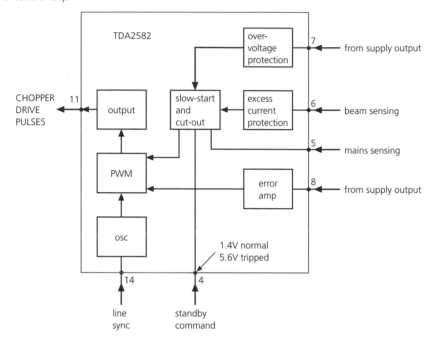

of the TDA 2582 control chip is shown in Fig. 13.16, in which the oscillator is driven by the line sync. pulse. The pulse-width modulator controls the space-to-mark ratio in such a way as to ensure a constant voltage output from the chopper supply. A remote standby command voltage to pin 4 in excess of 5.6 V will trip the control chip and switch off the chopper. The receiver will then rest in the standby mode. Excessive voltage and current protection as well as soft start are also provided. The trend in i.c. technology is to move to an ever greater degree of integration with a variety of functions being included in a single chip. The functions performed by the power control chip may be incorporated into a comprehensive power processor chip that includes line and field processing.

A complete SMPS circuit based on the Siemens TDA 4600 control chip is shown in Fig. 13.17, in which TR3 is a shunt type chopper transistor. The bridge rectifier produces a d.c. voltage of about 350 V from the mains input; this provides the supply voltage to the chopper transistor.

The start-up voltage is provided by thyristor SCR1 and associated circuitry. When the receiver is switched off, C87 is fully discharged and the cathode of SCR1 is at 0 V. At switch-on, a.c. mains is applied to the anode of the thyristor. Clipping zener D13 maintains SCR1 cathode at just under 5.6 V. D11 is reversed-biased and the thyristor is turned on, charging C87 during the positive half-cycles of the input. When the voltage across C87 reaches 6.5 V, D11 begins to conduct. SCR1 gate potential is now slightly lower than its cathode potential, so the thyristor is turned off. The voltage across C87, applied to the control chip at pin 9, is sufficient to bring the i.c. into operation. Short drive pulses are delivered to the chopper transistor to start the power supply operating. The pulses appearing across secondary winding 13–16 are rectified by D12, and the d.c. voltage

**Fig. 13.17** *Switched-mode power supply based on chip TDA 4600 (Ferguson TX9 chassis)*

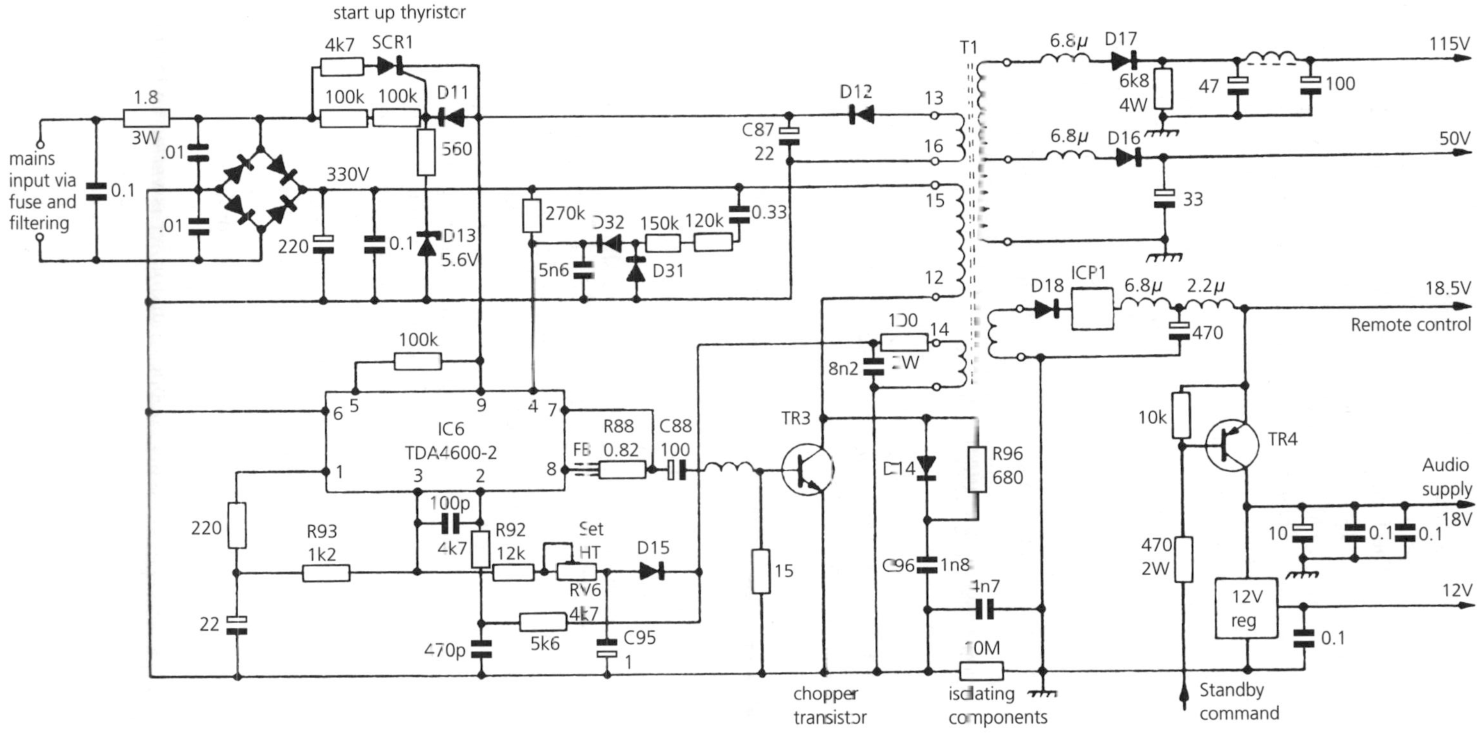

thus produced begins to charge C87. This provides a higher d.c. voltage to the chip, which in turn produces larger drive pulses and so on until the circuit reaches normal operation. Once the chopper circuit is running normally, D12 and C87 provide a stable 12 V supply for the chip. The feedback voltage is taken across a small winding (pin 14 on the transformer), rectified by D15 and fed into pin 3 of the chip via set h.t. resistor RV6.

The TDA 4600 control chip includes a free-running oscillator whose frequency is kept constant by components connected to pin 1 of the i.c. Each cycle is initiated by a zero-crossing or a zero-flux sensor incorporated within the i.c. at pin 2. When the a.c. voltage at pin 14 of the transformer crosses zero and attempts to reverse, the control chip switches on TR3 to commence the next cycle, thus ensuring minimum power loss.

Three rails are derived from the chopper transformer: 115 V for the line output stage, 50 V for the field output stage and 18 V for the remote control and the audio signal. This last rail also feeds a 12 V regulator chip which supplies the small-signal sections of the TV receiver.

Standby operation is obtained by turning off TR4 with a command signal to its base from the remote control unit. When its base goes high, a npn transistor TR4 switches off, thus disabling the audio circuit, the signal circuitry and the line and field oscillators. The power supply keeps working in the standby mode, delivering very small current and maintaining the supplies to the field and line output stages.

Chopper protection against excessive back e.m.f. is provided by D14, R96 and C96 connected across TR3.

## Syclops

Syclops, the synchronous converter and line output stage developed by Ferguson, combines the functions of the chopper and the line output into a single transistor. The basic elements of the Syclops circuit are shown in Fig. 13.18. The chopper/line output transistor TR1 feeds the normal chopper transformer and a standard line scan/line output transformer circuit via switching diodes D1 and D2 respectively. As described in Chapter 9, at time $t_1$ on the scan waveform (Fig. 13.19) TR1 is turned off to start the flyback. Its collector potential goes high, switching off D1. Tuned circuit C1/C2/L1 is pulsed into oscillation. At the end of the flyback, time $t_2$, the oscillation reaches its negative peak and current begins to change direction; D3 begins to conduct, which reverse-biases D2 and effectively places C2 across L1 to commence the first part of the scan. At time $t_3$, TR1 is turned on and its collector voltage drops, switching on D1. Current begins to flow into the primary winding of the chopper transformer from the mains rectifier, causing energy to be stored into T1. At time $t_4$, halfway along the scan, the current in the scan coils begins to reverse; D3 is turned off and D2, being forward-biased, is turned on. TR1 takes over from D3 to complete the second half of the scan. By varying the time at which TR1 is turned on, i.e. by advancing or delaying time instant $t_3$ (shaded area), the energy supplied to the chopper transformer is varied and with it the d.c. output. This has no effect on the scan waveform since D2 keeps TR1 isolated from the scan circuit until time $t_4$, when D3 turns off. Regulation is obtained by the Syclops control circuit comparing the output of the chopper transformer with a reference voltage and adjusting the mark-to-space ratio of the chopper drive pulse waveform accordingly.

**Fig. 13.18**  *Syclops power supply*

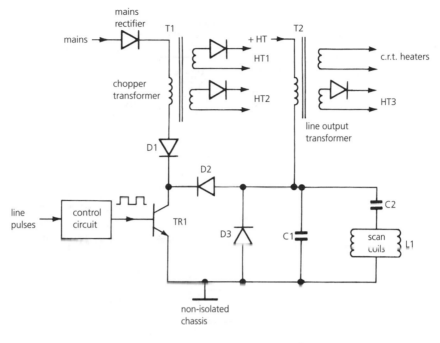

**Fig. 13.19**

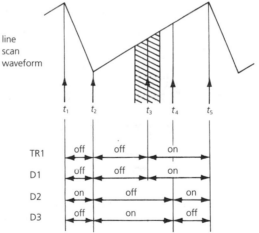

## Integrated power supply and line output

An improvement in the power dissipation of a power supply may be obtained by combining the functions of the chopper and line output transformers into a single transformer known as the *combitransformer*. The integrated power supply and line output (IPSALO) technique devised by Salora also saves on costs, space and weight of the receiver. Figure 13.20 shows the basic arrangements for an IPSALO circuit in which

**Fig. 13.20** *Integrated power supply and line output (IPSALO)*

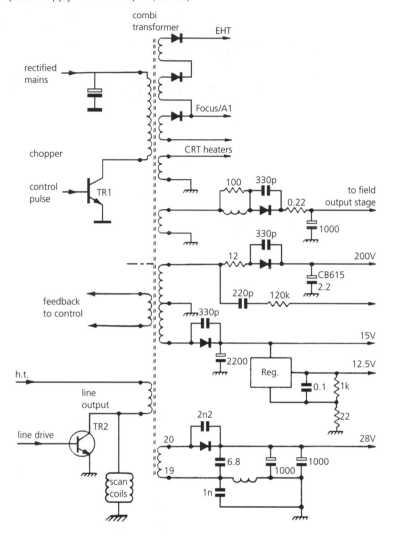

TR1 is the chopper and TR2 is the line output transistor. The combitransformer acts as the mains isolation tranformer and provides all d.c. rails, including the e.h.t. and the heaters to the tube.

## Self-oscillating power supply

The functions of chopper and oscillator may be combined in a single transistor to form a self-oscillating power supply (SOPS). A secondary winding of the chopper transformer is used to form a blocking oscillator arrangement (Fig. 13.21).

A different arrangement for a blocking oscillator is shown in Fig. 13.22. At switch-on the chopper transistor TR1 begins to conduct as a result of the forward-bias applied

**Fig. 13.21** *Blocking oscillator: one arrangement*

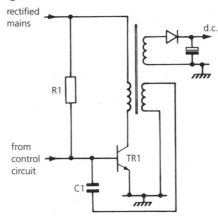

**Fig. 13.22** *Blocking oscillator: another arrangement*

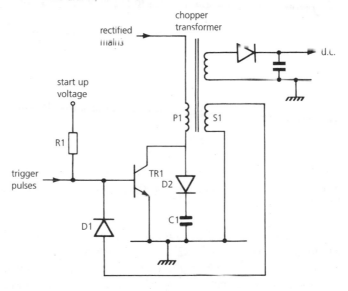

to its base via start-up resistor R1. The collector current increases, which induces a positive voltage across secondary winding S1. This forward-biases D1 and further increases TR1 current further. When saturation is reached, the increase in current ceases and a negative voltage is induced across S1; this reverse-biases D1, switching off TR1. At this point, the voltage across primary winding P1 reverses and D2 switches on. The tuned circuit formed by primary winding P1 and C1 begins to oscillate, transferring energy from P1 to C1. For the second half of the cycle, when energy begins to transfer back to P1, diode D2 is reverse-biased, oscillations stop and the current in P1 reverses; this causes a positive voltage to be induced across S1, turning on TR1, and so on. Trigger pulses from the control circuit are used to initiate each cycle to keep the d.c. output constant.

A self-oscillating chopper circuit based on a Panasonic chassis is shown in Fig. 13.23, in which TR3 is the chopper/oscillator transistor and C9 is the blocking capacitor, which

**Fig. 13.23** *Self-oscillating chopper power supply*

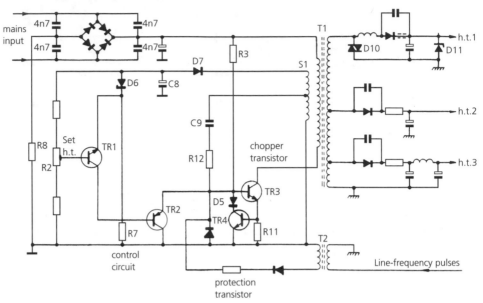

discharges through transistor TR2. Pulses derived from the line output stage are fed to the base of TR3, via isolating transformer T2, to turn on TR3 just before its natural turning-off point. Regulation is obtained by controlling the conduction of TR2 and thus the discharge of C9. Transistor TR2 is itself controlled by TR1, which senses any change in the voltage developed across C8 caused by changes in the loading of the chopper transformer as well as changes in the mains input. The voltage across feedback winding S1 is rectified by D7. The negative voltage thus produced is used to charge C8 to provide a measure of the loading of the transformer. This voltage is then used to provide TR1 emitter voltage via zener D6. Changes in the loading of the transformer, whether caused by changes in the loading or in the mains, cause the voltage at the emitter of TR1 to change, which determines the base bias of TR2, hence the discharging current of C9. The current through TR1 may also be varied by the 'set h.t.' control R2. TR4 provides a degree of protection against excessive current taken by TR3. A large current through R11 will cause TR4 to conduct, taking TR3 base to chassis via diode D5. Further protection is provided by D10 and D11 on the secondary side of the circuit. T1 and T2 provide mains isolation.

1.   State the advantages of a switched-mode power supply over a linear power supply.

2.   What is meant by soft start and how is it implemented?

3.   Explain why modern receivers have more than one power supply.

4.   Explain the reason for having two different chassis connections in modern TV receivers.

# 14 Television display tubes

Recall that in the c.r.t. the high-speed electrons in the form of a beam current are emitted by an electron gun, focused and accelerated by an electron lens and then directed towards a screen which acts as a positively charged anode. The screen is coated with phosphor and gives a visible glow when hit by high-speed electrons. The colour of the emitted light is determined by the type of phosphor used. For monochrome display, only one type of phosphor coating is used. For a colour display, three types of phosphor are used in order to obtain the three primary colours

## Extra high tension

A final anode voltage known as the extra high tension (e.h.t.) in the region of 15–30 kV is required by the tube to attract and accelerate the electrons. This e.h.t. is produced by a voltage multiplier network (Fig. 14.1). The a.c. voltage input is obtained from an overwind on the line output transformer. C1/D1 acts as a clamper which charges C1 to the peak of the input voltage $V_p$. This is then applied to the second clamping circuit D2/C2, charging C2 to $2V_p$, and so on. The elements of the voltage multiplier are contained in a single well-isolated package or capsule, known as a *tripler*.

## Monochrome tube

The monochrome display tube consists of a single electron gun, an anode assembly acting as the electron lens, and a viewing surface. The beam passes through a four-anode assembly (Fig. 14.2) which provides acceleration (A1), electrostatic focusing (A2/A3) and final e.h.t. anode (A4). As well as depending on the grid potential, the emission of electrons also depends upon the potential between the cathode and the first anode. An increase in the potential between these two electrodes causes more electrons to be emitted, and vice versa. The actual tube voltages depend on the size of the tube and its design, but here are some typical voltages for a monochrome receiver tube:

*Cathode*   *70 V*
*Grid*      *30 V*
*A1*        *300–400 V (accelerating anode)*

**Fig. 14.1** *Voltage multiplier network*

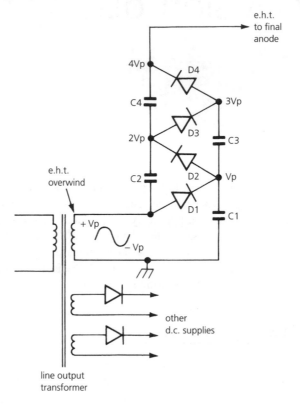

**Fig. 14.2** *Cathode-ray tube*

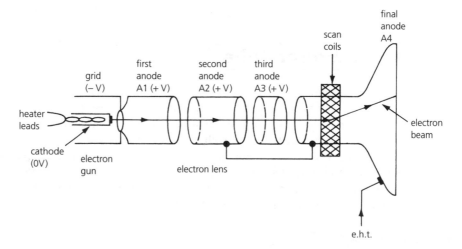

**Fig. 14.3** *Glass envelope of the display tube*

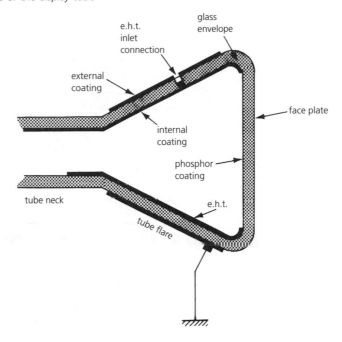

A2          *15–20 kV (connected to final anode)*
A3          *variable up to 500 V (focus anode)*
A4          *15–20 kV (final anode)*

The electron gun and the anode assembly are contained within a vacuumed thick glass envelope (Fig. 14.3). Access to the various electrodes is obtained via pin connections at the back of the neck of the tube, except for the final anode, which is accessed along the tube flare. The inside and outside of the tube flare are coated with a layer of graphite known as *Aquadag* coating. The outer coating is connected to chassis and the inner coating is connected to e.h.t. The glass separation between them forms a reservoir capacitor for the e.h.t. supply. This is why it is important the capacitance is fully discharged before handling the tube, otherwise a violent shock may be experienced.

In order to produce a display, the electron beam is deflected in the horizontal (line) and vertical (field) directions. Electromagnetic deflection is employed using two sets of coils (line and field), known as scan coils, placed along the neck of the tube (Fig. 14.4). The most common deflection angle is 90°, although 100° and 110° are widely used for large screens. The tube's angle of deflection forms an important specification of the c.r.t. It refers to the angle through which the beam is deflected along the diagonal of the screen. The actual angles of deflection in the horizontal and vertical directions are less than the specified value. For instance, a typical 110° tube may have a vertical deflection of 81° and a horizontal deflection of 98°. The deflection angle depends on the strength of the magnetic field created by the scan coils, the speed of the electron beam (which is a function of the e.h.t.) and the thickness of the neck of the tube. A narrow neck allows the scan coils to operate in close proximity to the beam, hence they exercise greater

**Fig. 14.4** *Beam deflection*

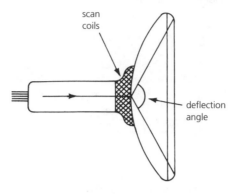

scan
coils

deflection
angle

**Fig. 14.5** *Screen size*

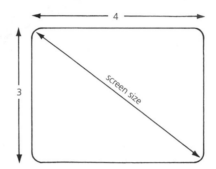

4

3

screen size

influence upon it. Modern receivers have deflection angles of 90° or 110°. Another specification of the c.r.t. is the size of the screen given in centimetres (Fig. 14.5). The centimetre figure refers to the diagonal measurement of the visible picture, whereas the traditional tube size, quoted in inches, refers to the overall diagonal measurement of the screen. With an aspect ratio (width:height) of 4:3, a 51 cm (21 inch) tube has a diagonal of 51 cm, a width of 41 cm and a height of 31 cm. The power required by the scan coils is a function of the size and geometry of the tube (screen size and neck diameter) as well as the deflection angle and the e.h.t. applied to the final anode.

## Raster geometry

Recall that, given a linear timebase waveform, the flat surface of the display tube causes non-linearity in the displayed picture. Equal angular deflections of the beam cause it to scan a smaller distance at the centre compared with the distance it scans at the two ends of the line. A similar effect is produced for the vertical scan. When both non-linearities are considered together, the raster produced on a flat screen shows what is known as *pincushion distortion*. Such distortion may be greatly minimised by using S-correction to change the shape of the scanning waveform to a non-linear sawtooth and by appropriate design of the scan coils. It can also be corrected by the use of small peripheral permanent magnets; the magnets are placed on the scan coils and manually manipulated to modify the shape of the magnetic field.

**Fig. 14.6** *Cathode modulation*

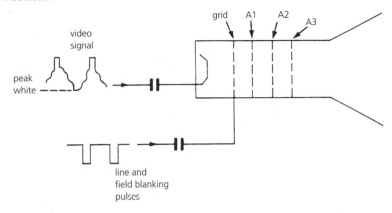

## Beam modulation

In order to produce an image on the tube, the brightness of the screen has to be varied as it scans the surface of the tube to recreate the picture information line by line. This is achieved by varying the intensity of the electron beam in accordance with the video signal, a process known as *beam modulation*. The beam is modulated by varying the potential between the cathode and the grid. There are two types of modulation. In *grid modulation* the cathode voltage is held constant and the grid voltage is varied by the video signal. In *cathode modulation* the voltage at the grid is fixed and the cathode voltage is varied with the video signal. With cathode modulation the video signal is negative-going (Fig. 14.6). Peak white is produced when the cathode is at its most negative potential. Its greater sensitivity means that cathode modulation is normally used, along with negative-going blanking pulses applied to the grid to cut off the beam during line and field flyback.

## Monochrome tube connections

Typical connections to a monochrome tube are shown in Fig. 14.7; W11 is the rectifier/doubler package providing the 11 kV e.h.t. for the final anode of the tube. Rectifier W13 provides the d.c. supply for the focus anode and the grid. Brightness is controlled by R98, which varies the voltage at the grid. Blanking pulses are applied to the grid via C108/R107 and clipped by diode W15.

# Colour receiver tubes

Recall that the cathode-ray tubes for colour display have three separate guns, one for each primary colour. The guns bombard a screen coated with phosphors arranged in triads. Each triad contains three different phosphors, one for each primary colour. Placed behind the coated screen, a steel *shadow mask* allows the three electron beams

**Fig. 14.7** *Monochrome tube connection (Ferguson 1613)*

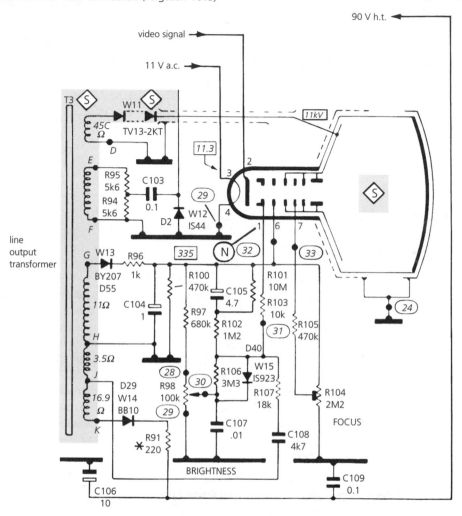

to converge and pass through slots before they strike their particular phosphor on the screen. Three primary colours are thus produced which, because they are very close to each other, are added by the human eye to create a sensation of colour.

Colour tubes require higher anode voltages and larger video drives than monochrome receivers. Typical voltages are as follows:

| | |
|---|---|
| *Cathode* | *100–150 V* |
| *Grid* | *20–30 V* |
| *A1* | *50–1000 V* |
| *Focus* | *2–7 kV* |
| *E.h.t.* | *25–30 kV (final anode)* |

**Fig. 14.8** *The three rasters*

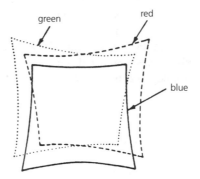

## Purity

For correct colour reproduction the red, green and blue beams must strike only their own particular phosphors and no other. This is known as *purity*. Purity adjustment involves changing the strength and direction of the magnetic field created by the scan coils to move the beams so that they strike the correct phosphor coating on the screen. This is achieved by using a pair of two-pole ring magnets placed along the neck of the tube.

## Convergence

The three beams form three separate rasters (Fig. 14.8), thus multiplying the problems of raster correction encountered with monochrome displays. The three rasters must not only be of the correct rectangular shape with no pincushion distortion, they must also coincide precisely. This is known as convergence. There are two types of convergence, static and dynamic. *Static convergence* involves the movement of the beams by permanent magnets placed inside or outside the tube to bring the three beams into coincidence in the central area of the screen. *Dynamic convergence* covers the rest of the screen, which involves the establishment of a continuously varying (dynamic) magnetic field to ensure convergence at the outlying areas and corners of the display. This requires electromagnetic waveforms that are a function of both the line and field frequencies.

## The delta-gun shadow mask tube

The first mass-produced colour tube was the delta-gun. It has three electron guns mounted at 120° to each other at the neck of the tube, each with its own electron lens. The guns are tilted by a small amount towards the central axis of the tube so that their electron beams converge and cross at the shadow mask and pass through carefully positioned holes to strike their correct phosphor dot. A large number of electrons miss their holes and are lost through hitting the mask, resulting in what is known as low *electron transparency*, low efficiency and low brightness. The delta arrangement of the three guns makes it difficult to achieve accurate convergence as the beams are made to scan the screen. This is the main disadvantage of the delta tube. Highly complex and expensive convergence circuits are necessary to overcome the two-dimensional distortion of the rasters and to maintain good convergence. Several presets have to be used in a complex sequence requiring highly skilled labour. Furthermore, the tendency for convergence to drift means frequent adjustments are necessary. For this reason, delta-gun tubes are no longer used for domestic TV receivers. However, they are still in production for use as monitors for advanced computer displays because of their high definition when fitted with a very fine-pitch shadow mask.

## The in-line colour tube

The in-line shadow mask tube has three guns placed side by side, and the phosphor coatings on the screen are in the form of striped triads. Each three-colour triad is arranged to coincide with a longitudinal grill or slot in the shadow mask. Having three beams in the same horizontal plane has two advantages. First, purity is unaffected by horizontal magnetic fields such as the earth's magnetic field. Second, the need for vertical convergence correction disappears because the three beams always travel in the same horizontal plane. Convergence is then reduced to the relatively easy task of deflecting the two outer beams slightly inwards to converge with the central beam. The first in-line tube was developed by Sony and is known as the *Trinitron*. This was followed by the Mullard *precision-in-line* (AX series) self-converging tube, which eliminated the need for dynamic convergence adjustment altogether.

## The Trinitron tube

The Trinitron uses a single in-line gun assembly and a single electron lens assembly (Fig. 14.9). The phosphors are arranged in vertical stripes forming three-colour strip triads. The shadow mask is replaced by a metal aperture grille with one vertical slit for each phosphor triad. Greater electron transparency is achieved as fewer electrons are lost by hitting the mask; this improves efficiency and brightness. The single electron gun employs three in-line cathodes. The three beams pass through a complicated anode arrangement; this bends the two outer (red and blue) beams so they seem to be emanating from the same source as the middle green beam. The Trinitron suffers from two basic disadvantages. The first is that the striped mask has very little stiffness in the vertical direction and has to be kept under considerable tension to prevent sagging or

**Fig. 14.9** *The Trinitron tube*

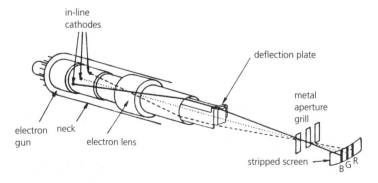

buckling. The second disadvantage is the need for some dynamic convergence adjustments, especially in wide-angle large-screen versions.

## The PIL tube

In the precision-in-line tube (PIL), three separate guns are mounted side by side on the same horizontal plane. The phosphors are arranged in vertical stripes on the screen and the shadow mask has the staggered slots shown in Fig. 14.10; staggered slots provide mechanical rigidity and high electron transparency, giving improved brightness. The main advantage of the PIL tube is the development of a special deflection yoke designed to produce a staggered magnetic field, known as an *astigmatic field*, which eliminates the need for dynamic convergence.

**Fig. 14.10** *The precision-in-line (PIL) tube*

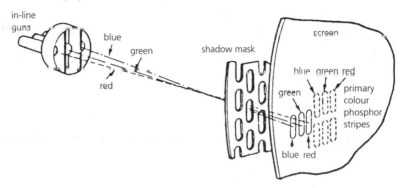

## Self-convergence

The angle of deflection applied to an electron beam is proportional to the strength of the magnetic field present along the path of the beam. For a uniform magnetic field, the three beams will be subjected to equal force and will thus converge at the centre of the screen at all points along a circular ring known as the image field (Fig. 14.11(a) ). However,

**Fig. 14.11** *Self-converging astigmatic field*

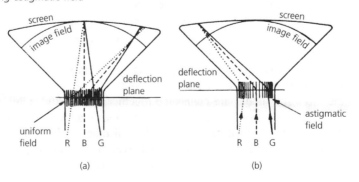

for a flatter screen, the three beams will diverge, resulting in misconvergence. This is overcome by the non-uniform or astigmatic field shown in Fig. 14.11(b), in which the two outer beams move along a path of varying magnetic strength. The centre beam passes through a relatively weak field in the middle of the deflection plane. The two outer beams will suffer different deflections as they move from a relatively strong field on the outside to a weaker one in the middle. As a result, the deflecting force acting on them is reduced and they turn through a smaller angle than the centre beam. Provided the astigmatic field is accurately designed, the three beams will converge at all points on a line across the flat screen. The same principles apply to the vertical deflection of the beams, resulting in two astigmatic fields established by the deflection yoke.

Later tube designs such as the Mullard 45AX tube incorporate a single-gun assembly with a narrow neck. Narrow-neck tubes require much less deflection energy and their closely spaced beams need less convergence correction.

## PIL tube connections

Typical connections to a PIL colour tube are shown in Fig. 14.12, in which the e.h.t. tripler capsule also provides the high voltage required by the focus electrode. Capacitor C730 is charged by the tripler, which is then used to provide the A1 voltage via preset resistor R721. The return path for the e.h.t. current is via current limiter diode W722. Diode W722 is forward-biased by R724/R725 with a forward current flowing in the opposite direction to the e.h.t. tube current. When the tube current exceeds the forward-bias current, W722 is cut off and the tube current is diverted via resistor chain R724/R725. This causes the voltage at the junction of the two resistors to increase, increasing the voltage at the grids and reducing the beam current. The change in voltage is also fed to the luminance amplifier to change its gain.

## Colour tube adjustments

Even with self-converging tubes, there remain a number of adjustments that have to be made to produce a fully linear display with the correct chrominance content. This involves two separate adjustments: grey scale and pincushion.

## Grey scale tracking

The purpose of grey scale tracking is to ensure that a monochrome display exhibits different shades of grey only with no traces of colour tint at all levels of brightness. Grey scale tracking is carried out on a staircase or grey scale display (Fig. 14.13) and involves two types of adjustments: lowlights (low brightness) and highlights (high brightness). The *lowlights* adjustment brings the cut-off points of the three guns into coincidence, thus ensuring that shaded areas of the picture have no coloured tint. In PIL tubes where the three A1 anodes are connected together, the lowlights adjustment is carried out by varying the d.c. potential of each cathode, a setting that is found at the video output stage. The *highlights* adjustment ensures that all other white levels are correctly reproduced without a coloured tinting. This is achieved by varying the video signal drive, a setting usually known as video gain control.

**Fig. 14.12** *PIL tube connections (Ferguson 9000 chassis)*

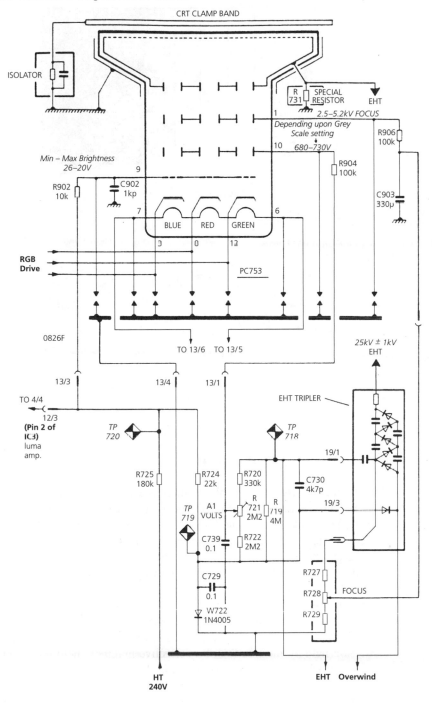

**Fig. 14.13** *Grey scale*

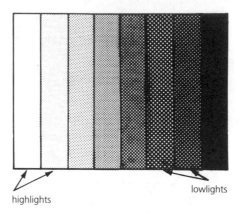

highlights

lowlights

**Fig. 14.14** *Pincushion distortion*

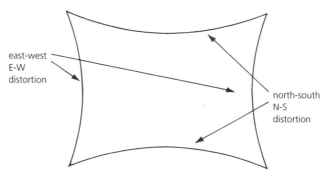

east-west
E-W
distortion

north-south
N-S
distortion

## Pincushion adjustment

Despite the S-correction described earlier, the screen display still suffers from pincushion distortion, especially in wide deflection tubes; it is caused by the flat surface of the screen. The joint effect of the line and field deflections produces a maximum angle of deflection along the diagonal, hence the stretched corners. Figure 14.14 shows a sketch of a raster with much exaggerated pincushion distortion. The bowing inward of the sides is called east–west (E–W) distortion; bowing at the top and bottom is called north–south (N–S) distortion. In E–W distortion the scan lines change in length as the beam is deflected vertically, i.e. field by field. E–W distortion is therefore a distortion of a line, the scan line, by the field frequency. Conversely, N–S distortion is a distortion of the field by the line frequency. The correction of both distortions involves modifying the deflection field by the creation of a corrective magnetic field that is equal and opposite to the field which created the distortion in the first place.

Correction along the E–W axis involves reducing the width of the line scan at the top and bottom of the picture until it is the same as in the middle of the picture

**Fig. 14.15**

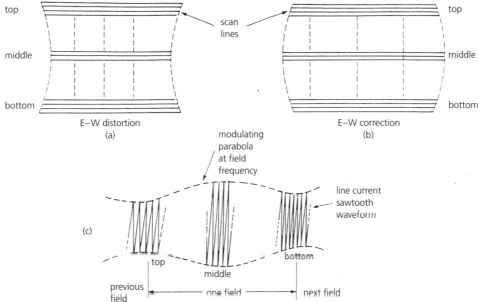

E–W distortion
(a)

scan
lines

E–W correction
(b)

modulating
parabola
at field
frequency

line current
sawtooth
waveform

(c)

previous
field

one field

next field

top

middle

bottom

**Fig. 14.16** *E–W correction circuit: block diagram*

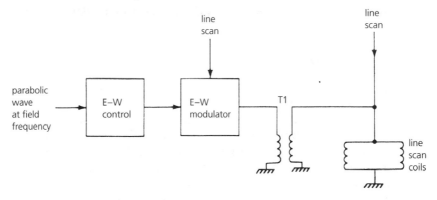

(Fig. 14.15). This is achieved by modulating the basic line scan sawtooth waveform by a parabolic waveform at the field frequency (Fig. 14.15(c)) to produce the necessary correction (Fig. 14.15(b)). A block diagram for an E–W correction circuit is shown in Fig. 14.16, in which the modulating parabolic waveform at field frequency may be changed in amplitude and phase by the E–W control. The modulated line waveform is then applied to the line scan coils via transformer coupling T1 to modify the basic current sawtooth scan waveform.

Correction along the N–S axis involves modulating the field sawtooth by small parabolic waveforms at line frequency (Fig. 14.17). The amplitude of the parabolas

**Fig. 14.17** *N–S correction*

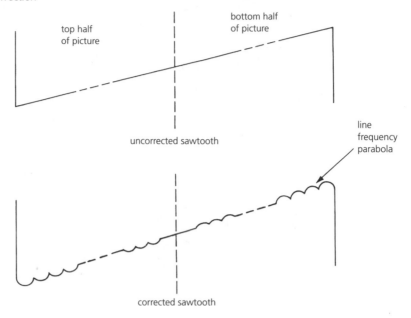

diminishes as the beam reaches the middle of the scan and reverses as it proceeds towards the bottom.

## Degaussing

The magnetic field applied to the three beams may be affected by the earth's magnetic field and any other stray field in the proximity of the receiver, including the effect of domestic appliances. The result is impurity in the form of colour patches on the display that cannot be removed by purity corrections. To avoid these effects, a magnetic shield is fixed over part of the cone of the tube. In addition, the shadow mask itself, together with other steel fittings near the tube face, must be demagnetised regularly to avoid impurity and misconvergence. Demagnetisation may be carried out manually using a degaussing coil; the coil is moved slowly and held parallel to the face of the screen. Energised by an a.c. supply such as the mains, the degaussing coil is moved slowly away, turned face downward and finally turned off when it is a few feet away. Everyday degaussing is carried out automatically every time the receiver is switched on. Two degaussing coils are fitted partly inside and partly outside the magnetic shield (Fig. 14.18(a)) and connected to the mains supply (Fig. 14.18(b)); R1 is a thermistor and R2 is a voltage-dependent resistor. At switch-on R1 and R2 have very small resistances, so they allow a large a.c. current to flow through the coils; this causes the temperature of R1 to rise, increasing its resistance. The voltage across R1 also increases, so the voltage across R2 falls, which increases its own resistance, and so on. The combined effect is a decaying current through the coils (Fig. 14.19) and this demagnetises the tube.

**Fig. 14.18** *(a) Colour tube magnetic shield and (b) degaussing circuit*

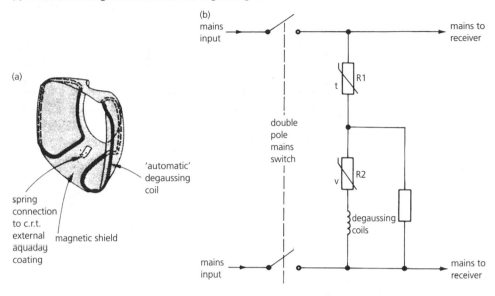

(b)
mains input

mains to receiver

R1

t

double pole mains switch

R2

v

degaussing coils

mains input

mains to receiver

(a)

'automatic' degaussing coil

spring connection to c.r.t. external aquadag coating

magnetic shield

**Fig. 14.19** *Degaussing current*

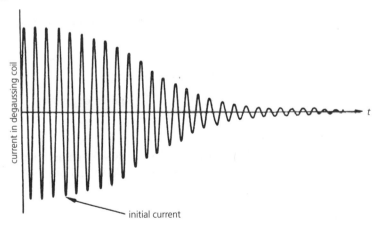

current in degaussing coil

t

initial current

## QUESTIONS

1. Explain each of the following, as used in a colour cathode-ray tube:
   (a) CRT isolator
   (b) electromagnetic deflection
   (c) purity
   (d) convergence

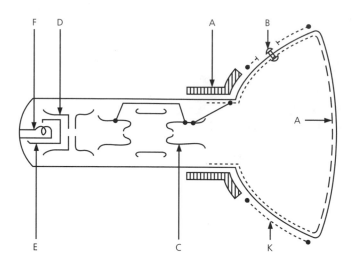

2.  Refer to Fig. Q14.1. Name the following parts:
    (a) part A
    (b) part B
    (c) part C
    (d) part F

3.  Refer to Fig. Q14.1. Explain the function of the following parts:
    (a) part D
    (b) part E

4.  Explain what is meant by pincushioning.

5.  Explain briefly why a colour CRT requires degaussing. State the visual effect of automatic degaussing circuit failure.

6.  Refer to Fig. Q14.2.
    (a) State, with reasons, the effect of R3751 going open circuit.
    (b) State the voltage between pin 8 (A19) and pin 7 (A20).

**Fig. Q14.2**

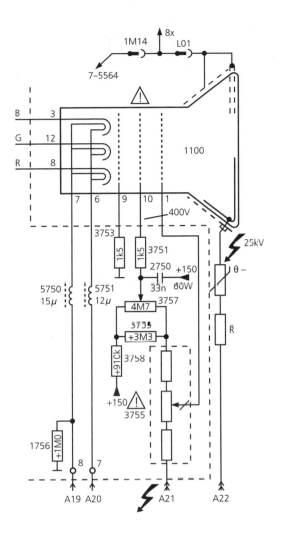

# **15** Digital processing

Physical matters that act upon our senses such as light or sound are of the analogue type; that is, they vary along a continuous curve. For this reason, electronic systems such as audio and TV systems have developed along analogue lines. Advances in digital technology and the manufacture of i.c.s have meant that sound, light and colour values can be expressed and manipulated in digital form, but at either end of the signal chain a transducer must be employed. At the transmitter, the transducer converts the analogue signal into a digital form. At the receiving end, the transducer transforms the digital expressions into an analogue form so they may be interpreted by our senses. The use of digital processing in TV transmission and reception provides a number of advantages and opens new opportunities. At the transmitter it is compatible with digital switching techniques used in TV studios. In the receiver it is compatible with teletex, remote control and NICAM sound transmission. With error detection and correction techniques, digital processing can provide improved transmission quality even in a noisy environment.

With current analogue public broadcasting, digital processing of sections of the receiver may be incorporated in a TV set. Such television sets are known as *analogue/digital receivers*. The major effect of this on the received picture is to produce an extremely steady display, which itself produces a subjective improvement in the picture quality. Flickering of the lines and fields may be eliminated with the use of memory devices. Memories may also be used to correct the shape of the raster and to reduce the effects of noise and interference on the tube image. One of the main advantages of digital processing is realised in the production process. Analogue/digital receivers require fewer adjustments at the manufacturing stage, reducing labour costs. It is also possible to manufacture multiple-standard TV receivers for use in different countries with minimal adjustments.

Digital processing in a TV receiver is amenable to microcomputer control and automatic adjustments such as grey scale tracking. It opens up wide possibilities such as digital storage, frame freeze and various forms of picture manipulation such as zoom and multiple displays. Analogue/digital receivers may also be used in conjunction with various other peripheral devices, such as computers and printers, which may be used to record text and images.

## **Digital transmission**

Unlike analogue signals, which are continuous and may take an infinite number of instantaneous values, a digital signal uses the binary system with two discrete levels: logic '0'

**TABLE 15.1**

| Denary | Binary columns | | |
|---|---|---|---|
| | C (4) | B (2) | A (1) |
| 0 | 0 | 0 | 0 |
| 1 | 0 | 0 | 1 |
| 2 | 0 | 1 | 0 |
| 3 | 0 | 1 | 1 |
| 4 | 1 | 0 | 0 |
| 5 | 1 | 0 | 1 |
| 6 | 1 | 1 | 0 |
| 7 | 1 | 1 | 1 |

is known as space and is represented by 0 V; logic '1' is known as mark and is represented by a certain voltage, normally 5 V. A single binary digit, known as a *bit*, provides basic YES or NO information. More information may be conveyed by grouping a number of bits together, e.g. 4, 8, 16. Such groupings are known as *words*. A word is a group of binary digits or bits which form the basic unit of information in a digital system. A 4-bit word, known as a *nibble*, can be used to represent $2^4 = 16$ different numbers from 0 to 15. An 8-bit word, known as a *byte*, can represent $2^8 = 256$ different numbers from 0 to 255, and so on.

In the same way as denary (decimal) columns represent increasing powers of 10, binary columns represent increasing powers of 2; the rightmost bit is known as the least significant bit (LSB) and has a value of $2^0 = 1$. The next column has a value of $2^1 = 2$, the next $2^2 = 4$, and so on (Table 15.1). In any binary word, the leftmost bit is known as the most significant bit (MSB).

## Hexadecimal

In order to avoid the use of long strings of binary digits, hexadecimal notation is used. Hexadecimal numbers have a base of 16, hence they have 16 distinct symbols:

0, 1, 2, 3, 4, 5, 6, 7, 8, 9, A, B, C, D, E, F

with A, B, C, D, E and F representing denary numbers 10, 11, 12, 13, 14 and 15 respectively. Each 4-bit binary number may thus be represented by a single hexadecimal digit (Table 15.2). An 8-bit binary number is represented by a 2-digit hexadecimal number and a 12-bit binary number by a 3-digit hexadecimal number (Fig. 15.1). To avoid confusion, the base (2 for binary and 16 for hexadecimal) may be shown as a subscript, e.g. $1001_2$ and $A3_{16}$. A more common way of distinguishing between the two types of numbering systems is to terminate binary numbers with a B and hexadecimal numbers with an H. For instance, binary 0110 is written as 0110B and hexadecimal 2F as 2FH.

| TABLE 15.2 | |
|---|---|
| Hexadecimal | Binary |
| 0 | 0000 |
| 1 | 0001 |
| 2 | 0010 |
| 3 | 0011 |
| 4 | 0100 |
| 5 | 0101 |
| 6 | 0110 |
| 7 | 0111 |
| 8 | 1000 |
| 9 | 1001 |
| A | 1010 |
| B | 1011 |
| C | 1100 |
| D | 1101 |
| E | 1110 |
| F | 1111 |

**Fig. 15.1**

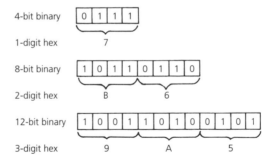

## Serial and parallel transmission

Digital transmission is the process of sending data in the form of a series of pulses down a line or by radio wave from one place to another, a process known as pulse code modulation (PCM). Digital data may be transmitted in one of two modes, serial and parallel (Fig. 15.2). In serial mode the bits are transferred in sequence, one after the other: b0, b1, b2, and so on. In parallel mode the bits are transferred simultaneously along a number of parallel lines, one line for each bit, in synchronism. Parallel transmission is fast, since complete words are sent at each clock pulse, but it requires as many lines as there are bits in the word. For this reason, serial transmission is used over medium and long distances. Another important consideration is the number of pins that must be made available on an i.c. to accommodate parallel data transmission. For large multibit systems the number of pins may be so large as to make construction impractical.

**Fig. 15.2** *Data transmission modes: (a) serial and (b) parallel*

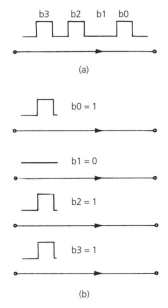

(a)

(b)

Digital transmission is synchronised by the system clock, which also determines the speed of transmission known as the *bit rate*. For instance, a 6-bit word transmitted serially on a single line has a bit rate given by

bit rate = no. of bits × clock frequency

and given a clock frequency of 32 kHz, then

bit rate = $6 \times 32 \times 10^3 = 192\,000 = 192$ kbit/s.

In serial mode a start bit and a stop bit are normally added at the ends of the word.

## Shift registers

A shift register is a temporary store of data which may then be sent out in a serial or parallel form. An 8-bit shift register is illustrated in Fig. 15.3, in which serial data is clocked into the register, bit by bit. When the register is full, the data stored in the register may then be clocked out serially, bit by bit. This type of shift register is known as a serial-in serial-out (SISO) shift register. Three other possible arrangements are possible: serial-in parallel-out (SIPO), parallel-in serial-out (PISO) and parallel-in parallel-out (PIPO).

**Fig. 15.3**

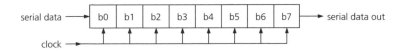

# Multiplexing

Multiplexing is a commonly used method to reduce the number of lines and therefore the number of pins on the chip. Two or more sets of data are made to share a single set of lines on a *time-sharing* basis, with each set of data occupying the lines for part of the time. The sets of data sharing the parallel lines may form part of a complete word. For instance, an 8-bit word may be transmitted along four lines by multiplexing one-half of the word, the most significant four bits, with the other half, the least significant four bits. At the receiving end, a demultiplexer is used to reconstruct the complete word. Multiplexing may also be used in the serial mode of transmission, in which different channels may be made to share a single line on a time-sharing basis, as in telephony and other data transmission systems.

# Error detection

Unlike analogue transmission, in which small errors in the received signal may not affect the operation of the system in a significant way, a data communication may suffer a fatal fault as a result of the smallest possible error, a change in the logic level of a single bit. The effect of errors on digital sound or video signals is to degrade the quality of reproduction. Digital transmission is prone to such errors; they are caused by noise and interference. Besides the purely random noise encountered on communication systems, transient noise can have a devastating effect on data transmission. A transient of say 10 ms duration can cause what is known as an *error burst*, which may obliterate a number of bits in a digital stream. For this reason, some form of error detection and correction system must be incorporated.

There are several methods of checking for errors. They all employ additional (or redundant) bits to detect the occurrence of errors. The simplest technique is the addition of a single *parity bit* at the end of the digital code to indicate whether the number of 1's in the digital coded word is even or odd. The parity bit may be set to 0 or 1 (Fig. 15.4). There are two types of parity checking: *even parity* (Fig. 15.4(a)) is when the complete code pattern (including the parity bit) contains an even number of 1's; *odd parity* (Fig. 15.4(b)) is when the complete code pattern contains an odd number of 1's. At the receiving end, the number of 1's is counted and checked against the parity bit; a difference indicates an error. The basic drawback of the simple parity check is that it can only detect a single bit error. An error affecting two bits will go undetected. Furthermore, there is no provision for determining which bit is actually in error. For these reasons, a more sophisticated system of error detection is normally used.

One method for detecting the actual wrong bits is based on the generation of a *parity word* for a group of digital codes arranged in a block of columns and rows. A parity bit for each row and column is generated to form a parity word. At the receiving end, the parity of each column and row is checked and if an error is indicated, the precise bit may be identified.

Another method is to *interleave* parity bits with bits carrying messages; one such technique is the Hamming code. Alternatively, the interleaved bits may be arranged in

**Fig. 15.4**

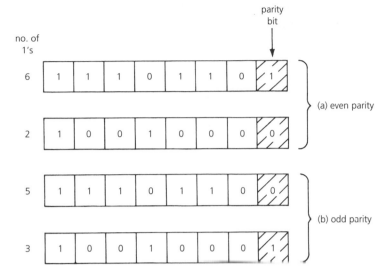

(a) even parity

(b) odd parity

a matrix consisting of a number of columns and rows. The bits within the block are then transmitted column by column. Interleaving is a simple and powerful error detection technique which is capable of detecting a number of simultaneous bits that may be in error. Interleaving may also be used to minimise the effect of error burst by separating adjacent bits, a technique employed very successfully in NICAM digital transmission.

## Digital coding

The conversion of a quantity or a number into a digital format may be carried out using one of a number of codes. The natural binary code is listed in Table 15.1, in which the columns of say a 4-bit binary number represent progressively increasing powers of 2, giving a count of 0 to 15. Such a count is not very appropriate for denary applications. A more appropriate coding system is *binary-coded decimal* (*BCD*), which converts each denary digit into a 4-bit binary number. A 2-digit denary number will thus result in two groups of 4-bit binary numbers (8 bits in total) as shown in Table 15.3.

Another popular coding technique is the Gray code; this ensures that only one bit changes state as the denary number is progressively incremented (Table 15.4). The Gray code avoids the problems of spurious transitional codes associated with the BCD technique.

### Twos complement

The coding techniques considered so far provide positive quantities of unsigned numbers. For instance, an 8-bit binary number may have values between

0000 0000 binary = 00 denary

and

**TABLE 15.3**

| Denary number | BCD |
|---|---|
| 00 | 0000 0000 |
| 01 | 0000 0001 |
| 02 | 0000 0010 |
| 03 | 0000 0011 |
| 04 | 0000 0100 |
| 05 | 0000 0101 |
| 06 | 0000 0110 |
| 07 | 0000 0111 |
| 08 | 0000 1000 |
| 09 | 0000 1001 |
| 10 | 0001 0000 |
| 11 | 0001 0001 |
| . | . . |
| . | |
| . | |
| 57 | 0101 0111 |
| . | . . |
| . | |
| . | |
| 83 | 1000 0011 |
| . | . . |
| . | |
| . | |
| 99 | 1001 1001 |

**TABLE 15.4**

| Denary | Binary code | Gray code |
|---|---|---|
| 0 | 0000 | 0000 |
| 1 | 0001 | 0001 |
| 2 | 0010 | 0011 |
| 3 | 0011 | 0010 |
| 4 | 0100 | 0110 |
| 5 | 0101 | 0111 |
| 6 | 0110 | 0101 |
| 7 | 0111 | 0100 |
| 8 | 1000 | 1100 |
| 9 | 1001 | 1101 |
| 10 | 1010 | 1111 |
| 11 | 1011 | 1110 |
| 12 | 1100 | 1010 |
| 13 | 1101 | 1011 |
| 14 | 1110 | 1001 |
| 15 | 1111 | 1000 |

1111 1111 binary = 255 denary

all of which are positive. In order to distinguish a negative number from a positive number, a − sign is used to precede a denary number, e.g. −25. In the binary system, the negative sign itself is given a binary code so that it may be recognised by a digital system. A binary bit known as the *sign bit* is introduced to the left of the MSB of the binary number; it is devoted entirely to indicating the sign of the number. Such numbers are often called *sign-and-magnitude*. When the sign bit is set to 0, the number is positive and when it is set to 1 the number is negative. It follows that a signed 8-bit number has its magnitude indicated by the first seven bits, 0–6, and its sign is indicated by bit 7. An 8-bit code will therefore lie in the range

[1] 111 1111 = −127
[0] 111 1111 = +127

where the bit in brackets is the sign bit.

The sign-and-magnitude system of representation is not conducive to arithmetic operations using positive and negative numbers. To overcome this limitation, the twos complement representation of negative numbers is used, and this automatically produces the correct sign following arithmetic operations.

For positive binary numbers, the twos complement is identical to the sign-and-magnitude system in that it also uses the MSB as the sign bit, leaving the remaining bits to indicate the magnitude.

For negative numbers, the twos complement is obtained by first producing the complement of the original binary number (with a positive sign) and then adding a 1 to the least significant bit (LSB). The complement of binary 1 is 0 and the complement of binary 0 is 1. Hence the complement of a binary number is obtained by inverting all its digits, i.e. changing all 1's to 0's and all 0's to 1's.

For example, the twos complement for +54 is given as

+54 = [0] 0110110

However, to find the twos compliment for −54, start with its positive equivalent:

```
              sign
              bit
   +54      = 0 0110110
complement = 1 1001001
                    +1
    twos
complement = 1 1001010
```

Table 15.5 shows the conversion of denary numbers into 8-bit twos complement codes.

# Digital memory chips

A memory chip consists of a number of memory cells into which data bits may be stored (or written). The stored data may then be retrieved (or read) from the device. These

**TABLE 15.5**

| Denary number | Twos complement |
|---|---|
| +127 | 0 111 1111 |
| +126 | 0 111 1110 |
| +2 | 0 000 0010 |
| +1 | 0 000 0001 |
| zero | 0 000 0000 |
| −1 | 1 111 1111 |
| −2 | 1 111 1110 |
| −126 | 1 000 0010 |
| −127 | 1 000 0001 |
| −128 | 1 000 0000 |

**Fig. 15.5** *Eight 4-bit memory locations*

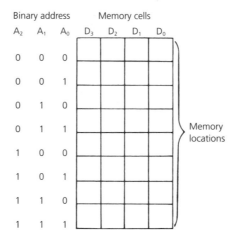

memory cells are grouped together to form a memory location (e.g. a 1-bit, 2-bit, 4-bit or 8-bit memory location). For the purposes of identification, each location is given a unique binary code known as an address.

Figure 15.5 represents the arrangement for eight 4-bit memory locations. Each location has four memory cells, D0, D1, D2 and D3, into which a 4-bit word may be stored. Each location is identified by a unique 3-digit (A0, A1, A2) binary *address*; 000, 001, 010 . . . 110, 111. A 3-digit binary number can address up to $2^3 = 8$ locations. In order to be able to address a bigger memory store with a larger number of locations, a binary address of a higher order must be used. For example, to address 16 (= $2^4$) different locations, four digits (A0, A1, A2, A3) are necessary and to address 32 locations five digits are required, and so on. A memory store with $2^{10} = 1024$ locations is known as having a 1K memory. Such a memory device requires a 10-digit address (A0, A1, . . . , A9). The actual size of the memory is determined by the number of bits in each location as well as the number of locations available on the device. Hence a memory chip with 1024 locations, with each location consisting of two memory cells or bits, has a total memory capacity of $1K \times 2 = 1024 \times 2 = 2048$ bits in total.

## Read-Only Memory

A read-only memory (ROM) is a non-volatile memory used for storing data permanently. The data stored can only be read by the user, hence its name, and no new data can be written into the device. It is programmed by the manufacturer in accordance with the user specifications.

## Programmable Read-Only Memory

The fact that ROMs are programmed by the manufacturer means that they can be expensive unless they are produced in large quantities. Furthermore, subsequent changes to the program once it has been written into ROM are very costly. To avoid this, programmable read-only memory (PROM) chips are used. PROMs fulfil the same basic function as ROMs except they may be programmed by the user instead of by the manufacturer. Once programmed, the data stored in a PROM cannot be altered.

## Erasable programmable read-only memory

The main disadvantage of a PROM is the fact it cannot be reprogrammed. Mistakes in programming cannot be corrected. The erasable PROM (EPROM) overcomes this by allowing the user to delete or erase the stored data and thus change the program. The stored program in an EPROM may be erased by exposing the memory cells to ultra-violet light through a window on the i.c. package.

## Electrically erasable programmable read-only memory

An electrically erasable programmable read-only memory (EEPROM) overcomes the disadvantages of an ordinary EPROM. EEPROMs can be programmed and erased in circuit by the application of suitable electrical signals. Furthermore, individual locations may be erased and programmed without interfering with the rest of the data pattern.

## Random access memory

Random access memory (RAM) is a volatile, i.e. non-permanent memory chip, which the user may read from and write into, hence it is also known as read/write memory. Locations may be accessed at random by placing the address of the selected location onto the address lines. RAMs are divided into two major categories according to the type of storage technique. *Dynamic* RAMs (DRAMs) store information in the form of a charge on a capacitor. However, due to leakage, the charge is lost and has to be restored, a process known as refreshing the cell. Dynamic RAMs have the advantage of higher density and lower power consumption. *Static* RAMs employ flip-flops as the basic cell, hence they require no refreshing. Static RAMs with hold data as long as d.c. power is applied to the device.

## Access time

An important property of a memory chip is its access time. *Access time* is the period between an address appearing on the address bus and valid data appearing on the data bus (reading mode) or data being written into the chip (write mode). Dynamic RAM (DRAM) chips have a longer access time (typically 60 ns) compared with static RAM (SRAM) chips (typically 20 ns). Memory chips with a short access time can operate at a faster clock pulse rate, providing faster read/write cycles. Typical access times for different memory chips are shown in Table 15.6.

### TABLE 15.6

| Type | Access time (ns) |
| --- | --- |
| DRAM | 60 |
| SRAM | 20 |
| ROM | 200 |

# Digital processing in a TV receiver

Figure 15.6 shows the principles of digital processing in a TV set receiving analogue broadcast signals. The u.h.f. analogue signal is received in the normal way by the aerial. Following i.f. amplification and detection by the video demodulator, the composite video is fed to the analogue-to-digital converter (ADC). The ADC acts as a pulse code modulator (PCM) encoder, translating the analogue input into a coded multi-line digital stream of pulses. Binary-coded decimal, Gray code or other coding techniques may be used. The digitised composite video is then fed to the digital processing unit. A number of chips are used to perform a variety of functions, including chrominance decoding, matrixing, clamping, blanking, horizontal and vertical deflection and other controls such as contrast and brightness. Luminance and chrominance delay is implemented in a RAM chip. Data bits of digitised signals are written into successive memory cells of a RAM

**Fig. 15.6** *Digital signal processing*

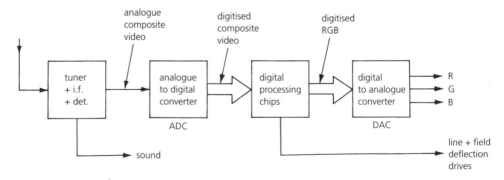

chip. The chip is then put in the read mode and the bits are retrieved, resulting in time delay. The amount of delay is determined by the number of memory cells used for each bit of data.

The digitised RGB signals from the digital processor are then fed into a digital-to-analogue converter (DAC) to translate the digitised primary colour signals back into an analogue form before going into the c.r.t. via appropriate output stages. The digital-to-analogue and analogue-to-digital converters may be incorporated into a single i.c. known as a *code–decode* or *codec* chip. Furthermore, the digital processor carries out the necessary sync. separation and timebase generation to produce the required drives for the line and field output stages. Sound may also be digitally processed using NICAM technique, as will be explained later.

Digital processing may take place after the separation of the composite video into its separate individual components: luminance, Y and the two colour difference signals, $B' - Y'$ and $R' - Y'$. The three separate components are then digitised by an analogue-to-digital converter before going into the digital processor. This technique is known as component coding as opposed to the composite coding shown in Fig. 15.6. Composite coding will be used for the rest of this chapter as this is the most popular technique employed by TV manufacturers.

## Analogue-to-digital converter

The analogue-to-digital converter (ADC) is a pulse code modulator. It takes the analogue input, samples it then converts the amplitude of each sample to a digital code (Fig. 15.7). The output is a number of parallel digital bits (four in Fig. 15.7) whose simultaneous logic states represent the amplitude of each sample in turn. A variety of codes may be used for this representation, including the natural binary code.

### Sampling

The Shannon theory of sampling states that, for satisfactory results, the analogue signal must be sampled at a rate which is at least twice the highest frequency of the baseband of the original analogue input. This sampling rate is known as the *Nyquist rate*. When the samples are reproduced and the dots are joined together, the reconstructed waveform contains all the information of the original analogue waveform. Figure 15.8 shows the sampling and reconstruction of an analogue waveform. If the sampling rate were lower than the Nyquist rate, i.e. comparable to the highest frequency of the analogue

**Fig. 15.7** *Analogue-to-digital conversion*

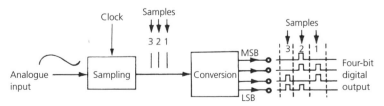

**Fig. 15.8** *Reconstruction of a sampled waveform*

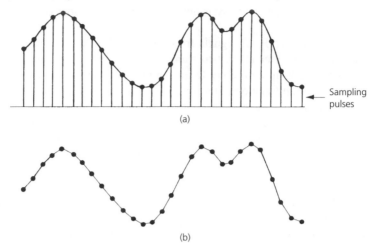

Sampling pulses

(a)

(b)

signal, an overlap would occur between the sidebands produced by the sampling pulses. This creates an effect known as aliasing, which makes it impossible to recover the original signal without distortion.

For digital processing in a TV receiver, the highest composite signal frequency is 5.5 MHz. Theoretically, therefore, a sampling frequency of $2 \times 5.5 = 11$ MHz will be adequate. However, in practice a whole multiple of the colour subcarrier frequency is used to avoid intermodulation between the sampling clock pulse and the subcarrier frequency caused by the non-linearity of the analogue-to-digital conversion process. Sampling at a whole multiple of the subcarrier frequency has a number of other operational advantages, including simplified composite processing and a more convenient conversion process in the case of component coding. The lowest multiple that gives a sampling frequency greater than the Nyquist rate is three times the subcarrier frequency, i.e. $3 \times 4.434 = 13.3$ MHz. Sampling at four times the subcarrier frequency, $4 \times 4.434 = 17.7$ MHz, is more commonly used, giving further operational benefits in terms of a simplified colour decoding technique.

## Quantising

The final stage of the ADC is the conversion process itself. A number of levels are established, say 0.25, 0.5, 0.75, 1.0, etc., and each level is given a binary code. This is known as *quantising*. The number of these quantum steps or discrete levels is determined by the number of bits at the converter output. For instance, in a 3-bit ADC the binary output can have a coded value from 000 to 111, a total of eight levels. Suppose that a scale or quantum of 250 mV is used, then Table 15.7 will apply.

## Quantising noise

With the input being analogue (i.e. continuous), sample voltages will invariably fall between the quantum levels. Hence there is always an element of uncertainty or ambiguity in

**TABLE 15.7**

| Level sample | Voltage (V) | Binary code | | |
| | | MSB | | LSB |
| --- | --- | --- | --- | --- |
| 0 | 0 | 0 | 0 | 0 |
| 1 | 0.25 | 0 | 0 | 1 |
| 2 | 0.50 | 0 | 1 | 0 |
| 3 | 0.75 | 0 | 1 | 1 |
| 4 | 1.00 | 1 | 0 | 0 |
| 5 | 1.25 | 1 | 0 | 1 |
| 6 | 1.50 | 1 | 1 | 0 |
| 7 | 1.75 | 1 | 1 | 1 |

terms of the value of the least significant bit (LSB). This uncertainty gives rise to a quantising (or ±1) error, an error that is inherent in any digital coding of analogue values. The quantising error gives rise to quantising distortion, known as quantising noise. The effect of noise is that some of the binary digits are received incorrectly and distort the reconstructed waveform.

Quantising noise has a constant quantity equal to one-half the quantum level. The effect of quantising noise is therefore more noticeable at low analogue signal voltages, resulting in a poor signal-to-noise (S/N) ratio. For instance, using Table 15.7, with a quantum level of 250 mV, a binary code of 110 may represent a voltage from 1.375 V to 1.625 V, i.e.

$$\text{quantised level} = 1.500 \text{ V} \pm 1/2 \text{ quantum}$$
$$= 1.500 \text{ V} \pm 1/2 \times 0.250 \text{ V}$$
$$= 1.500 \pm 0.125 \text{ V}.$$

This gives a signal-to-noise ratio of

$$\text{S/N} = 1.5/0.125 = 12 \text{ or } 21.6 \text{ dB } (20 \log 12).$$

For a smaller analogue signal level, e.g. 0.25 V, the signal-to-noise ratio is

$$\text{S/N} = 0.25/0.125 = 2 \text{ or } 6 \text{ dB } (20 \log 2).$$

## Companding

Although the quantising error cannot be wholly avoided, it can be minimised by improving the resolution of the converter through increasing the number of bits used and thus reducing the quantum level, the quantising error and the quantising noise. However, this still leaves weak signals with a poor signal-to-noise ratio. To overcome this, non-linear quantising may be used, in which the quantum level for weak signals is decreased and the quantum level for strong signals is increased. At the receiving end, a complementary non-linear digital-to-analogue conversion is employed to reproduce the original analogue signal. This non-linear coding/decoding technique, called companding, tends

**Fig. 15.9** *A 7-bit analogue-to-digital converter (ADC)*

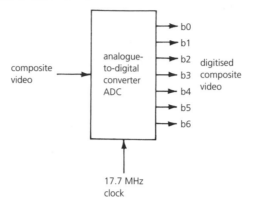

to equalise the signal-to-noise ratio over the range of sample amplitudes generated by the analogue signal. Another companding technique which accomplishes the same result is to use an analogue voltage compressor to precede a linear encoder. The companded analogue voltage from the voltage compressor gives prominence to weaker signal levels. After the decoding stage, a complementary expander is used to restore the original signal. Companding may also be used to reduce the number of output bits for the same number of quantum levels, a highly desirable outcome with limited bandwidths.

## Video encoding

Digital processing of video signals requires a digital code with a minimum of 6 bits, giving $2^6 = 64$ different levels. Assuming a peak-to-peak input voltage of 1 V, a quantum step of 1 V/63 = 16 mV is obtained. A 6-bit code will enable an acceptable picture to be reproduced. In practice, however, 7-bit or 8-bit codes are used for a better quality picture. Figure 15.9 shows a 7-bit ADC driven by a 17.7 MHz sampling clock.

A digital broadcast system requires a minimum of 256 quantum levels, i.e. an 8-bit code. To that must be added the redundant bits for error detection. For our purposes, the coding and decoding are carried out within the receiver with little possibility of the noise and the interference associated with TV broadcasting. Hence there is no need for error detection, rendering the system far simpler than those required if the broadcast system itself were digitally coded.

## Flash ADC

There are many types of analogue-to-digital converters on the market. For TV processing applications, a fast and accurate ADC such as the flash (also known as simultaneous) converter is necessary. A typical arrangement for a 6-bit flash type ADC chip is shown in Fig. 15.10. It consists of 63 comparators, one for each quantum step above zero. One input of each comparator is connected to a tapping point along a potential divider chain consisting of 64 separate resistors connected to a reference voltage $V_{ref}$. Each resistor provides one quantum step. The resistor chain thus provides each comparator with a constant reference voltage that is progressively increasing in steps of one

**Fig. 15.10** *A 6-bit flash ADC chip*

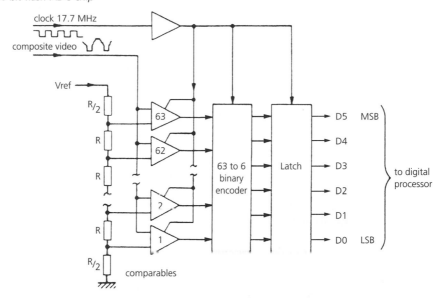

quantum. The value of the quantum step is determined by the reference voltage. For instance, given a reference voltage of 0.96 V, the quantum step is

$$V_{ref}/64 = 0.96/64 = 15 \text{ mV}$$

The composite video input is simultaneously applied to the second sampling inputs of all 63 comparators. Comparisons thus take place between the instantaneous value of the input signal and 63 different voltage levels. When the instantaneous value of the input is higher than the portion of the reference voltage fed into the particular comparator, the comparator conducts and a logic 1 is obtained at its output. Thus an input level equal to two quanta results in the bottom two comparators conducting; an input of three quanta results in the bottom three comparators conducting, and so on. The input level is therefore quantised into 64 steps. At each positive (or negative) edge of the 17.7 MHz sampling clock, the comparators are enabled, feeding one quantised sample into the binary encoder which converts the 63-line quantised output from the comparators into a 6-bit binary code. At the second clock edge, a second sample is encoded, and so on. The output from the 63-to-6 binary encoder is latched to hold the 6-bit code long enough for the digital processor to accurately capture the information. The latch is cleared regularly by the sampling clock, ready to store the next coded sample, and so on.

## Digital-to-analogue converter

The digital-to-analogue converter receives a parallel digital input and converts it back to a voltage (or a current) value that is represented by the binary input. If this is repeated for successive coded inputs, an analogue waveform may be reconstructed. For instance, assuming a 3-bit binary input, then 000 will be represented by a zero output and 111

**Fig. 15.11** *Digital-to-analogue converter (DAC)*

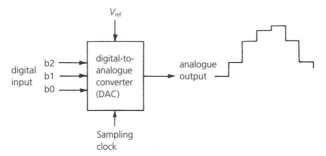

**Fig. 15.12** *Digital-to-analogue converter (DAC) chip: block diagram*

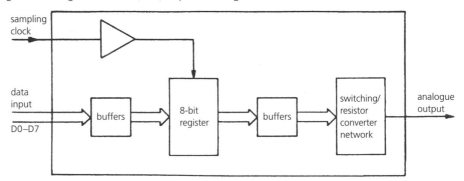

by a maximum voltage output determined by the reference voltage $V_{ref}$ in Fig. 15.11. Other inputs are reproduced as a proportion of $V_{ref}$, e.g. 001 as $\frac{1}{8}V_{ref}$, 011 as $\frac{3}{8}V_{ref}$ and 110 as $\frac{6}{8}V_{ref}$ (or $\frac{3}{4}V_{ref}$). Each bit of the binary input is reproduced in accordance with its weighting, giving the following general formula:

output level $= V_{ref}$ (b2/2 + b1/4 + b0/8)

A typical block diagram for a DAC chip is shown in Fig. 15.12. The 8-bit input data is fed into a temporary store via a buffering stage. The store is in the form of an 8-bit register which samples the input data at regular intervals determined by the clock frequency. It holds each 8-bit item of coded data long enough for the switching/resistor converter network to capture and decode the input. At each clock pulse, the register updates its contents with the next incoming 8-bit code, and so on.

The switching/resistor converter network is normally of the R-2R ladder type, which converts the succession of 8-bit coded data into an analogue waveform. A common arrangement for an R-2R ladder converter is shown in Fig. 15.13; it contains eight electronic switches, one for each bit. The position of the switches is determined by the logic state of the relevant bit. When all the bits are 0, all the switches are downconnected to chassis and the output is zero. When one bit has a logic 1, the relevant switch is connected to the reference voltage line $V_{ref}$. A portion of the reference voltage then appears at the output depending on the weighting of the bit. For instance, if the most significant bit (MSB) is logic 1 and the other bits are at logic 0, then an output of $\frac{1}{2}V_{ref}$ will be produced, representing the correct weighting of the most significant bit, and so on.

**Fig. 15.13** *R-2R ladder DAC*

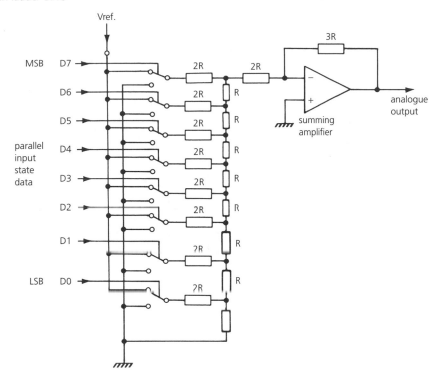

## 'Digital' TV chassis

A block diagram for a TV receiver employing digital video processing and based on a Fidelity chassis is shown in Fig. 15.14. The master clock chip provides a 17.7 MHz sampling pulse waveform, which is also used to synchronise the operation of the other units. The clock frequency which is four times the colour subcarrier is locked to the colour burst at the video processor by a phase-locked loop (PLL). Composite video from the vision detector is fed to an analogue-to-digital converter, which usually forms one-half of a codec chip. Sampling is carried out at a frequency of 17.7 MHz to produce a 7-bit digital output encoded in BCD or Gray code. The digitised composite video is then passed via a 7-bit bus to the teletex decoder, video processor and deflection processor chips.

The teletex decoder produces appropriate outputs for the display memory and four output lines, RGB and fast blanking for the c.r.t. display.

The video processor chip carries out all necessary luminance and chrominance processing functions, including the separation of luminance and chrominance components, luminance processing (delay, contrast, etc.), chrominance delay and decoding of the PAL signal into colour difference components. The output of the video processor consists of a 4-bit multiplexed colour difference signal and an 8-bit luminance signal. Eight bits are used for the luminance to allow for 256 shades of grey. The two colour difference signals are demultiplexed before being matrixed with the luminance signal to produce

**Fig. 15.14** *A 'digital' TV chassis*

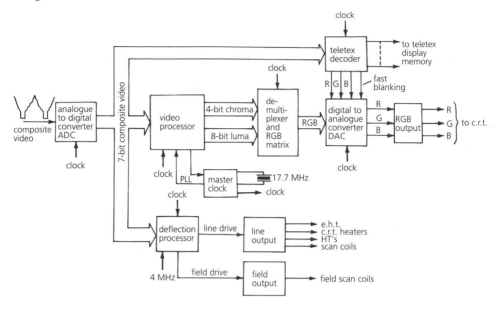

three coded primary colour signals, R, G and B, ready for conversion back into an analogue format by the digital-to-analogue converter.

The deflection processor chip provides timebase synchronisation and line and field drive waveforms, which feed into the line and field output stages respectively. The deflection processor decodes the 7-bit input and separates the line and field sync. pulses. The process of decoding is clocked by the 17.7 MHz pulse. The processor also receives a 4 MHz reference frequency from a central control unit (not shown). The line frequency is obtained by dividing this frequency by 256, resulting in a very steady 15.625 kHz waveform which is then locked to the line sync. pulse. E.h.t. and other d.c. supplies are produced in the normal way by the line output stage.

The deflection processor also provides a pulse-width modulated (PWM) field drive output which is locked to the field sync. pulse. A sawtooth drive is then obtained by integrating the PWM waveform using an RC network before going into the field output chip.

### QUESTIONS

1. Convert the following binary numbers into hexadecimal:
   (a) 00101101
   (b) 101010101010
   (c) 1111000011000011

2. Explain briefly the meaning of the following terms:
   (a) parity
   (b) BCD
   (c) interleaving

3.  Give a typical access time for
    (a)  a dynamic RAM (DRAM) chip
    (b)  a static RAM (SRAM) chip

4.  In a digital system, state the purpose of
    (a)  ADC
    (b)  DAC

5.  It is required to digitise an analogue signal with a bandwidth of 0–0.5 MHz; state
    (a)  the Nyquist sampling rate
    (b)  a practical sampling frequency

# 16 NICAM digital stereo and Dolby sound

The system of transmitting sound used in British 625-line TV broadcasting employs a sound frequency modulated carrier placed just outside the video spectrum. It can be used to provide a good quality sound given good quality sound amplification at the receiver. However, it is incapable of producing hi-fi quality and unable to carry stereophonic sound. Stereo sound has been successfully transmitted by v.h.f. radio broadcasting using analogue modulation of an f.m. carrier. Such a system does not readily commend itself for TV broadcasting because of its bandwidth requirements. It is not possible to add a second sound carrier between 6 and 8 MHz to the video frequency spectrum without causing unacceptable interference to either the vision or the primary 6 MHz sound carrier. To avoid this, compromises would have to be reached, and this would defeat the original aim of stereo hi-fi sound transmission.

After years of research and development BBC engineers came out with a radically new sound system for TV broadcasting using state-of-the-art technology. The system became known as NICAM 728 or NICAM for short. NICAM stands for near instantaneous companded audio multiplex and 728 refers to the data rate of 728 kbit/s. It provides two completely independent sound channels so that dual-language sound tracks may be transmitted as well as stereophonic sound. It can carry data in one or both channels and is completely separate and independent of the existing f.m. monophonic sound channel.

The NICAM system is a classic example of how a digitised waveform can have its coded data manipulated in a variety of ways (scattered, compressed, companded, interleaved and scrambled) and yet can be reconstituted without any loss of the original information. It heavily relies on modern digital technology and on the ability of i.c. manufacturers to make very complex digital chips at low cost. NICAM has been accepted by the UK government as the standard for stereo broadcasting and is recommended by the European Broadcasting Union as the digital standard for terrestrial, i.e. land-based, stereo broadcasting in Europe. It is now being considered as a possible international standard by the International Standards Radio Consultative Committee (CCIR).

## NICAM system outline

In the NICAM system, digitised analogue sound signals are grouped into *data blocks* of 704 bits each. The data blocks are organised in a 1 ms *frame* structure (Fig. 16.1). Each data block is preceded by a *frame alignment* word (FAW) to inform the receiver

**Fig. 16.1** *NICAM 1 ms frame structure*

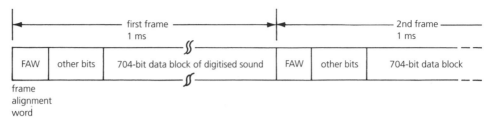

of the start of each frame. The total frame is then used to modulate a 6.552 MHz carrier, which falls just outside the normal 6 MHz f.m. sound but remains within the total TV channel bandwidth of 8 MHz. For stereo transmission, the two sound channels are multiplexed, digitised and transmitted in turn.

The basic outline of the NICAM stereo system is shown in Fig. 16.2. The two analogue sound channels, A and B, are pre-emphasised before going into an analogue-to-digital converter. Groups of 32 samples of each channel are grouped together to form the basic *data segment* of the system. They are then digitised segment by segment and sample by sample into 14-bit codes using a sampling rate of 32 kHz. This is followed by a 14-to-10 companding network which compresses the 14-bit codes into 10 bits without any significant loss of quality. The error detection parity bit is then added, resulting in 11-bit samples. Next the channel data segments are organised into data blocks. Each data block consists of two 32 × 11-bit sample segments, one from each channel, a total of

$$2 \times 32 \times 11 = 704 \text{ bits}$$

These 704-bit chunks of data form the basic block (sound + parity) of the NICAM broadcast data frame. Each frame consists of a total of 728 bits as follows:

704 bits for 64 × 11-bit samples
24 bits for frame alignment and operation control

The time duration of each frame is 1 ms, resulting in a bit rate of

728/1 ms = 728 kbit/s.

Framing is followed by interleaving and scrambling. Interleaving is necessary to ensure that error bursts are distributed among several samples which are far apart. Scrambling avoids the uneven distribution of energy which follows the process of modulation. The companded, interleaved and scrambled data frame is then used to modulate a subcarrier that is 6.552 MHz above the vision carrier (Fig. 16.3). The differential quadrature phase shift keying (DQPSK) modulator is very economical in bandwidth requirements; this is the key to NICAM's ability to squeeze yet another signal in the tightly packed 8 MHz bandwidth allocated for each TV channel. Before transmission the modulated NICAM carrier is passed through a sharp cut-off, low-frequency filter to ensure the NICAM frequency spectrum does not overlap with the analogue f.m. carrier, a process known as spectrum shaping. The analogue 6 MHz f.m. sound carrier is retained for compatibility, and both sound carriers are added to the video signal for u.h.f. transmission.

We shall now describe the system in detail.

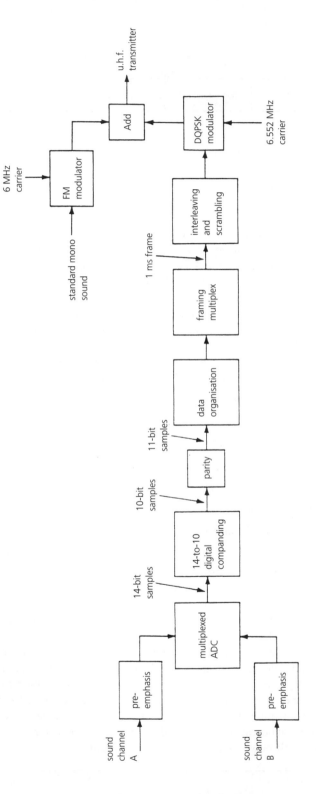

**Fig. 16.2** *Basic outline of a NICAM transmitter*

**Fig. 16.3** *Frequency response with a NICAM subcarrier at 6.552 MHz above the vision carrier*

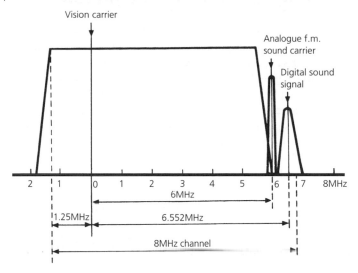

**Fig. 16.4** *Pre-emphasis and de-emphasis*

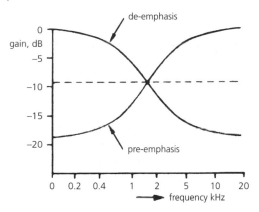

## Pre-emphasis

Pre-emphasis is applied to each audio channel in accordance with international standards to give a boost to the high-frequency components of the signal (Fig. 16.4). The purpose of pre-emphasis is to reduce the noise level, which resides mainly at the high-frequency end. Pre-emphasis is applied either while the sound signal is in analogue form or by means of digital filters while it is in a digitised form. At the receiving end, the balance between the strength of the low and high frequencies is restored by a complementary de-emphasis of the signal, which reduces the amplitude of the high-frequency components.

# Analogue-to-digital conversion

Studies have shown that an audio bandwidth of 15 kHz is quite adequate for good quality broadcasting to the home. Sampling a 15 kHz audio signal therefore requires a minimum sampling frequency of $15 \times 2 = 30$ kHz (the Nyquist rate). However, to prevent aliasing and its consequential distortion which occurs when the lower sideband of the sampling frequency (30 kHz/2 = 15 kHz) overlaps with the upper end of the audio frequency (also 15 kHz), a higher sampling rate, namely 32 kHz, is used. For this purpose, a filter with a very sharp cut-off at 15 kHz is inserted in the audio path before the sampling process.

Sampling of the two channels is carried out simultaneously. A group of 32 samples of one channel is then converted sample by sample into a 14-bit coded word followed by another group of 32 samples of the second channel, and so on. The twos complement method of representing binary numbers is used, being the most convenient to represent positive and negative excursions. The output of the analogue-to-digital converter thus consists of segments of data representing groups of 32 samples of one channel followed by a second segment representing 32 samples of the other channel, and so on.

A 14-bit digitiser provides 16 384 quantum levels, which is adequate for high-quality sound reproduction. If fewer than 14 bits are used, the quantising error can become audible in the form of a 'gritty' quality for low-level signals, an effect known as *granular distortion*.

# 14-to-10 digital companding

The use of 32 kHz sampling with a coding accuracy of 14 bits per sample would require a data bit rate of approximately 1 Mbit/s, and consequently a very large bandwidth which could not be accommodated within a single TV channel. For this reason, near-instantaneous digital companding is used, which enables the number of bits per sample to be reduced from 14 to 10 with virtually no degradation in the quality of sound reproduction. Consequently, the data bit rate of the system is markedly reduced.

Unlike the analogue companding described earlier, which has the aim of improving the signal-to-noise ratio, the purpose of digital companding is to reduce the number of bits per sample and hence the data bit rate. Furthermore, because all the operations of digital companding are performed in digital form, the compressor at the transmitting coding stage and the expander at the encoder receiving stage can be matched precisely, without the mistracking that is associated with analogue companding.

The companding technique used in NICAM is based on the fact that the significance of each bit of a binary code depends on the sound level which the particular sample code represents. For instance, assuming a peak analogue input of 1 V, then with a 14-bit code the quantum step is given by

quantum step = 1 V/quantum levels = 1 V/16 384 = 61 $\mu$V

This is the value of the least significant bit. The second least significant bit has a value of $2 \times 16 = 32$ $\mu$V, and so on. It can readily be seen that for a loud sound, i.e. a high-amplitude sample, say 500 mV or over, the effect of the three or four least significant

**Fig. 16.5** *Coding of companded sound signals*

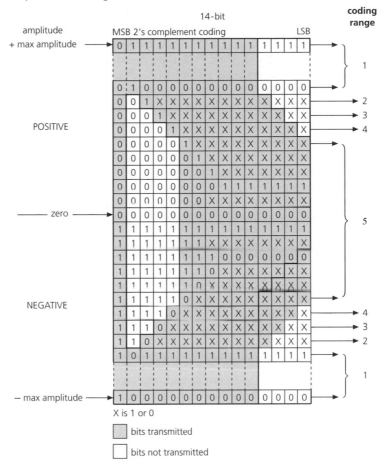

bits (LSBs) is imperceptible and may be neglected. However, for delicate or quiet passages with sample amplitudes in the region of a few hundred microvolts, the LSBs are all important.

NICAM companding reduces the 14-bit sample codes to 10-bit codes in such a way that for low-level signals the receiver is able to recreate the original 14-bit samples, and for high-level signals it is possible to discard between one and four LSBs as irrelevant.

Each segment of 32 successive audio samples is investigated to find the largest sample in that segment. The amplitude of this sample is then used to indicate the audio strength of the whole segment in five coding ranges (Fig. 16.5). Coding range 1 represents a segment where the largest sample falls between the maximum amplitude and one-half the maximum amplitude; range 2 is from one-half to one-quarter maximum amplitude; range 3 is from one-quarter to one-eighth; range 4 is from one-eighth to one-sixteenth; and range 5 represents one-sixteenth of maximum amplitude to zero or silence. The shaded bits in Fig. 16.5 show the bits actually transmitted for each range. In each case the most significant bit, the 14th, is the sign bit and is therefore retained to indicate a positive or negative value. The 13th bit is discarded if it is the same as the 14th; the 12th is likewise discarded if it is the same as the 13th and 14th, and similarly with the 11th and 10th bits. Where high-order bits are discarded, NICAM provides a labelling technique known as *scale factor coding*, which enables the missing bits to be reconstituted at the receiver. When the discarding of the high-order bits is completed, coded samples of between 10 and 14 bits are left, depending on the sequence of the high-order bits. Where a code has more than 10 bits, a sufficient number of bits are removed to reduce the size of the code to 10 bits; the first bit to be removed is the least significant bit and successive bits are removed by working upwards.

It follows, therefore, that for a segment of 32 samples falling in the largest amplitude range, range 1, the four LSBs of each sample are discarded and lost for ever. In the case of segments falling in range 2, the bit next to the most significant bit, the 13th bit of each sample, is discarded along with the three LSBs. Although the three LSBs are lost, the 13th bit is reconstituted at the receiver, since it always has the same value as the most significant bit, and so on for ranges 3, 4 and 5.

## Parity, protection and scale factors

The next stage is the addition of the parity bit to each sample code, resulting in an 11-bit word. One parity bit is added to the 10-bit sample to check the six most significant bits (MSBs) for the presence of errors. The remaining bits, the five least significant bits, are transmitted without a parity check. Even parity is used for the group formed by the six most significant bits. Subsequently, the parity bits are modified to introduce greater error protection and correction as well as coding range information.

The decoder at the receiving end needs to know the number of high-order bits that have been discarded, so they may be reinserted. This is carried out by labelling each coding range with a code known as the scale factor. The scale factor is a 3-bit code which informs the decoder of the number of discarded high-order bits. To save on bandwidth, this information is conveyed without the use of additional bits. Instead, the information is inserted by modifying the parity bits, a technique known as *signalling-in-parity*.

Signalling-in-parity takes a group of nine samples within a basic 32-sample segment and uses it to indicate one bit of the scale factor. Two other groups of nine samples within the same basic segment are used to indicate the other two bits of the scale factor. If a scale factor bit needs to be set to 0, the group of nine samples are allocated even parity. Conversely, odd parity is used to set the scale factor bit to 1. For instance, assuming the number of missing MSBs is 4, then a scale factor of 011 is necessary. The first group of nine samples are given odd parity to represent the LSB (logic 1) of the scale factor, and the parity bit of each sample in the group is then chosen accordingly. The second group of nine samples are also given odd parity to represent the second bit (logic 1) of the scale factor. The third group of nine samples is given even parity, representing the MSB (logic 0) of the scale factor. At the receiving end, the decoder checks each sample for parity in the normal way, compares the results with the transmitted parity bit of each group of nine samples and extracts the 3-bit scale factor. This process also restores the original parity bit for the six MSBs. Assuming there are no errors due to transmission, the decoder deduces the type of parity, even or odd, used by the transmitter for each

**TABLE 16.1**

| Coding range | Scale factor | Protection range | No. of MSBs same |
|---|---|---|---|
| 1 | 1 1 1 | 1 | 0 |
| 2 | 1 1 0 | 2 | 2 |
| 3 | 1 0 1 | 3 | 3 |
| 4 | 0 1 1 | 4 | 4 |
|   | 1 0 0 | 5 | 5 |
| 5 | 0 1 0 | 6 | 6 |
|   | 0 0 1 or | 7 | 7 or |
|   | 0 0 0 |   | over |

group of nine samples and hence the relevant scale bit. In cases of bit error in the coded samples, the decoder uses what is known as majority decision logic, in which it accepts the parity indication of the majority of the group of nine samples and disregards the minority. This technique is very effective because a mistake in a scale factor bit can take place only if more than four of the group of nine samples suffer errors simultaneously, something that is highly unlikely under normal reception conditions.

Table 16.1 shows the scale factor for each coding range. Notice that coding range 5 is divided into three different protection ranges. This is because the scale factor codes are used to represent seven protection ranges, also shown in the table. An eighth protection range is not employed, in order to keep NICAM compatible with MAC/packet systems for satellite transmission. While the coding range informs the receiver of the number of high-order bits that have been compressed and not transmitted, the protection range provides information on the number of high-order bits that are the same. For example, in protection range 3, corresponding to coding range 3, the three MSBs should all be the same (Fig. 16.5), and in protection range 6 the six MSBs are the same, and so on. This makes it possible for the receiver to identify errors and correct them, even if the parity check indicates no error.

## Framing multiplex

The data emerging from the analogue-to-digital converter is in the form of 352-bit segments. Each segment consists of 32 × 11-bit samples: A1–A32 for channel A and B1–B32 for channel B (Fig. 16.6). Before framing occurs the data stream is organised into blocks, each of which is composed of two segments, one from each channel; and in the case of stereo sound broadcasting, the blocks are multiplexed as shown in Fig. 16.7. Each sound block is then preceded by an additional 24 bits for identification and control, to give a total of 704 + 24 = 728 bits for each frame (Fig. 16.8). The 24 additional bits are divided as follows:

*the first 8 bits for the frame alignment word (FAW)*
*the next 5 bits for control information*
*the last 11 bits for additional data for future use*

**Fig. 16.6** *NICAM segment composition*

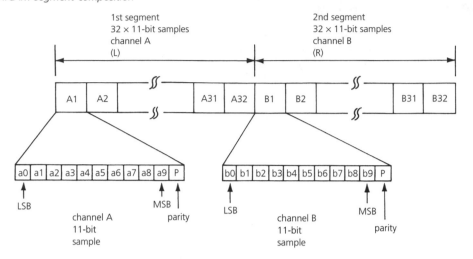

**Fig. 16.7** *NICAM block composition*

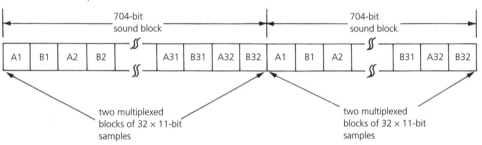

**Fig. 16.8** *Structure of a 728-bit NICAM frame*

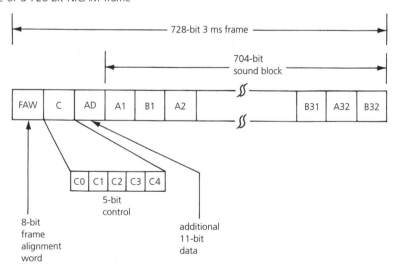

**TABLE 16.2**

| Control bits | | | Application |
|---|---|---|---|
| C1 | C2 | C3 | |
| 0 | 0 | 0 | Stereo signal with multiplexed samples |
| 0 | 1 | 0 | Two independent mono signals (M1 and M2) in alternate frames, e.g. dual language |
| 1 | 0 | 0 | One mono sound and one data channel sent in alternate frames |
| 1 | 1 | 0 | One data channel |

The frame alignment word synchronises and sets up the decoder in the receiver. It is always set to 01001110. The application control bits, C0–C4, are used for decoder control and switching. C0 is known as the frame flag; it is set to 0 for eight successive frames and 1 for the next eight frames, and so on, in order to define a 16-frame sequence used to synchronise changes in the type of information sent. Bits C1, C2 and C3 provide application control information (Table 16.2) and C4 is set to 1 when NICAM carries the same audio information as the f.m. analogue sound carrier.

## Bit interleaving

Bit interleaving is applied to the 704-bit sound block in order to minimise the effect of multiple bit error, known as error burst, caused by transient noise; this may corrupt a number of adjacent bits with devastating effect on sound quality. NICAM bit interleaving separates adjacent bits so that when the data stream is finally transmitted (following the scrambling process) they are at least 16 clock cycles apart (i.e. a minimum of 15 other bits occur between them). Thus, provided an error burst spans fewer than 16 bits, it will spread as single bit errors in different samples.

Bit interleaving is achieved by writing the 704-bit sound block into memory locations of a RAM chip and then reading them out in a different sequence which separates adjacent bits by a predetermined bit space. The readout order is stored in a read-only memory known as a ROM *address-sequencer*, which is also used in the decoding stage to restore the original bit pattern.

Bit interleaving ensures that the most likely errors are single bit errors. Errors affecting the six most significant bits of any sample are detected and corrected by the parity bit. Errors in the remaining five least significant bits will not be detected, but should they occur they will cause very small annoyance to the listener.

## Scrambling

Before modulation, the bit stream is scrambled to make the signal appear like random sound, thus dispersing the energy, and to reduce further the likelihood of interference with the analogue f.m. sound or video signal. The frame alignment word (FAW) is not scrambled since this is needed to synchronise the transmitter with the receiver. Total random scattering is not possible because there is no way of descrambling the bits back into their original order at the receiver. However, a pseudo-random scattering can be achieved using a *pseudo-random sequence generator* (*PRSG*) which produces the same result as total random distribution. The output of the PRSG is predictable and may be repeated at the start of each frame. At the receiving end, a reciprocal process takes place which descrambles the data bits back into their original form.

## DQPSK modulation

A digital signal has only two states, 1 and 0, and when it is used to modulate a carrier, only two states of the carrier amplitude, frequency or phase, are necessary to convey the digital information. In terms of bandwidth, the most economical form of modulation is phase modulation known as phase shift keying (PSK), in which the carrier frequency remains constant while its phase changes in discrete phase states in accordance with the logic state of the data bit. Binary PSK is a two-phase modulation technique in which the carrier is transmitted with a reference phase of $0°$ to indicate a logic 1 and a phase of $180°$ to indicate logic 0.

Quadrature phase shift keying (QPSK), also known as 4-phase PSK, has four phase settings. The four phases are produced by employing two carriers in quadrature ($90°$): $I$ (in-phase) and $Q$ (quadrature) as shown in Fig. 16.9. The frequency of the two carriers is constant, but their phase may change by $180°$ to produce four resultant phasors: $45°$, $135°$, $225°$ and $315°$ (Fig. 16.10). Each phasor may be used to represent two bits of data (Table 16.3). The advantage of this type of modulation is its ability to send twice as much information as the binary PSK for the same bandwidth.

Differential phase shift keying (DPSK) has no specific reference phase. The phase shift indicates whether the current bit is different from the previous bit. The phase reference is therefore the previously transmitted signal phase. The advantage of this is that the receiver as well as the transmitter does not have to maintain an absolute phase reference with which the phase of the received signal is compared.

**Fig. 16.9**

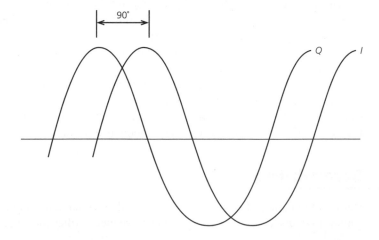

**Fig. 16.10** *Phase shift keying (PSK)*

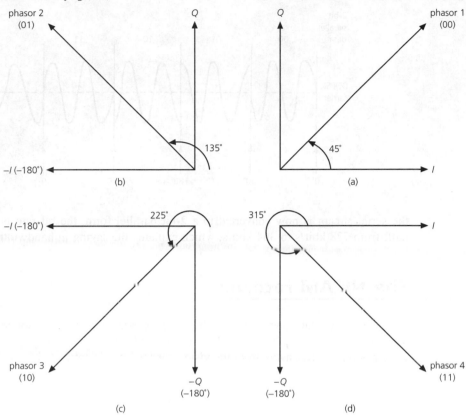

**TABLE 16.3**

| Input bits | | Phase change |
|---|---|---|
| 0 | 0 | no change |
| 0 | 1 | −90° |
| 1 | 0 | −180° |
| 1 | 1 | −270° |

The differential quadrature phase shift keying (DQPSK) technique used in NICAM combines the advantages of QPSK and DPSK. Using a serial-to-parallel converter, the serial bit stream is first converted into a 2-bit parallel format. The instantaneous states of each pair of bits, known as dibits, can take one of four combinations: 00, 01, 10 and 11. Each of these combinations changes the phase of the carrier from its previous setting by a different angle (Table 16.3). The four 2-bit data combinations are thus represented by four different phase changes. For example, data 00 is represented by no change in the phase of the carrier and 11 is represented by a 270° phase change (Fig. 16.11). Since

**Fig. 16.11** *DQPSK carrier waveform*

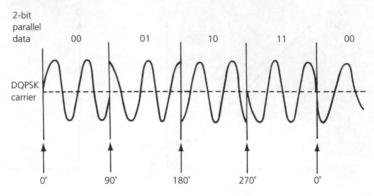

2-bit parallel data

00      01      10      11      00

DQPSK carrier

0°      90°      180°      270°      0°

the serial stream is now converted to a 2-bit parallel form, the bit rate is reduced by half, from 728 kbit/s to 364 kbit/s, which explains the saving in bandwidth.

## The NICAM receiver

At the receiver, the tuner converts the vision carrier and the f.m. sound intercarrier to an i.f. of 39.5 MHz and 33.5 MHz, respectively, in the normal way. The NICAM carrier is 6.552 MHz away from the vision carrier, so it is converted to an intermediate frequency of

$$39.5 - 6.552 = 32.948 \text{ MHz or } 32.95 \text{ MHz}$$

This is demodulated by a DQPSK detector and applied to the NICAM decoder, which reverses the processes carried out at the transmitter to recreate the 14-bit sample code words for each channel. This is then followed by a digital-to-analogue converter, which reproduces the original analogue two-channel, left and right, sound waveforms.

The basic elements of NICAM sound reception in a TV receiver are shown in Fig. 16.12. Following the tuner, a special surface acoustic wave (SAW) filter provides separate vision and sound i.f. outputs. A sharp cut-off removes the two sound i.f.s, 33.5 MHz for mono and 32.95 MHz for NICAM, from the 39.5 MHz vision carrier. The SAW filter provides a separate path for the f.m. and NICAM carrier i.f.s. It also provides for a very narrow peak at 39.5 MHz. In the sound i.f. demodulator, the 39.5 MHz pilot frequency is used to beat with the f.m. sound i.f. and with the NICAM i.f. to produce 6 MHz f.m. and 6.552 MHz DQPSK carriers. Sharply tuned filters are then used to separate the two sound carriers. The f.m. carrier goes to a conventional f.m. processing channel for mono sound and the 6.552 MHz NICAM phase modulated carrier goes to the NICAM processing section. This consists of three basic parts. The DQPSK decoder recovers the 728 kbit/s serial data stream from the 6.552 MHz carrier. The NICAM decoder descrambles, deinterleaves, corrects and expands the data stream back into 14-bit sample code words. Finally, the digital-to-analogue converter reproduces the original analogue signals for each channel.

**Fig. 16.12** *NICAM sound receiver: basic elements*

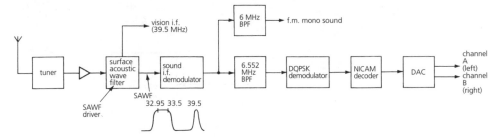

**Fig. 16.13** *DQPSK demodulator: main elements*

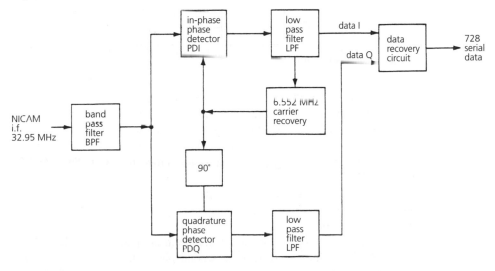

## DQPSK decoder

The phase demodulator or detector works on the same principles as an f.m. detector in which a variation in phase (or frequency) produces a variation in the d.c. output. In the case of two-phase modulation, the d.c. output of the detector has two distinct values, representing logic 1 and logic 0. However, in the case of quadrature, i.e. 4-phase modulation, the output of the detector is ambiguous. The same output for a 90° phase shift is obtained as for a phase shift of 270°, and similarly for phase shifts of 0° and 180°. In order to resolve the ambiguities, a second phase detector operating in quadrature (90°) is used. Figure 16.13 shows the main elements of a DQPSK demodulator using an in-phase phase detector (PDI) and a quadrature phase detector (PDQ). The outputs from the two phase detectors, data I and Q, are fed into a data recovery circuit which reproduces the original 728-bit serial data stream. The 6.552 MHz reference carrier frequency is generated by a carrier recovery circuit which includes a crystal-tuned voltage-controlled oscillator and a phase-locked loop.

**Fig. 16.14** *Data recovery circuit*

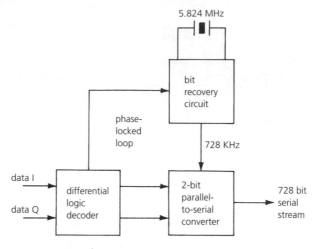

The main elements of a data recovery circuit are shown in Fig. 16.14, in which the bit recovery circuit includes a second PLL locked to the bit rate of 728 kHz. In order to ensure a 'clean' bit rate clock, a master system clock which is a multiple of the bit rate is used. In this case a clock frequency of 5.824 MHz is used. The bit rate is then retrieved by dividing the master system clock by eight. The I and Q data streams from the DQPSK detector are fed into a differential logic decoder, which produces the correponding 2-bit parallel data. The pairs of parallel data are then fed into a parallel-to-serial converter, which reconstructs the original 728-bit serial stream before going to the NICAM decoder.

Practical DQPSK decoders are available in i.c. packages such as the TA8662N and the TDA 8732. They incorporate a second phase detector driven by the Q data; this generates a muting signal to turn the f.m. mono sound on and off depending upon the presence or otherwise of the 6.552 MHz NICAM carrier. A practical DQPSK decoder circuit using the TA8662 chip is shown in Fig. 16.15, in which two chassis levels are available, a D chassis for digital signals and an A chassis for analogue signals. Transistor TS01 is a filter driver and TS02/TS03 form a two-stage NICAM i.f. amplifier. The 6.552 MHz carrier enters at pin 4 and the outputs are at pin 29 (728-bit data stream), pin 21 (728 kHz clock) and pin 26 (5.824 MHz system clock). The 5.824 MHz system clock crystal QS01 and the 6.552 MHz crystal QS02 form part of the bit and carrier recovery circuits respectively. The I and Q low-pass filters are connected to pins 11 and 10 respectively.

## NICAM decoder

The NICAM decoder, sometimes known as the NICAM multiplexer, descrambles, deinterleaves and reconstitutes the original 14-bit words. It provides data, ident and clock signals to the digital-to-analogue converter. A simplified block diagram of the

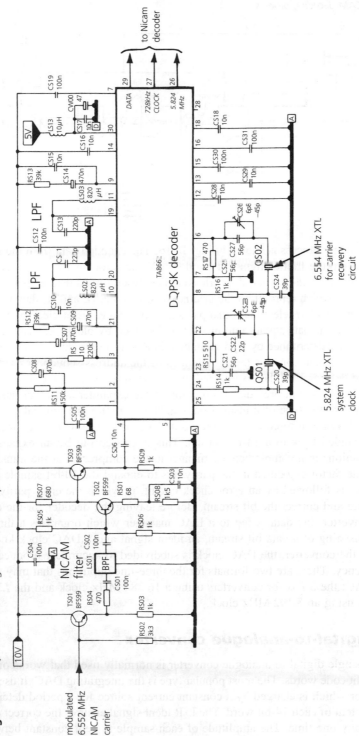

**Fig. 16.15** DQPSK decoder using demodulator chip TA8662

**Fig. 16.16** *NICAM decoding process*

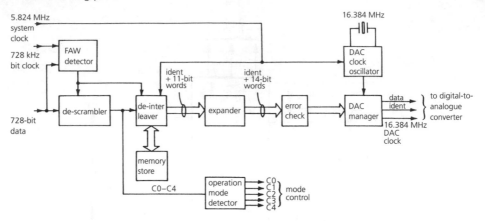

decoding process is shown in Fig. 16.16. The encoded data from the DQPSK detector is fed into the frame alignment word (FAW) detector for frame recognition and resetting of the descrambler and deinterleaver. The descrambled data is then fed into the deinterleaver, which reproduces the original dual-channel (L and R) data together with an L/R ident signal to select the signal paths as appropriate. Deinterleaving is carried out by first writing the data stream into memory cells block by block. The cells are then read out in an order determined by a program held in ROM to reproduce the correct order of the bits. The program contains the complement of the address-sequencer used at the transmitter. The descrambled data is also fed to the operation mode detector which decodes control bits C0–C4 and provides information to the expander and other parts of the system in terms of the type of transmission, e.g. stereo, mono or bilingual.

Having restored each 11-bit word (10 + parity) to its correct order, they have to be expanded back to a 14-bit format. This is carried out by an expansion circuit, which functions in a complementary manner to the compressor at the transmitter but uses the scale factor embodied in the parity bits to expand the 10-bit sample codes into 14 bits. This is followed by an error check circuit, in which the error parity is used to investigate and correct the bit stream. Before leaving the decoder for the digital-to-analogue converter, the data is fed to a DAC manager which organises a three-line bus output consisting of a data bit stream, an ident signal and a DAC clock known as DACOSC. At the converter, the DAC clock is subdivided to accurately produce the sampling frequency. There are two formats for the three-line ouput bus that may be used to feed the DAC: the *S-bus* for converters using a 16.384 MHz clock and the *I2S-bus* for converters using an 8.192 MHz clock.

## Digital-to-analogue converter

A single digital-to-analogue converter is normally used that works on alternate left and right code words. The most popular type is the integrating DAC. It uses a precision capacitor which is charged by a constant current source for a period determined by the data content of each 14-bit word. The L/R ident signal ensures the correct channel is selected at any one time. The amplitude of each sample is kept constant between samples by a

**Fig. 16.17** *NICAM audio signal processing*

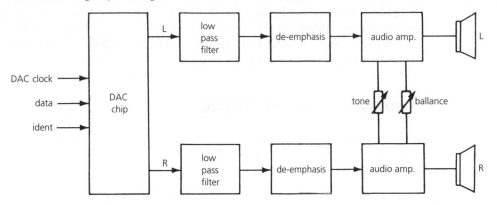

hold circuit. The 32 kHz sampling frequency is derived by dividing the DAC clock from the decoder by 512 in the case of a 16,384 MHz clock (S-bus) or by 256 in the case of an 8.192 MHz clock (I2S-bus).

## Switching arrangements

All modern TV receivers have some complex switching arrangements to implement changes between NICAM stereo, mono or bilingual, f.m. sound and auxiliary sound or video inputs as well as making the sound and video signals available for outside use. Control of the switching network is carried out by the NICAM decoder and other control chips such as a microprocessor unit.

## Audio signal processing

Following the DAC, the analogue signals known as the baseband audio signals first pass through a low-pass filter with a sharp 15 kHz cut-off frequency in order to smooth out the quantising steps and minimise quantising noise in the reconstructed audio signal (Fig. 16.17). This is followed by a de-emphasis network which re-establishes the correct response for each channel and restores the signals to their original audio form, ready for tone control and amplification. Two separate hi-fi audio amplifiers available in i.c. packages are used for the L and R channels.

# Dolby sound

The Dolby sound system was first introduced for sound reproduction in cinemas and auditoriums. It was found that traditional stereo sound was only effective for those seated on a central axis between the left and right speakers. For those seated off-axis, the output from one of the loudspeakers predominates, giving an imbalance in the stereo impact or image. This central area, sometimes known as the sweet spot, in which balanced and realistic stereo is obtained may be extended if a central channel is added, primarily for dialogue. Such an arrangement, which places a third loudspeaker behind the screen, gave acceptable results even for those well off the central axis. Further improvements may be obtained by adding a fourth surround speaker at the rear. Sound is thus given a three-dimensional quality, which boosts the stereo effect. More than one surround speaker may be used for that purpose.

## Dolby surround sound

In sound reproduction systems such as 35 mm films and NICAM, where only two sound channels are available, it is necessary to encode the four Dolby channels into two, which may later be decoded back into the original four channels. These two encoded channels are known as *right total (RT)* and *left total (LT)*. This is known as Dolby surround sound.

A Dolby surround sound encoder is shown in Fig. 16.18. Encoding involves adding the left (L) channel to the centre channel; the sum thus produced is then added to the surround channel, which has been shifted by +90° to produce the left total (LT) channel. The same process is repeated for the right sound channel, except the surround channel is now phase shifted by −90°. Thus the resulting LT and the RT channels contain all four elements of the original sound. At the receiving end, a decoder is used to recover the original four sound channels. For domestic use such as TV receivers, where the area is small, the centre channel is considered optional. Its effect may be simulated by the left and right loudspeakers to produce what is known as a *phantom centre channel*. This method significantly enlarges the sweet spot, which is adequate in a domestic situation. Further enhancement is obtained by an *audio delay*, which introduces an artificial lag in the surround channel to give the impression of a large auditorium. For an average living-room, Dolby recommends an audio delay of between 15–25 ms.

**Fig. 16.18** *Dolby surround sound detector*

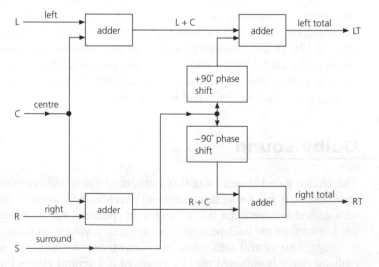

**Fig. 16.19** *Adaptive mixing*

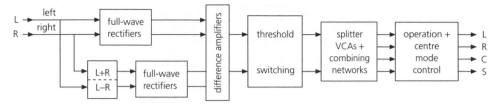

# Dolby Pro Logic sound

Dolby Pro Logic is an active surround sound system as compared with the simple Dolby surround sound, which is merely passive. In Dolby Pro Logic systems the encoded sound channels, LT and RT, are continuously monitored to evaluate the direction and intensity of the dominant sound. *Adaptive mixers* are then used to represent this effect in the reproduction of the four elements of Dolby sound. Adaptive mixing (Fig. 16.19) involves assessing the relative intensities of the four channels. This is carried out by full-wave rectification of the four signals and using operational amplifiers producing left–right and centre–surround difference signals. Two axes (L–R and C–S) are created and used to obtain a vector that represents the intensity and direction of sound effects. However, when a vector exceeds a certain threshold, the decoder switches to 'fast' mode operation, producing four control voltages representing the four Pro Logic signals: VL, VR, VC and VS. Following amplification by eight voltage-controlled amplifiers (VCAs), the two left and right components are then weighted and combined in a network (combining network). The result is then fed into the operation and centre mode control, which provides the output signals.

Dolby Pro Logic processing was originally carried out by a customised analogue chip (ASIC). Today modern TV receivers employ a *digital sound processor* (*DSP*).

A generalised block diagram of an analogue sound processor is shown in Fig. 16.20. The LT and RT signals from the NICAM decoder or a VCR are first amplified before pink noise is added for channel balance adjustment. For normal stereo reception, the *pink noise* input is disconnected. The two channels are then split, with two feeds going directly into the processor and a further two going via two low-pass filters. Processing Dolby Pro Logic sound employs adaptive mixing and noise reduction techniques to extract the original four audio channels: L, R, C and S. A 2 MHz crystal-controlled oscillator provides the necessary clock pulse for the processor. A 16 or 32K RAM is used for audio delay if required. Two separate volume control chips are used, one for the L and R channels and one for the C and S channels. The processor may be configured to operate in different Dolby modes, which are set by connecting the various control pins to the reference voltage, $V_{ref}$. In order to avoid digital noise appearing on analogue signals, two separate analogue and digital chassis connections are used which are linked together at one point.

**Fig. 16.20** *Analogue sound processor*

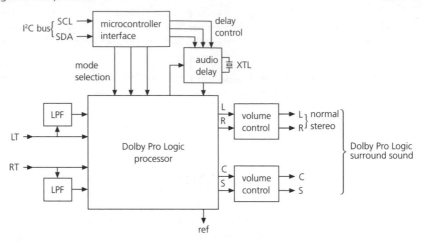

## Practical decoder

Figure 16.21 shows a circuit diagram of an analogue Pro Logic sound decoder NJM2177L used in a Ferguson GLF 2.20 chassis. Digital audio delay is provided by NJU9701. LT and RT signals enter the decoder at pins 15 and 22 respectively. Dolby surround channels L, R, C and S are produced at pins 32, 33, 38 and 29 respectively. Two other audio signals are also provided, L + R and L – R (pins 34 and 35); they are not required for Pro Logic but may be useful for other applications. A further output for audio delay is provided at pin 39, which returns back at pin 42. If a delay chip is not used, pins 39 and 42 are shorted. The delay chip contains a 16Kb fast SRAM, delay control, difference amplifiers, a 2 MHz crystal-controlled oscillator and a control unit. The analogue input enters the chip at pin 23 and is converted to a digital format. Following processing, the signal is converted back into an analogue format before exiting the chip at pin 13. According to the diagram, the delay time is set manually by earthing pins 4, 5 and 6. These same pins may be used for microcontrolled delay time. Microprocessor control is provided via an interface at pins 23, 24 and 25 (pink noise control), 31 (Pro Logic/ normal control) and 36 (phantom control) with LT and RT inputs at pins 15 and 22 respectively.

The *digital sound processor* (Fig. 16.22) carries out the same function as the analogue audio processor. Analogue audio signals LT and RT are fed into the processor directly and via appropriate low-pass filters as shown. Pink noise may be added for channel balance. At the DSP, the signals are converted into digital format, processed for Pro Logic and converted back into an analogue format. An external RAM chip is used for audio delay and the process is controlled by an external 12 MHz crystal-controlled oscillator. Six outputs are produced by the DSP chip; they are fed into a multi-switching chip to obtain the Pro Logic signals R, L (internal and external), C and S.

**Fig. 16.21** Analogue Pro Logic sound decoder circuit (Ferguson GLF 2.2)

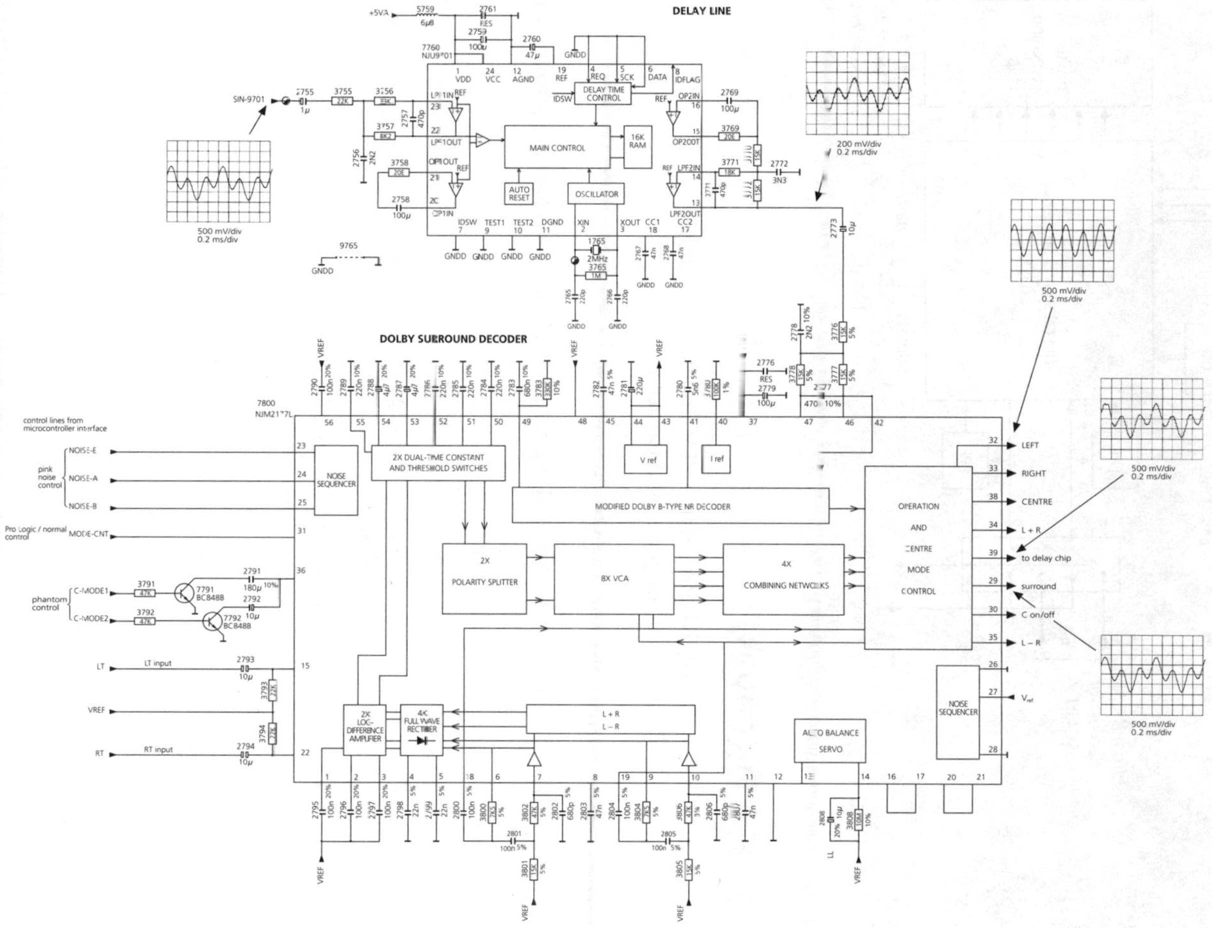

**Fig. 16.22** *Digital sound processor*

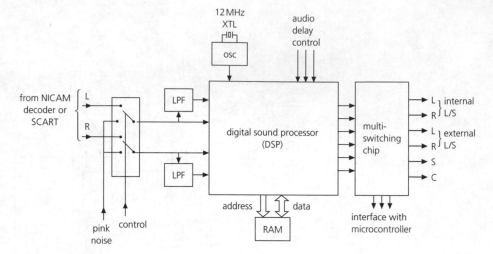

| | | QUESTIONS |

1. For British PAL colour TV broadcasting, state
   (a) the analogue sound carrier frequency
   (b) the NICAM sound carrier frequency
   (c) the NICAM bit rate

2. In relation to NICAM digital sound, briefly explain the purpose of
   (a) companding
   (b) pre-emphasis
   (c) scrambling

3. Explain the difference between phase shift keying (PSK) and quadrature phase shift keying (QPSK).

4. In relation to a NICAM receiver, state
   (a) the analogue sound i.f.
   (b) the NICAM i.f.

5. State the four components of Dolby surround sound.

# 17 Microprocessor-controlled receivers

The microcomputer, also known as a microcontroller, is a complete microprocessor-based system constructed on a single chip. The basic structure of a microprocessor-based system is shown in Fig. 17.1. It consists of the following elements:

- microprocessor unit (MPU) or central processing unit (CPU)
- memory chips (RAM and ROM)
- parallel input/output ports (PIO)
- serial communications chip (UART)
- programmable devices
- a bus structure.

The MPU chip contains all the necessary circuitry to interpret and execute program instructions in terms of data manipulation, logic and arithmetic operations and timing and control. The capacity or bit size of a microprocessor chip is determined by the number of data bits it can handle. An 8-bit processor has an 8-bit data width, a 16-bit processor has a 16-bit data width, and so on. RAM and ROM are two types of memory chips that are normally used. Other types such as PROM, EPROM, EEPROM or FLASH may also be used.

The parallel input/output (I/O) ports provide a link to and from the system peripheral devices such as keyboards, VDU, and transducers or drive circuitry for stepper motors, LEDs and relays. Parallel ports consist of a number of bidirectional lines, each

**Fig. 17.1** *Microprocessor system: general block diagram*

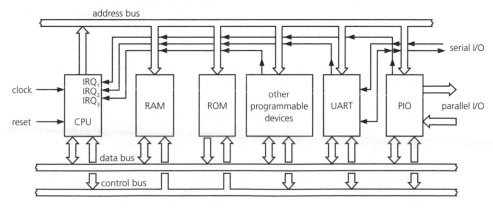

**Fig. 17.2** *RS-232 connector ports: 9-pin and 25-pin*

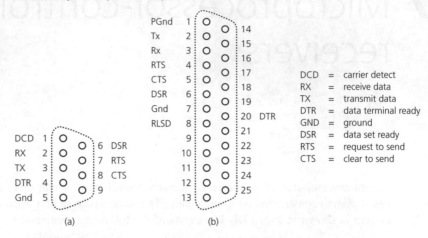

of which may be designated as input or output. The *UART* (*universal asynchronous receiver transmitter*) provides serial communication with external devices such as modems and printers. The UART provides a two-line (data in and data out) serial interface together with all necessary control lines, known as handshake lines. A standard connector for the UART is the 9-pin or the 25-pin RS-232 D-type connector or port (Fig. 17.2). Both the PIO (*programmable I/O*) and the UART are programmable devices. Programmable devices have a number of internal registers that are accessible by the CPU. Programming them involves the CPU entering the appropriate data into the registers to set their operational parameters. For the PIO and the UART this involves setting the communication parameters such as the speed or baud rate, the data width and the type of parity, if any. Other programmable devices such as a modem or a video decoder may also be provided.

The various elements of the system are interconnected by a bus structure. The number of tracks or lines used in a bus is determined by the capacity and complexity of the microprocessor chip. There are three types of buses: the data bus, the address bus and the control bus. The *data bus* is used to transfer data between the MPU and other elements in the system. The *address bus* is used to carry the address of memory locations from which data may be retrieved, i.e. read, or stored, i.e. written, into memory locations. It is also used to address other elements in the system, such as the I/O ports. The *control bus* carries the control signals of the CPU such as the *clock*, reset (*RES*), read (*RD*) and write (*WR*). The number of control lines depends on the microprocessor used and the design of the system. The processor also provides a number of interrupt request lines (IRQ1–IRQ3). When a device such as a video decoder requires a service from the processor, it enables an IRQ control line. Upon receiving this request, the processor halts its current program and initiates the software routine requested by the device. When that is finished, the processor returns to its original program. Where two or more IRQs are provided, they are dealt with on a predetermined priority basis.

In order to reduce the number of tracks in a bus, *multiplexing* may be used either between groups of lines within a bus or between one bus and another. This technique also reduces the number of pins on the silicon chip.

**Fig. 17.3** *Power-on reset (POR) circuit*

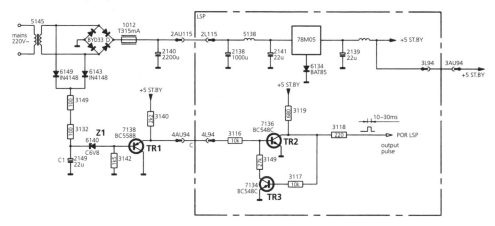

## Resetting and initialising

All programmable IC chips, including the CPU, must have their registers set to an initial setting, which determines the start-up parameters. This is known as *initialisation*. Initialisation involves taking the RES control line from GND to +5 V. Following power-on, a reset pulse known as power-on reset (POR) is generated; this is then used to reset and initialise all programmable chips. Resetting must only take place when the various d.c. voltages have been established. Thus a delay of 20–60 ms has to be introduced before the POR pulse is made active.

A POR circuit is shown in Fig. 17.3. After switching on the television set, the 5 V standby voltage becomes available. This provides the necessary d.c. power for the circuit. At the point of switching on, capacitor C1 is fully discharged and zener Z1 is open-circuit. TR1 is on and its emmitter goes low causing TR2 to cut off (TR3 on) with its collector going high. The POR pulse output thus goes high. It remains high until C1 charges up to just over 6.8 V (Z1 breakdown voltage), upon which TR1 turns off and TR2 turns on, taking the POR output low. Initialisation of the CPU, the power supply control chip and the other ICs starts at the descending edge of the POR pulse. The width of the POR pulse is 20–60 ms, depending on the mains input voltage. Following this initialisation the set itself starts up under the control of the microprocessor.

## CPU architecture

Microprocessor chips have complex architectures; they vary from one manufacturer to another, but all have the following features in common:

- arithmetic and logic unit
- timing and control logic
- accumulator and other registers
- instruction decoder
- internal bus

# General operation of the system

The heart of the system, the microprocessor chip, operates on a *fetch and execute* cycle synchronised by the system clock. During the fetch phase, the CPU receives the instruction from the memory location where the program is held and stores it into an internal register known as the instruction register. During the execute phase, the MPU having received the instruction will then decode it and execute it. This is carried out by the MPU generating the necessary timing and control signals for the execution of that particular instruction. The execute phase may involve a simple arithmetic operation, e.g. add or subtract, or a more complex data transfer to or from a memory chip or a peripheral device. When the instruction is completed, the microprocessor then fetches the next instruction, and so on.

# The instruction set

The microprocessor performs its tasks in a predetermined sequence known as the program. The program is a series of instructions which breaks down each operation into a number of individual tasks. These instructions are fed into the microprocessor chip in the form of binary digits. An instruction consists of two parts: an *operator* and an *operand*. Each instruction, such as ADD or MOVE DATA, is represented by a binary number known as the machine code or *opcode* (operational code) of the particular microprocessor. This is the operator part of the instruction. The data items that the opcode is to operate upon, i.e. the two numbers to be added or the data to be moved, form the second part of the instruction, the operand. Assuming an 8-bit system, we will have an 8-bit operator and one or more 8-bit operands. An instruction with many operands takes longer to complete than an instruction with fewer operands. Each make of microprocessor has its own set of machine codes, known as the instruction set.

Writing programs directly in machine code is a very lengthy and tedious process. Programs are normally written in a language which uses letters and words. This is then translated into the appropriate series of opcodes. The simplest form of translation is the assembler, which employs an *assembly programming* language. In the assembly language each opcode is given a mnemonic such as EN for enable, MOV for move, ANL for logical AND, and INC for increment.

# Microcomputers

Microcomputers or microcontrollers are complete microprocessor systems on a single chip. They contain the elements of the microprocessor itself as well as RAM, ROM or other memory devices and I/O ports. A variety of microcomputers are available from various manufacturers (Intel 8048/49 and 8051 series, Motorola 6805 and 146805, Texas TMS1000 and Ziloc Z80 series) for use as dedicated computer systems in such applications as car engines, washing machines, VCRs and of course TV receivers. The difference between one type of microprocessor and another lies in the type and size of memory, instruction set, operating speed, number of available input and output lines and data length, e.g. 4, 8 or 16 bits. Where microcomputers are customised for specific use, they are called *application-specific integrated circuits (ASICs)*. In the majority of cases, microcontrollers have their program stored permanently into an internal ROM at the manufacturing stage, a process known as mask programming. Some chips have an internal EPROM available for user programming.

# Microcontroller internal architecture

The basic architecture of an 8-bit microcontroller or minicontroller is shown in Fig. 17.4. The program is held in ROM with a small RAM of 1–4 K available for data and other external control signals. The timer/counter may be loaded, started, stopped or read by software commands. In TV applications it is used to keep track of the sequence of lines and fields, and to prompt the controller to carry out certain operations at specific times. Two parallel 8-bit ports are shown, ports A and B, which may be assigned as inputs or outputs. A serial input/output port may be established by using two lines of the parallel ports, one to receive and the other to transmit serial data. The ALU carries out arithmetic operations such as adding two numbers, or it performs a logic function such as NAND or NOR on two numbers. The ALU therefore has two inputs, one input for each number. When the ALU operation is completed, the result is stored in the accumulator. The timing and control unit provides the necessary synchronisation of the system through the clock and other control signals. Interconnection between the various units is provided by a single multiplexed 8-bit bus.

**Fig. 17.4** *Microcontroller: internal architecture*

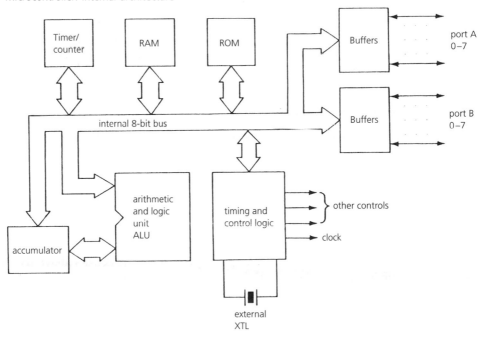

**Fig. 17.5** _Minicontroller: pin connections for the 8048_

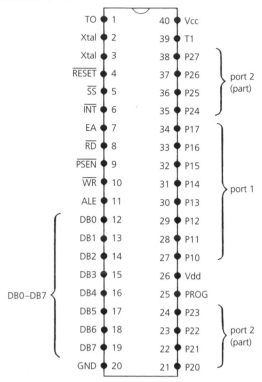

Television minicontrollers are based on one or other of the main types of microcomputer. Each TV manufacturer has custom-built microcontrollers such as the MAB8400 (Mullard), CCU2000 (ITT), CCU7070 (Fidelity), MC6805 (Grundig), FERG01 (Ferguson) and PCF84C640 (Hitachi).

## 8048 minicontroller

The pin connections for the 8048 series microcomputer chip are shown in Fig. 17.5 in which

- P10–P17 and P20–P27 are two bidirectional ports.
- DB0–DB7 provide access to the internal data bus and may also be used as a third 8-bit port.
- T0 and T1 are two input lines.
- Pins 2 and 3 are for clock crystal connections.
- INT and RESET are interrupt and reset inputs. (The bars above INT and RESET indicate these lines are active low, i.e. they become active when at logic 0.)
- Pins 8 and 10 provide active-low read and write output control.
- SS (pin 5) allows single stepping of the program for debugging purposes.
- PSEN allows external or additional program storage.

- PROG drives an output expansion chip.
- ALE (pin 11) is the address latch enable, used in conjunction with an external address latch.

The 8048 has a powerful set of 96 instructions, a 1 K ROM and a RAM capacity of 64 bytes (64 × 8 bits). It may be programmed using assembler language. For memory efficiency, each instruction (operator and operand) consists of only one or two bytes.

When used for any particular application, e.g. in a TV receiver, a program is implanted in the microprocessor, which assigns each pin a particular function. Micro-controllers may also be modified to include other functions such as tuning and analogue-to-digital conversion. Additional memory space may be added in the form of electrically erasable programmable read-only memory (EEPROM).

## TV microcontrollers

A variety of functions can be performed by a microcomputer in a TV receiver, including generating test signals and storing parameters and customer adjustments. The microcontroller may be used to perform some or all of the following functions:

- front panel keyboard scanning
- LED display drive
- volume and tone control
- audio muting
- brightness and saturation control
- remote control decoding
- switching and control of luma and chroma processing
- switching between video, teletext and auxiliary inputs
- test signal generation and automatic adjustments
- onscreen graphic generation and control
- u.h.f. tuner control
- analogue-to-digital decoding

A functional block diagram for a TV microcontroller, usually known as a *central control unit* (*CCU*), is shown in Fig. 17.6. The microcomputer core circuit controls the other units by an internal bus in accordance with the program stored in the internal ROM. The keyboard on the receiver front panel is scanned, i.e. read by a number of I/O port bits designated as input bits. Other bits are assigned to control a multi-digit LED display. A further port bit receives remote control signals and feeds them into a remote control decoder. Channel selection or other adjustments (contrast, saturation, etc.) from either the keyboard or the remote control handset are fed via the internal bus into the microcomputer core circuit, which sends the appropriate command signals, including those to the multi-digit LED display unit. Control of the receiver may be effected by assigning the various port bits a dedicated function such as keyboard scanning or volume, contrast or teletext control. It may also be operated through a serial bus which connects to other compatible chips in the receiver. A 4 MHz crystal is used to generate the microcontroller clock pulse. Connection to the other units of the receiver is realised

**Fig. 17.6** *TV microcontroller: chip architecture*

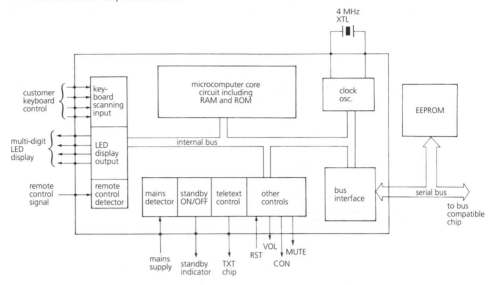

by a two-line or three-line serial bus. This also connects an EEPROM to the system, which provides a non-volatile memory store. Unlike the internal RAM, which loses its contents when power is switched off, the EEPROM chip retains its data regardless of the power supply. Furthermore, selected cells may be cleared and new data written into the chip. At the manufacturing stage, the EEPROM is used to store the data necessary to customise the receiver for its particular range of facilities, teletext, NICAM etc.; for the type of transmission system (PAL, NTSC, SECAM); and for other normal settings such as height, vertical sync. and saturation. Customer's preferences as to contrast, saturation, etc., may also be stored into EEPROM at home. Casual variations on these settings are stored in the internal RAM within the central control unit.

## RAM backup battery

Where a volatile RAM is used to store receiver settings, it is necessary to provide d.c. supply to the memory when the receiver is switched off, in order to maintain the stored data intact. An external backup battery maintains the supply to the chip. Where a rechargeable battery is employed, a charging circuit must be incorporated to maintain the d.c. level. If the battery is changed, it is then necessary to reprogram the microprocessor with all the receiver settings.

## The CCU serial bus

Two main types of serial bus are used in conjunction with TV central control units: the two-line *inter i.c.* (*IIC or I²C*) bus and the three-line *intermetall* (*IM*) bus.

The I²C bus has two bidirectional lines: a serial clock (SCL) and serial data (SDA). Any unit connected to the bus may send and receive data. Data is transmitted in 8-bit words or bytes (Fig. 17.7). The first byte contains the 7-bit address of the device for

**Fig. 17.7** *I²C bus data construction*

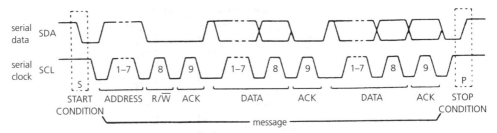

**Fig. 17.8** *Intermetall bus lines*

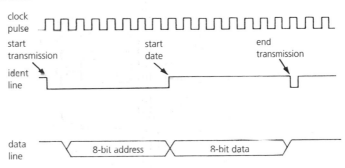

which the information is intended, and the eighth bit is a read/write bit to signify whether the data is required from, or being sent to, the device. A number of data bytes follow; the total number in a message depends on the nature of the information being transferred. Each data byte is terminated by an acknowledge (ACK) bit. Like all other bits, the ACK bit has a related clock pulse on the clock line as shown. The first byte of any data transfer is preceded by a start condition and is terminated by a stop condition. To ensure that two devices do not use the bus simultaneously, an arbitration logic system is used. The clock, which operates only when data is transferred, has a variable speed. Data may then be sent at a slow or a fast rate of up to 100 kbit/s.

The Intermetall bus has three lines: ident (I), clock (C) and data (D). Both the ident and the clock lines are unidirectional between the microcontroller and the other peripheral devices. The data line is bidirectional. The start of transmission is indicated by the ident line going low (Fig. 17.8). An 8-bit address is sent along the data line. At the end of eight clock cycles, the I line goes high, indicating the start of data transmission. Data is then transmitted along the D line for 8 (or 16) clock cycles for an 8-bit (or 16-bit) data word, at the end of which the I line goes low again, indicating the end of data transmission.

Both buses may be used simultaneously in a single receiver to provide connections with different sections of the receiver. Ferguson's ICC5 chassis uses four different buses, IM, I²C, tuning and Thomson for channel selection and analogue control, teletext and graphics, u.h.f. tuner and video processing respectively. Manufacturers have available a number of peripheral chips, including tuner interfaces, EEPROMs, and ADCs and TV processing chips for operation with I²C or IM buses.

**Fig. 17.9** *Pin connections for the SAA1293 microcontroller*

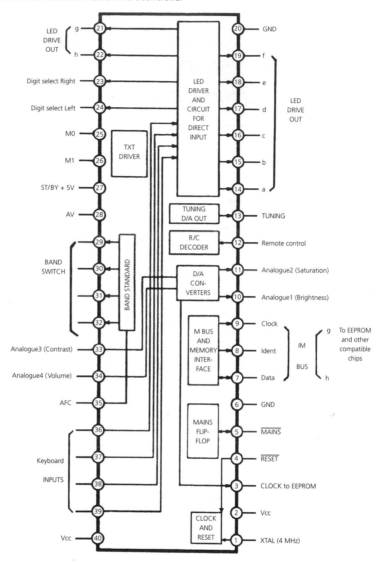

The pin connections for a practical microcontroller chip (SAA1293) are shown in Fig. 17.9. A two-digit eight-segment front display is driven by pins 14–19 and pins 21–24. The two digits are multiplexed, with pins 23 and 24 selecting each digit in turn. The eight segments a–h of each digit are then driven by pins 14–19 and pins 21 and 22. The input to the LED drive circuit is derived directly from the keyboard (pins 36–39) on the front panel or via remote control (pin 12). Four analogue output control signals are provided at pins 10, 11, 33 and 34 for brightness, colour saturation, contrast and volume. Pins 7, 8 and 9 provide the data, ident and clock lines for an IM bus to interface with an external EEPROM and other compatible chips. Pins 29–32 provide a facility for

**Fig. 17.10** *Microcomputer-controlled TV receiver*

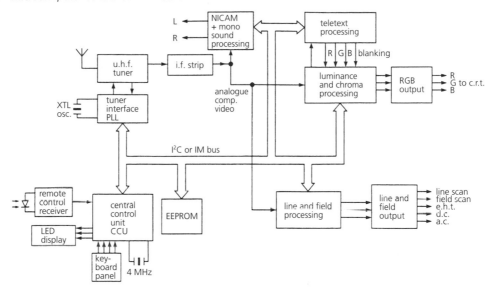

selecting different u.h.f. bands and transmission standards to customise the receiver for use in different countries. Such customisation is carried out by software programming of the EEPROM using parameters stored in the microcontroller's internal ROM.

Figure 17.10 shows a block diagram of a microcomputer-controlled TV receiver employing a single-bus structure. The CCU controls a number of processing chips via an I²C or an IM bus; it also receives and decodes signals from the remote control and the keyboard panel and drives an LED display. Each processing unit is a specialised system-on-chip ASIC dedicated to a particular function such as video decoding or sound processing. Chip manufacturers provide a complete set of processors, known as the *chip set*, which provide all the necessary functions for a TV receiver system.

## I²C bus-controlled front end

Figure 17.11 shows the front end combining both the tuner and the i.f. section used in the Ferguson GLF2.2 chassis. The tuner section contains the tuning stage and a combined oscillator–mixer. Tuning is achieved by sending a frequency to the PLL in the tuner. The tuning voltage of approximately 40 V is derived by zener D6304. A part of this voltage is fed to pin 11 of the front end to change the tuning frequency. The circuit around TR1 and TR2 provides sound suppression when no video signal is being received. TR1 emitter voltage is set to 3.1 V by resistor chain R1 and R2. When a CVBS signal is present on pin 23, TR1 saturates for a small part of the waveform, producing short-duration negative-going pulses which are fed into the base of pnp transistor TR2. Transistor TR2 will therefore only conduct for the short duration of the pulse. The average voltage at the collector is thus kept low; it is fed to the microcontroller along control line STR-FE. Without a CBVS signal present on pin 21, the average voltage on

**Fig. 17.11** *Bus-controlled front end*

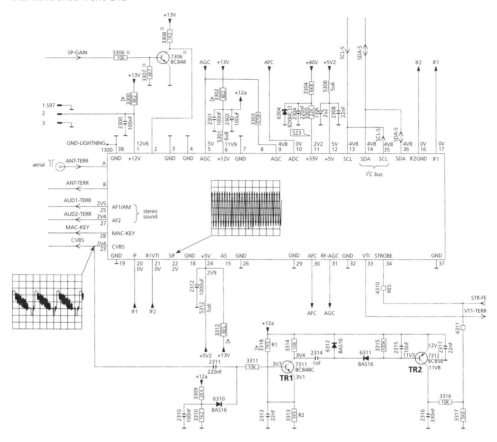

TR2 collector is high, hence the STR-FE line is high. The microcontroller receiving a high voltage on the STR-FE line will suppress the sound channel. The outputs from the front end are the CVBS baseband composite video signal (pin 23), the audio signals which feed the stereo decoder (pins 25 and 27) and the i.f. signals for the NICAM decoder (pin 22). The front end is fully controlled by the microcontroller via the I²C bus (pins 35 and 36).

## The TDA 94141 video and sync. processor

The simplified functional diagram for the TDA 94141 video processor chip is shown in Fig. 17.12. The luminance Y (CVBS) and the chrominance C signals (pins 26 and 25) are fed into the video decoder, which converts them into Y, R − Y and R − B signals (pins 12, 14 and 13). The composite video also goes to a sync. separator, which provides line (H) and field (V) sync. pulses and a sandcastle pulse (pins 17, 11 and 10). External RGB signals may be fed directly into the TDA 9414 as shown.

**Fig. 17.12** *Video processor chip*

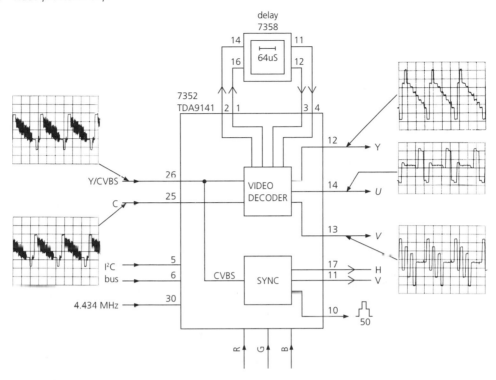

# The TDA 9155 digital deflection processor

The TDA 9155 (Fig. 17.13) is *a digital deflection processor* (*DDP*) with in-built CPU, RAM and ROM. The DDP carries out the following functions: synchronisation, frame drive, line drive, east–west drive and protection. All geometry adjustments are performed by the microcontroller via the I²C bus (pins 1 and 2) and stored in an EEPROM chip. Horizontal and vertical drive pulses are fed into pins 4 and 3 with control flyback pulses going into pins 16 and 17. Line output is obtained at pin 2, and differential vertical deflection drives are provided at pins 23 and 24 from the vertical digital-to-analogue converter (DAC). Full E–W correction (picture width, trapezium, parabola or pincushion and corner correction) is provided at pin 19 from the E–W DAC, and the sandcastle pulse is provided at pin 13. The DDP provides protection facilities against horizontal, vertical, E–W and audio failure. All failures are reported to the microcontroller.

## Onscreen display

The purpose of onscreen display (OSD) is to give visual indication on the TV screen when any function is requested, and to give its state. The display contains characters and numbers arranged in rows along the screen. Each row occupies a number of lines

**Fig. 17.13** *TDA 9155 digital deflection processor*

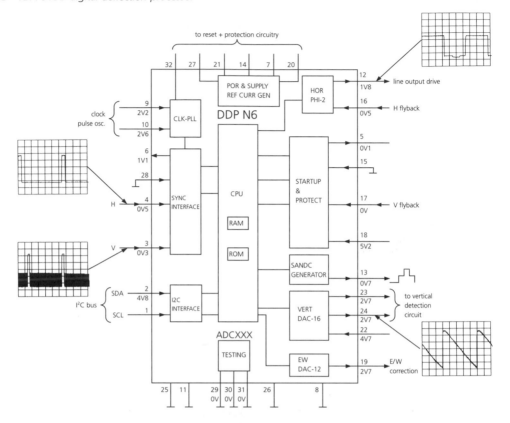

depending on the size of the displayed characters. Assuming a 5 × 7 dot format, each display line of characters will occupy seven scanning lines (Fig. 17.14). Assuming a non-interlaced system, the process of scanning involves the electron beam sweeping across the first dot matrix row of all the characters in the first display line, displaying the appropriate dots along that line scan. This is then followed by the second matrix row, and so on. The video signal thus consists of the first row of matrix dots for each successive character, followed by the next row and the next, up to the seventh row. The video signal of such a display has two levels only, white and black, representing the presence or absence of the dots along each scan line.

The principal elements of an OSD system are shown in Fig. 17.15. Character generation is carried out by dedicated chip IC1, under the control of a microcomputer. The character generator contains a code for each character that may be displayed. To display a character, the code is recalled by the microcomputer and placed in an appropriate location in a memory map within the chip itself. The precise location is determined by the position of the character on the screen. For each field, the microcomputer causes the character codes to be retrieved from the memory map in the correct order. Each code generates a set of black and white level pulses which correspond to the luminance content of each scan line. These pulses are then fed into black and white level

**Fig. 17.14** *Character display on a TV screen*

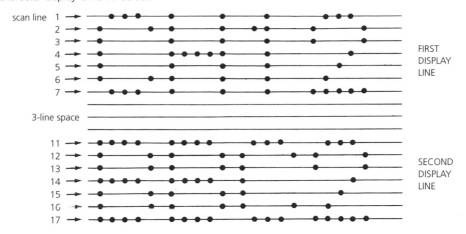

**Fig. 17.15** *Onscreen display (OSD) system*

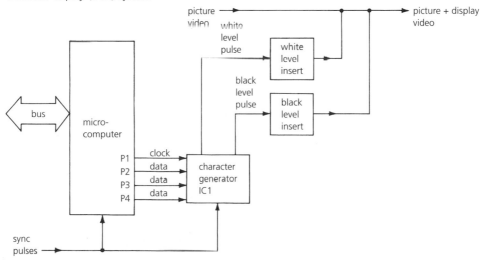

insert switches which connect the video line to the appropriate d.c. level to superimpose the characters on a plain raster or an existing picture.

## Practical configuration of a bus-driven chassis

An I²C bus configured Philips 3A chassis is shown in Fig. 17.16, in which two micro-computers are employed to control the various elements via a single bus. The address of each device is shown in hexadecimal on the diagram, including the two microprocessors, which have addresses 52 and 50 respectively.

**Fig. 17.16** *Microcomputer-controlled TV receiver using the I²C bus (Philips 3A chassis)*

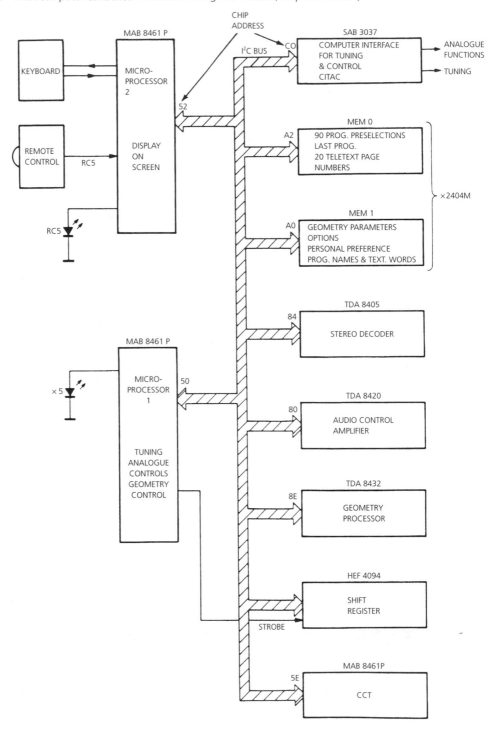

Microprocessor 2 receives commands from the remote control system and the keyboard and passes them on to microprocessor 1 via the bus. It also controls the onscreen display (OSD).

Microprocessor 1 controls all analogue controls and channel tuning. When a program selection command is given, it obtains the relevant channel number from memory MEM 0, converts it to frequency-related information and sends it via the bus to the CITAC, the computer interface for tuning and analogue control. The CITAC proceeds to tune the receiver and informs the microprocessor. Microprocessor 1 also sends data related to four other analogue controls (brightness, colour, contrast and hue) to the CITAC, which converts them to analogue voltages. The memory devices, MEM 0 and MEM 1, are EEPROMs containing details of tuning preselection and teletext page numbers. MEM 1 contains details of picture geometry parameters, TV standards for other countries and personal preferences as to volume level, etc.

A bus controlled geometry processor chip, TDA 8432, is used to provide S-correction, linearity, E–W correction and compensation for e.h.t. variations as well as height, shift (line and field) and line hold controls. The geometry processor chip receives information from microprocessor 1 via the I²C bus; this enables it to carry out twelve picture geometry functions.

# Synthesised tuning

Synthesis is the process of combining or adding up incremental amounts to obtain a certain quantity of, say, a voltage or a frequency.

In the *frequency synthesised tuner* (*FST*) the tuning voltage is obtained from a programmable phase-locked loop composed of the tuner's local oscillator, a controlled prescaler, a phase discriminator and a low-pass filter (Fig. 17.17). A sample of the tuner's local oscillator frequency is fed back to the phase discriminator via a controlled divider, $\div n$, known as a prescaler. The value of $n$ is set by the channel select input; it determines

**Fig. 17.17** *Synthesised tuner*

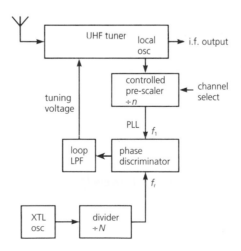

the frequency $f_1$ going into the phase discriminator. The discriminator then compares $f_1$ with reference frequency $f_r$. A d.c. output is produced that reflects the difference between the two frequencies. After filtering, this d.c. voltage is then used to tune the tuner to the selected channel.

Synthesised PLL-controlled tuners have extremely stable output with practically no drift, thus removing the need for an a.f.c. circuit. The stability of the tuner output depends to a large extent on the stability of the reference frequency. For this reason, frequency divider $\div N$ is used to divide the frequency of a crystal-controlled reference oscillator by a large factor; this improves stability by the same factor $N$.

## Voltage synthesised tuning

The principle of *voltage synthesised tuning* (*VST*) is shown in Fig. 17.18. The process starts by loading the tuning register with a binary number. The contents of the register are then fed into a digital-to-analogue converter to produce the required tuning voltage, which is amplified and filtered before going to the tuner. While a channel is being tuned, the a.f.c. loop is opened by the a.f.c. switch, which switches over to the a.f.c. level detector. When a tuning voltage is established by the DAC, the i.f. detector checks for a correct intermediate frequency. If a drift in frequency is recognised, the a.f.c. level detector will increment or decrement the tuning register for fine tuning. When the correct intermediate frequency is established, the a.f.c. loop is closed by the a.f.c. switch.

The register may be loaded manually by incrementing (+) or decrementing (−) its contents, or by using preset channel selection. With preset channel selection, the setting (i.e. the number) for each channel is stored in a RAM chip which is offloaded

**Fig. 17.18**

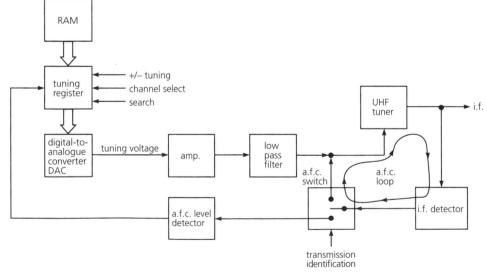

**Fig. 17.19** *Circuit based on the PLL chip TD6316 AP (Ferguson ICC5 chassis)*

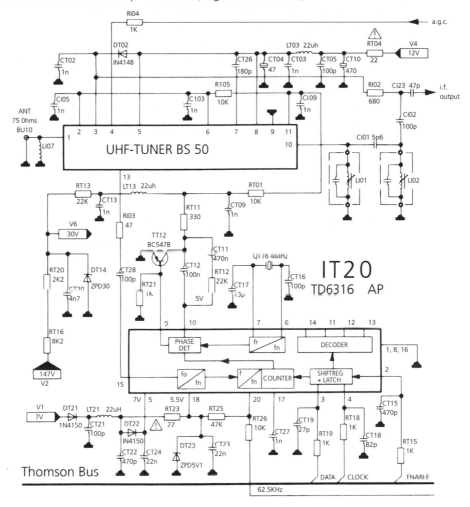

into the register when a particular channel is selected. In the search mode, the register continuously increments its contents, gradually increasing the tuning voltage. When a channel is recognised, a transmitter identification signal is sent to the register to freeze its contents.

## PLL-Controlled FST

Figure 17.19 shows a circuit diagram of programmable PLL chip TD6316 AP, used by Ferguson for frequency synthesised tuning. The chip is controlled by the central control unit via a three-line Thomson bus, data/clock/enable. An 18-bit data word is entered whenever the enable line (pin 2) is held high, which occurs whenever there is a channel selection. The data bits are stored in the shift register latch and subsequently decoded to set the division ratio of programmable divider f/fn. The resulting frequency is then fed into the phase detector, where it is compared with the constant frequency derived from a 4 MHz crystal. A factor of 512 is used to divide the 4 MHz signal to produce a very stable frequency of 4 MHz/512 = 7812.5 Hz. The output from the discriminator is used to control transistor TT12, which functions as a variable load resistor for the 30 V tuning voltage produced by diode DT14. A tuning voltage of 0.5 V to 30 V is thus made available to the tuner. The tuner frequency at pin 13 of the tuner is fed back to the frequency divider f/fn via prescaler fo/fn to complete the loop.

## QUESTIONS

1. Name the three types of bus in a microprocessor-based system.

2. Name the two signals carried by the I²C bus.

3. State the functions of a microcontroller in a TV receiver.

4. Explain the difference between RAM, ROM and EPROM devices.

5. Explain the reason for a delay in the reset circuit of a microprocessor.

6. What is a UART and what is it used for?

# **18** Remote control

The use of remote control is now common to almost all TV receivers. Very early remote control systems used long cables to connect to the receiver. This was followed by the use of visible light from a torch, which activated a sensor in the receiver. Next came the ultrasonic system, which provided additional facilities and improved performance. Remote control came into its own by the introduction of infrared systems, and today they are beginning to resemble data transfer links, employing microcomputers at both the transmitter and receiver.

## Remote control system

The basic block diagram of a remote control (RC) system is shown in Fig. 18.1. The keyboard consists of a number of press-button contacts, one for each command. When one key is operated, the encoder generates a coded signal which represents the particular selected control key. There are two types of encoder, static and scanning. The *static encoder* produces an output when two i.c. pins are shorted by pressing a key. The *scanning encoder* contains a program which interrogates the keyboard to identify the closed key. The encoded signals are transmitted as ultrasonic or infrared waves via a transducer diode (DT). The encoded signal is received by a sensing diode (DS); it is decoded and the appropriate control is then activated.

**Fig. 18.1** *Remote control system: basic blocks*

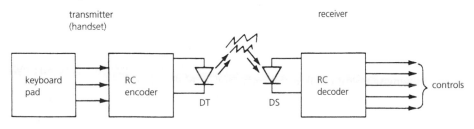

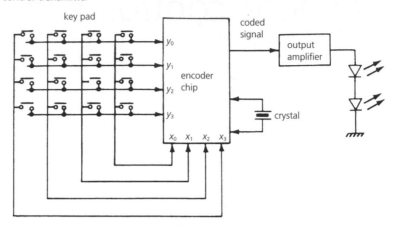

**Fig. 18.2** *Remote control transmitter*

# Remote control transmitter

The basic arrangement for a remote control transmitter (handset) using a static encoder is shown in Fig. 18.2, in which a 4 × 4 matrix keypad is used. The keypad consists of four vertical wires ($x_0$–$x_3$) and four horizontal wires ($y_0$–$y_3$). At each of the 16 cross-points is a push-button keyswitch which, when pressed, makes a connection between a vertical line and the corresponding horizontal line. Two i.c. pins are thus shorted, causing a unique coded signal to be generated; the signal is amplified then transmitted by the diode transducer. The number of control keys may be increased by using a larger matrix, e.g. 4 × 8 (32 keys) or 8 × 8 (64 keys).

# Ultrasonic encoding

The ultrasonic encoder contains a crystal-controlled oscillator and a variable frequency divider. When a key is selected, it generates a multibit word that determines the setting of the variable divider and hence the frequency of the transmitted ultrasonic wave. The frequency range of ultrasonic transmission is limited, so the number of commands is also limited. Ultrasonic systems have a number of other drawbacks, such as spurious responses to rattling keys and coins, and even chattering budgerigars. Reflections from walls and other close objects create confusion at the receiving end, and ultimately they cause malfunction of the system. For these reasons, modern remote control systems employ infrared transmission; its high velocity removes the problems associated with reflections. Furthermore, because it employs digital coded transmission, infrared remote control provides a far greater number of functions.

**Fig. 18.3** *Remote control data word construction*

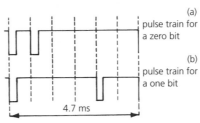

start   control   system   command
bits    bit       bits     bits

**Fig. 18.4** *Pulse-position modulation for the SAA5000*

(a)
pulse train for
a zero bit

(b)
pulse train for
a one bit

4.7 ms

# Infrared encoding

The main function of the infrared (IR) encoder chip is to convert a specific $x$–$y$ matrix selection into a corresponding serial data word to drive the IR light-emitting diode. The data word is continuously repeated while the button of the keypad remains pressed. The data word consists of four basic parts (Fig. 18.3). It begins with constant time reference *start bits* which are used to synchronise the receiver to the transmitter. This is followed by the *control bit* which changes its logic level, i.e. toggles with each new switching contact of the keypad. This informs the receiver whether a button has been held down or pressed a second time, important in commands where the same digit is repeated, e.g. channel 11 or teletext page 222. The control bit is followed by the *system bits* which define the device being addressed, e.g. a TV set or a video cassette recorder. This is usually carried out by a two-position VCR/TV switch on the keypad. Finally the *command bits* are transmitted. They instruct the receiver as to the setting of various controls. The coding technique and the bit size of the component parts of the data word differ from one manufacturer to another. In remote control systems that control a TV receiver only, the system bits may be dispensed with and, where the number of channels does not exceed nine, the control bits may also be dispensed with, leaving start (also known as framing) and command bits only.

Before transmission the data word has to be modulated. This is necessary because of the characteristic of the infrared transducer diode which, given a stream of similar pulses, would be continuously on and off, resulting in a very heavy current drain on the battery and effectively turning the transmitter off. Two types of modulation are used: pulse position and amplitude modulation. The SAA5000 remote control chip uses pulse position modulation (PPM) which separates the pulses, thus improving the duty cycle of the IR diode and reducing power consumption. Each logic level is given a 5-bit code: a logic 0 bit is given the code shown in Fig. 18.4(a) and a logic 1 bit is given the code shown in Fig. 18.4(b). Modern encoders use a constant frequency (in the region of 36 kHz) to amplitude modulate the pulse train. The result is the series of high-frequency bursts shown in Fig. 18.5, with each burst representing a logic 1 bit.

**Fig. 18.5** *Amplitude modulation of infrared data*

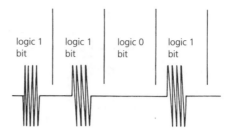

## Scanning encoders

Scanning is performed by first taking vertical lines $x_0$–$x_3$ (Fig. 18.2) to logic 1 and then sequentially turning horizontal lines $y_0$–$y_3$ on and off. When a closed key is detected, the appropriate bits of the data word are generated. The scanning process commences as soon as a key is pressed and before the start bits are initiated.

## Practical infrared transmitter

A practical remote IR transmitter circuit using a scanning encoder is shown in Fig. 18.6. The coded pulse pattern is derived from the system clock, generated by a crystal-controlled oscillator operating at a frequency of 72 kHz. The keyboard consists of two matrices: the X–Y matrix consists of two groups of eight terminations on the encoder, X0 to X7 and Y0 to Y7; the Z–Y matrix is labelled Z0 to Z3 and Y0 to Y7. The Z–Y matrix and switch SK1 determine the system bits (TV or VCR).

The total number of possible commands is 64 for the X–Y matrix and 32 for the Z–Y matrix, giving a theoretical maximum of $64 \times 32 = 2048$. However, far fewer commands are required and in the circuit under consideration only 49 keys are used. Scanning is carried out twice, the first time the Z–Y matrix is scanned to determine the system bits of the data word and the second for the X–Y matrix to generate the command bits. The timing and coding unit completes the data word pulse stream by adding the start and control bits. The whole data word is then modulated by half the oscillator frequency ($72/2 = 36$ kHz) and this signal is used to drive the output stage; transistors 7002 and 7003 with diodes 6001 and 6002 provide signal limiting.

## Universal RC handset

The increase in the number of remotely controlled items of equipment in the home has meant a proliferation of the number of transmitter handsets. This is exacerbated by the fact that manufacturers use different RC techniques. To overcome this, universal handsets are available which can store the commands codes of the various manufacturers in memory. This may be done by programming an EEPROM with the required set of commands or by holding a complete set of all command codes in ROM.

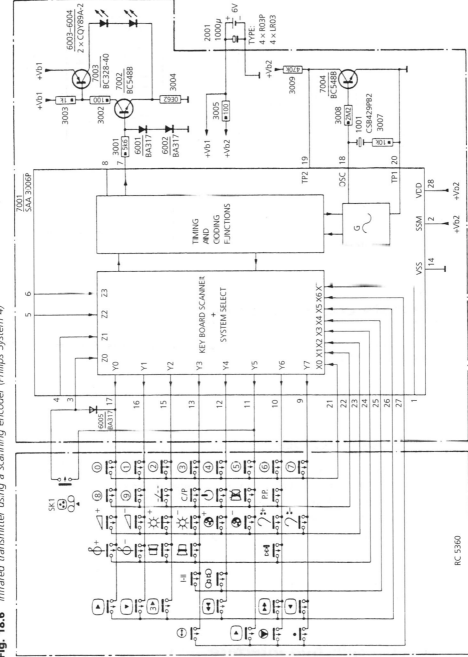

**Fig. 18.6** Infrared transmitter using a scanning encoder (Philips System 4)

**Fig. 18.7** *Remote control receiver: generalised block diagram*

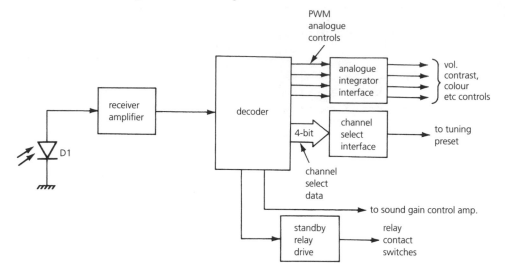

# Remote control receiver

A generalised block diagram for a remote control receiver is shown in Fig. 18.7; D1 is a transducer which converts the ultrasonic or infrared waves into electrical signals. After amplification the coded signal is fed into a decoder, which translates the command word into a control output signal. Each output is fed into an appropriate interface circuit, which carries out the control instruction.

## Analogue interface

When an analogue command is detected, the decoder produces a pulse-width modulated pulse train. The mark-to-space ratio of the pulse is used to change the setting of the control. With a high mark-to-space ratio, the mean d.c. level is high. This d.c. voltage is used to move the control in one direction. Conversely, when the mark-to-space ratio is low, the d.c. level is also low, moving the control in the other direction. A typical interface circuit is shown in Fig. 18.8, in which R1/C1 is an integrator. Capacitor C1 charges up towards the mean d.c. voltage of the pulse-width modulated input, changing the base voltage of TR1 and with it the voltage across emitter resistor R2. This changed emitter voltage is then fed into the appropriate gain control amplifier to vary the setting of the selected control.

## Channel selection interface

Programme selection commands are translated into a 4-bit parallel code, which provides $2^4 = 16$ different channel selections. The coded data is then fed into a 4-to-16 demultiplexer, which activates one and only one of its 16 outputs for each combination

**Fig. 18.8** *Interface circuit for analogue controls*

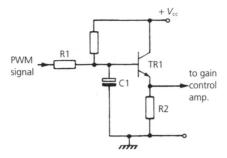

**Fig. 18.9** *Channel selection interface*

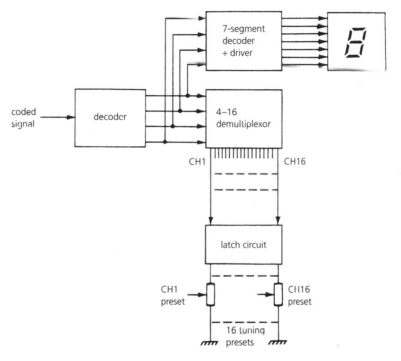

of the input data (Fig. 18.9). The enabled output drives a latch which selects one of the 16 tuning presets. The 4-bit data is also fed to a seven-segment decoder/driver to drive a seven-segment LED display.

## Standby interface

The purpose of the standby control is to disconnect the main h.t. lines from the receiver, leaving subsidiary d.c. voltages such as those that feed the remote control receiver itself. The standby interface is essentially a relay drive circuit; in Fig. 18.10 RL1 is the relay armature operating contact switches S1 and S2. Under normal conditions, relay driver TR1 is off with no current flowing through relay armature RL1. Contact switches S1 and

**Fig. 18.10**  *Standby control interface*

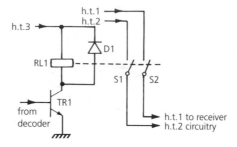

**Fig. 18.11**  *Practical standby interface circuit*

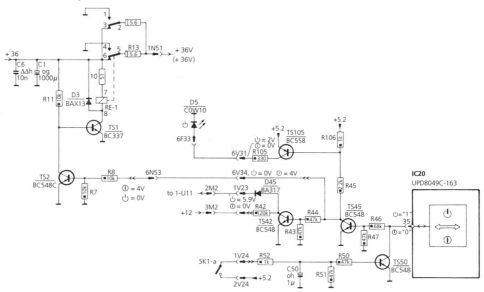

S2 are thus closed, providing normal h.t. to the receiver. When a standby instruction is received, the decoder standby output pin goes high (logic 1) causing TR1 to conduct. TR1 current energises relay RL1, causing S1 and S2 to open, disconnecting h.t.1 and h.t.2 and turning the receiver off. Diode D1 is the normal protection diode; it protects the relay driver transistor from any overvoltage resulting from the back e.m.f. induced in the relay armature when TR1 switches off.

A practical standby interface circuit is shown in Fig. 18.11; TS1 is the relay driver with D3 its protection diode, RE-1 is the relay armature and IC20 is the remote control decoding part of microcomputer chip UPD8049C. When a standby command has been received, pin 35 of IC20 switches from low to high, turning on TS45. The low collector voltage of TS45 turns off TS2 and therefore turns on TS1, energising relay RE-1. Contact is made between relay contacts 2 and 1, also 4 and 5, and this switches off the 36 V supply. The low voltage of TS45 collector also turns off TS42 so that D45 conducts via R42, which applies a high voltage (5.9 V) to pin 1 of main power supply unit U11 (not shown), switching off the main h.t. line. The circuit provides for an LED to indicate a standby mode. When the receiver is in the standby mode, with TS45 turned on, TS105 is turned on, driving current through LED indicator D5. When the receiver is switched on again, pin 35 of IC20 goes from high to low. As a result, T45 turns off, TS2 turns on, TS1 turns off and relay RE-1 is de-energised, Contacts 2 and 3, and contacts 5 and 6 make, restoring the 36 V supply. Furthermore, pin 1 of main power unit U11 goes low, enabling the h.t. supply, and TS105 and D5 turn off.

# The decoder chip

A block diagram for an infrared remote control decoder chip is shown in Fig. 18.12. Remote control of the analogue settings (volume, brightness and colour) is achieved by digital-to-analogue conversion of the outputs of separate 6-bit counters driven by clock pulses from the oscillator. When an analogue up or down command is recognised by the decoder, the appropriate counter is incremented or decremented by the clock pulses as necessary. The counter's 6-bit binary output is then converted into an analogue value by the digital-to-analogue converter. This analogue value is then used to control the pulse width of a square wave derived from the clock oscillator. A 6-bit counter provides a maximum count of $2^6 = 64$. The analogue control voltage therefore varies in steps totalling 64. Given a command repetition frequency of say 100 ms, it will take $64 \times 100$ ms = 6.4 s to adjust the level from minimum to maximum or vice versa.

**Fig. 18.12**  *Decoder chip for infrared remote control*

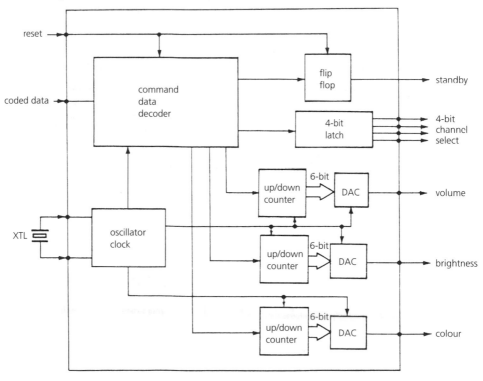

When a programme change command is received, the decoder converts the coded instruction into a 4-bit binary number and stores it into a 4-bit parallel latch. The decoder also determines the logic state of the standby output pin via the flip-flop. When the receiver is in standby mode, the decoder sets the flip-flop to produce a high and vice versa for normal operation. On switch-on from the off or standby modes, a pulse is fed into the chip to reset the decoder and the standby for normal operation.

## Microcomputer decoding system

Remote control decoding using a microcomputer offers a wider range of possible functions. Apart from the infrared detector and amplifier, the decoding and control functions are all carried out by the microcomputer. A generalised block diagram for a microcomputer decoding system is shown in Fig. 18.13, in which microcomputer chip IC1 decodes commands from the remote control receiver as well as those from the local control keyboard matrix. The local control keyboard usually provides a number of functions, including a synthesised tuning facility. The bus system allows IC1 to communicate with and control the other chips. IC1 sets the division ratio of the dividers within the programmable phase-locked loop IC4, thus selecting the channel. The final tuning data is fed back to the microcomputer via the bus and stored into memory. The system ensures accurate tuning without the need for a.f.c. Any drift due to tuner ageing is compensated by the memory being continually refreshed. When the microcomputer detects an analogue control instruction, it sends the appropriate instruction to the analogue control chip IC2 via

**Fig. 18.13** *Microcomputer decoding system*

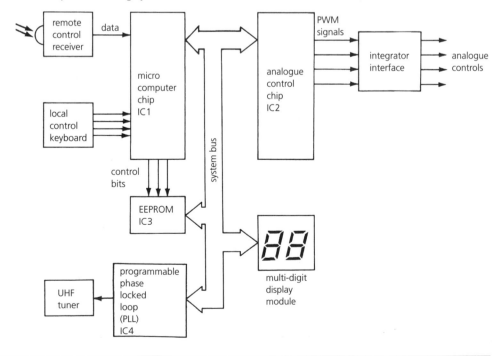

the bus. IC2 converts the instruction into a pulse-width modulated waveform which, when integrated by the interface, is used to vary the analogue setting. The system also provides for a multidigit LED display controlled by the bus. Finally, memory chip IC3 stores all the relevant information with regard to the customer preferred analogue settings, channel frequencies and other settings at the manufacturing stage. The EEPROM is controlled by the microcomputer using a 3-bit parallel code which instructs the chip to write, read, erase, etc.

We shall now describe the various parts of the computerised system.

# Receiver for microcomputer remote control systems

Apart from infrared detection by a transducer diode, the remote control receiver provides narrowband and selective amplification, demodulation and pulse shaping before feeding the data into the microcomputer. A practical circuit is shown in Fig. 18.14, in which IC1 provides the necessary preamplification after detection by D1. IC1 incorporates automatic gain control which ensures that TS3 receives a constant amplitude signal. The signal level is determined by the setting of preset R11. Demodulation is carried out by TS3, TS4 and IC2, which detects the envelope of the modulating 36 kHz frequency. TS3 is biased so that it conducts at the positive peaks of the 36 kHz modulating signal. The resulting negative-going pulses at the collector of TS3 drive the SET input of IC2 (pin 2) so that the signal on output pin 3 becomes high. Resetting occurs when the RESET input (pin 6) exceeds a certain trigger level. The RESET input is controlled by TS4. When a remote control signal is received, TS4 is turned on (via D4) by the negative-going voltage at the collector of TS3, keeping the RESET input below its threshold level. At the end of the 36 kHz burst, TS4 turns off; this allows C8 to charge via R9, which raises the RESET input above the trigger level, pulling pin 6 to zero. Pins 7 and 6 are also pulled to zero by the action of the NOT gate in IC2, discharging C8. The sequence is then repeated for the next burst of carrier frequency, and so on, to reproduce the original coded pulse stream.

**Fig. 18.14** *Receiver for microcomputer remote control system*

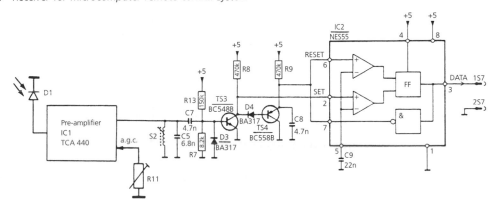

**Fig. 18.15** *A two-digit bus-controlled display module (Philips)*

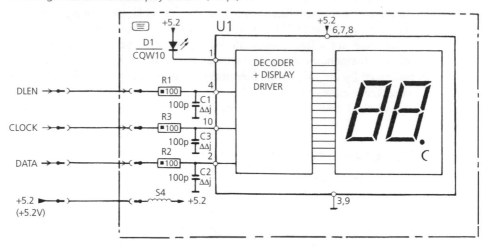

## LED display module

A circuit of a two-digit bus-controlled display is shown in Fig. 18.15 with an additional LED which, when illuminated, appears as a decimal point to indicate that a channel number is being displayed. There is a further LED (D1) which lights up when the teletext mode is selected. The decoder and display driver chip is controlled via a three-line DBUS system by the CCU. The data signal comprises 18 bits; it includes information for the teletext LED and the decimal point, the first and last bits providing the address of the converter chip. The chip converts the binary-coded programme or channel number sent along the DATA line into a seven-segment code to drive the two digits.

### QUESTIONS

1. State three advantages of an ultrasonic remote controller over an infrared remote controller.

2. What is meant by the universal RC transmitter?

3. Figure Q18.1 is the circuit diagram of a remote control transmitter.
   (a) State the circuit function of **each** of the following:
      (i) D101
      (ii) R1 and C1
      (iii) C3
      (iv) C5
   (b) Explain why **each** of the following single component faults will render the transmitter inoperative:
      (i) C4 short circuit
      (ii) R6 open circuit
   (c) What would be the effect on the transmitter output if one of the diodes marked D were to become open circuit?

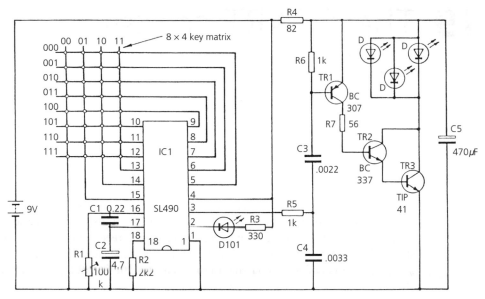

# 19 The SCART socket

Modern TV receivers are designed to receive direct audio/video (AV) input from external sources such as video recorders, camcorders and satellite decoders. They also provide direct AV signals to peripherals such as video recorders and audio systems. The standard outlet for such facilities is the SCART socket connection (Fig. 19.1). Also known as Peritel (peripheral television) and Euro-Connector, the SCART connector allows direct RGB access to the receiver as well as bidirectional composite video known as CVBS (composite video, blanking and sync.) together with independent stereo sound channel connections. The function of each pin is listed in Table 19.1, along with the expected type of signal. Two types of video input may be processed by the receiver – composite video (pin 20) and RGB (pins 15, 11 and 7) – but notice that only composite video (pin 19) is available as an output signal for external devices. Pins 10 and 12 are provided for intercommunications between devices connected to the SCART socket.

**Fig. 19.1**  *SCART socket pin-out*

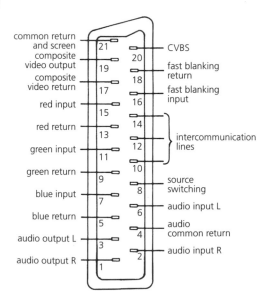

**TABLE 19.1**

| Pin no. | Function specification | Signal |
|---|---|---|
| 1 | Right channel audio out | 0.5 V into 1k |
| 2 | Right channel audio in | 0.5 V into 10k |
| 3 | Left channel audio out | 0.5 V into 1k |
| 4 | Audio earth | |
| 5 | Blue earth | |
| 6 | Left channel audio in | 0.5 V into 10k |
| 7 | Blue in | 0.7 V into 75$\Omega$ |
| 8 | Source switching (9–12 V) | not specific but usually max 12 V into 10k |
| 9 | Green earth | |
| 10 | Intercommunication line | |
| 11 | Green in | 0.7 V into 75$\Omega$ |
| 12 | Intercommunication line | |
| 13 | Red earth | |
| 14 | Intercommunication line earth | |
| 15 | Red in | 0.7 V into 75$\Omega$ |
| 16 | Fast RGB blanking | varies (1 3 V) |
| 17 | Composite video, blanking and sync. earth | |
| 18 | Fast blanking earth | |
| 19 | Composite video, blanking and sync. out | 1 V into 75$\Omega$ |
| 20 | Composite video, blanking and sync. in | 1 V into 75$\Omega$ |
| 21 | Socket earth | |

The SCART connector is a 21-pin non-reversible device. There are 20 pins available for connections; pin 21 is connected to the skirt, hence the chassis provides the overall screening for cable communication.

## The SCART interface

Figure 19.2 shows a standard SCART interface. The interface provides the switching, amplification and impedance matching necessary for routing the signals in and out of the receiver. The interface has two modes of operation: TV mode and AV (monitor) mode. In TV mode the receiver processes the off-air signals in the normal way. In AV mode the receiver blanks out these signals in order to process incoming signals from an external device connected via the SCART socket. Switching between the two modes may take place in one of two ways:

- The external source asserts SCART pin 8 (source switching) by applying a high voltage of 9–12 V to the pin.
- A TV/AV select signal from the receiver to the interface is produced when the AV button on the receiver is pressed.

**Fig. 19.2** *SCART interface*

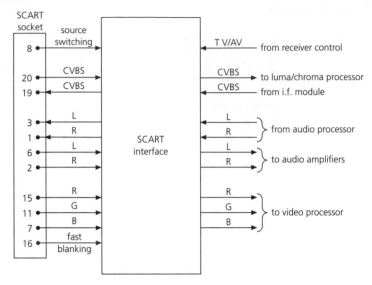

**Fig. 19.3** *Simple routing circuit*

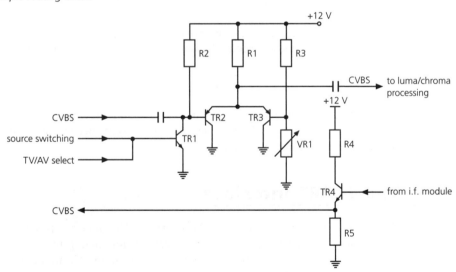

## Routing composite video

Figure 19.3 shows a simplified circuit for routing an incoming composite video (CVBS) signal from an external source into the video processing stage of a TV receiver. The external CVBS signal at pin 20 of the SCART socket is fed into the base of TR2. When the interface is in the TV mode, the voltage at the base of TR1 (SCART pin 8) is low, turning off TR1; this turns off TR2, which turns on TR3. An incoming CVBS signal is thus blocked. When pin 8 goes high, either by a switching signal from an external source or by pressing the AV button on the receiver, TR1 base goes high, turning it on. TR1's collector voltage drops, forcing TR2 to turn on and TR3 to turn off. With TR2 on, its input, the incoming CVBS signal, appears across common-emitter resistor R1, and it is fed into the video processing stage of the receiver. The same switching signal at SCART pin 8 is used for two other functions:

- To activate a separate audio routing circuit to connect the left and right sound channels directly to the audio amplifier stage of the receiver.
- To blank out the off-air produced RGB signals from the video processing stage; this prevents the video and chrominance signals generated within the set from interacting with the signals fed to the SCART socket.

The circuit in Fig. 19.3 also provides a composite video output on SCART pin 19, regardless of the mode of operation. Emitter follower TR4 acts as a buffer stage between the receiver and the external device receiving the CVBS signal.

## RGB routing

The RGB inputs to SCART pins 15, 11 and 7 are routed into the video processing stage by a switching arrangement such as Fig. 19.4. In the TV mode, S1, S2 and S3 in the RGB routing network are in the positions shown, with the internally generated RGB

**Fig. 19.4** *RGB routing*

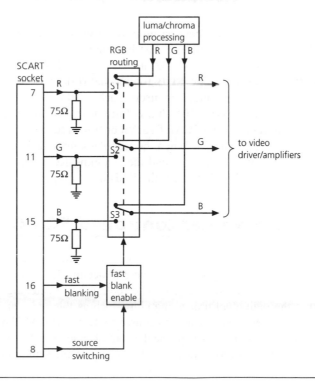

**Fig. 19.5**

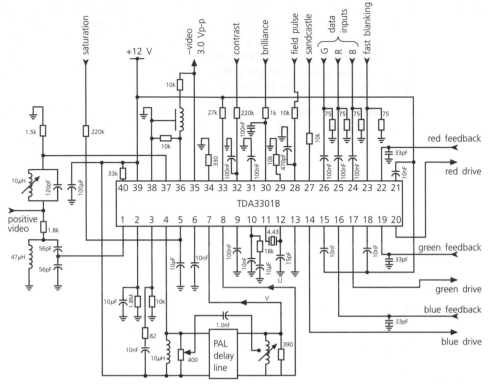

signals from the video/chroma processor routed to the RGB amplifiers. When source switching pin 8 is taken high, fast blanking is enabled and fed into the RGB routing network, switching over S1, S2 and S3. External RGB signals are now connected to the RGB amplifiers. The RGB routing network is normally incorporated within the video/chroma processing chip itself, such as the TDA 3301 colour processing chip (Fig. 19.5). (The TDA 3301 chip is fully described in Chapter 12). RGB signals from an external source via a SCART socket or from a teletext decoder are connected to RGB input pins 25, 26 and 24. When the fast blanking signal (pin 23) is enabled, the off-air RGB signals are blanked out and the incoming RGB signals at pins 25, 26 and 24 are directed to pins 20, 17 and 14, the RGB drive outputs.

## Microcomputer-controlled SCART interface

In a microcomputer-controlled receiver, the routing of composite video, RGB and the audio channels is supervised by the microcontroller chip. Figure 19.6 shows a typical SCART interface arrangement. The rooting and switching networks are incorporated within the individual processing chips such as the audio and luma/chroma processors. When an external device asserts source switching (SCART pin 8), a status signal is sent to the controller chip. In response the controller sends a TV/AV select signal to the CVBS switching network to switch it to AV mode. At the same time, the TV/AV select signal

Fig. 19.6

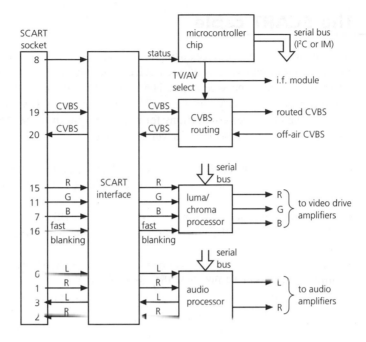

is sent to the AV/tuner switch in the i.f. module to blank out the off-air composite video signal. Audio AV switching and RGB selections are carried out under the control of the microcontroller chip via the serial bus.

Where *separate luminance and chroma* (*S-Y/C*) and stereo audio inputs are provided at sockets at the front of the receiver, these signals are also selected by the TV/AV signal from the controller. With the routing and switching being carried out by the microcontroller chip, the SCART interface is reduced to a few simple switching and buffering circuits.

# Dual-SCART receivers

Modern receivers provide two or more SCART sockets, labelled AV1, AV2, and so on. Where two sockets are provided, AV1 is fully wired for all input and output signals. AV2, on the other hand, may be wired as an output or a video link only. Audio and video loops through the SCART sockets are provided to ensure that signals may be routed from one external device such as a camera to another external device such as a VCR. The type of SCART socket is indicated by the arrow notation shown in Fig. 19.7.

**Fig. 19.7** *SCART sockets: (a) input only; (b) input/output*

# The SCART cable

When connecting two devices using a SCART cable, care must be taken to ensure the pins carrying output signals of one device are connected to the respective input pins of the second device, and vice versa. The normal SCART cable has a male plug at either end to connect two devices such as a TV receiver and a VCR via their respective SCART sockets. The crossover connections for such a plug-to-plug SCART cable are shown in Fig. 19.8. The cable consists of several sections:

- The audio section contains two independent input and output stereo channels. Screening for the four audio wires is provided by the audio earth connection at pin 4.
- The composite video section consists of two CVBS channels (input and output). A 75 Ω coax, with the video earth providing the outer screen, is used for each video channel.
- The intercommunication data channels consist of two wires in a balanced pair arrangement; pin 14 provides the earth connection.
- Source or function switching consists of a single insulated wire connecting pin 8 at one end of the cable to pin 8 at the other end.
- The RGB section consists of three signal wires (RGB) and three wires for their respective earths, together with a fast blanking.

**Fig. 19.8** *Plug-to-plug SCART connection*

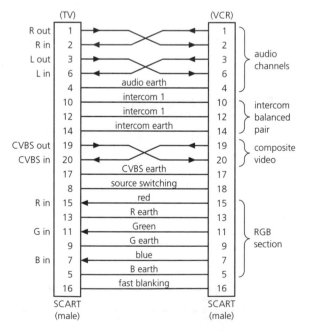

1. State three reasons for including a SCART connector to a TV receiver.

2. In relation to a SCART connector, state the purpose of the following pins:
   (a) CVBS in
   (b) CBVS out
   (c) fast blanking

3. Explain the reason why more than one SCART connector is normally available in TV receivers.

# 20 Teletext

Teletext is a system of information broadcasting in which pages of text and graphics are transmitted in coded format which may be retrieved by a special decoder and displayed by a television receiver. The viewer may select particular pages and display them on the television screen. The page may be displayed in place of, or added to, the normal television picture. Teletext also offers newsflash and subtitles services that may be inset in the picture.

A typical page contains up to 24 rows with each row containing up to 40 columns in the form of alphabetic characters, numbers or graphic symbols. Each television channel is capable of carrying up to eight groups of pages called magazines. Each magazine contains up to 100 pages, giving a total of 800 pages of information. The graphic symbols may be joined together to display simple pictures such as weather maps, graphs and charts.

## System overview

At the transmitting end, teletext information is encoded row by row and page by page then added to the normal television composite video signal, which is modulated and transmitted in the normal way.

At the receiving end and following the demodulation stage, a decoder is used to retrieve and display the information one page at a time. Figure 20.1 shows the basic components

**Fig. 20.1** *Basic components of a teletext decoder*

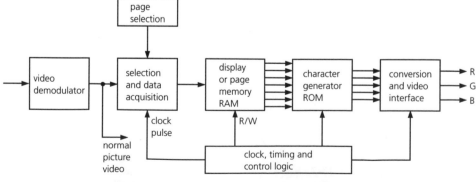

of a teletext decoder. The teletext information is continuously monitored by the selection and acquisition unit. When the number of the transmitted page corresponds to the number selected by the viewer, the contents of that page are captured. The coded information is then loaded into the *display or page RAM* memory chip, row by row and character by character. Each row is identified by its own number, known as a *row address*. The size of the memory must be large enough to store a minimum of one full page of coded information. The contents of display memory are then sent, character by character, to a character generator. The *character generator* is a ROM chip which translates the coded information into RGB drive signals. The whole system is synchronised and controlled by the clock, timing and control logic.

## Teletext broadcasting

Teletext uses the normal broadcast television signal to carry the additional information without interfering with the normal TV reception. This is achieved by employing some of the unused scanning lines at the start of each field. Recall that, during the field blanking period, line scan signals continue to be transmitted to ensure the line oscillator remains in sync. with incoming signals. The field flyback lasts for 25 complete lines, comprising 3 lines of the current field and 22 lines of the following field (Fig. 20.2). At the end of the odd field, flyback starts at line 623 and ends at line 23 of the subsequent field. At the end of the following even field, flyback begins at line 311 and ends at line 336.

Of the 25 'flyback' lines, only 9.5 are utilised by normal television broadcasting:

- 2.5 lines for the first five equalising pulses
- 2.5 lines for the five broad pulses
- 2.5 lines for the second five equalising pulses
- 2 lines for test signals: 19 and 20 for the first or even field, and 332 and 333 for the second or odd field

The remaining lines are unused for normal picture transmission and may thus be utilised for teletext data transmission. Initially only lines 17 and 18 (even field) and 330 and

**Fig. 20.2** *Teletext data lines: ITS = insertion test signal*

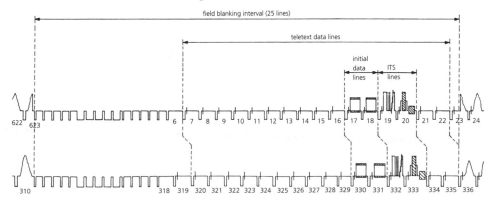

331 (odd field) were used. However, today all 12 available lines (7–18 and 320–331) are utilised, which among other things provides faster page selection. Each teletext line carries enough information for one complete row of the display. Thus 24 lines are necessary for a full page of teletext. The time it takes to transmit one full page may be calculated as follows:

transmission time for one field = 1/50 = 0.02 s

With one field carrying 12 lines of teletext, two fields are required to transmit a full page. Therefore

transmission time for a full page = 2 × 0.02 = 0.04 s

A full 100-page magazine will therefore take

0.04 × 100 = 4 s

## The Teletext data line

A teletext data line carries data in *packets*; each packet contains 360 bits of digital information (Fig. 20.3). The logic state of each bit is represented by the following voltage levels:

*logic 1    66% (∓6%) of peak white*
*logic 0    black level ( ∓2% of peak white)*

For a standard 1 V peak-to-peak video signal, logic 1 will be represented by approximately *330 mV*.

Data pulses are filtered before transmission to remove the very high harmonics. At the receiving end, they are further rounded by the receiver's i.f. and demodulator circuits, resulting in almost sinusoidal waveforms by the time they reach the decoder. For this reason, a *slicer* is used at the input of the decoder to return the pulses back to a square shape.

**Fig. 20.3**  *A single teletext data line*

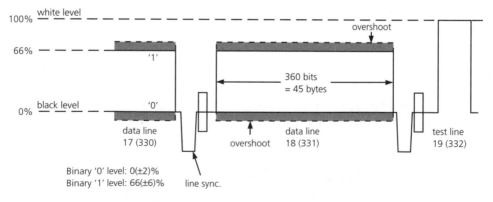

**Fig. 20.4**

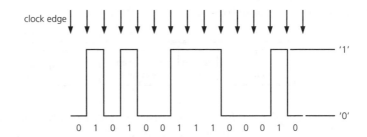

clock edge

'1'

'0'

0 1 0 1 0 0 1 1 1 0 0 0 1 0

## Non-Return to Zero format

Teletext data bits are transmitted in series using what is known as a *non-return to zero* (*NRZ*) format. In this format, a change in the logic state of the signal occurs only if the preceding state is different. When several logic high bits follow each other, the signal voltage remains the same at logic high for the duration of the sequence of 1's; the same goes for a sequence of logic 0's (Fig. 20.4). The NRZ format requires fewer changes in the signal's logic state to convey the same amount of information in the form of bits. Figure 20.4 shows how 14 data bits are represented by eight changes in the logic level of the signal. At the receiving end, a clock pulse running in synchronism with the transmitter clock is required to establish the correct timing of the data bits encoded within the NRZ format. For this reason, a sample of the encoding clock is transmitted along with the teletext data.

## The bit rate

The bit rate, i.e. the rate at which the data bits are transmitted, is determined by the number of bits and the time available for their transmission. In the case of teletext, 360 bits must be transmitted during the line scan. In the 625-line system, the total period of a line scan is 64 $\mu$s, of which 12.05 $\mu$s is occupied by the front porch, the line sync. and the back porch. The remaining 51.95 $\mu$s is available for picture or teletext information. The minimum data bit rate is therefore

360/51.95 = 6.9297 Mbits/s (megabits per second).

Since it is preferable for the bit rate to be a harmonic of the line frequency, a bit rate of

*6.9375 Mbit/s* (±25 ppm)

is used; this is 444 times the line frequency. With two successive bits of opposite values (0 and 1) forming a single cycle, the actual frequency of the clock waveform is 1/2 × bit rate = *3.46875 MHz*.

## Data line construction

The 360-bit packets of the teletext data line are divided into groups of 45 bytes, where each byte contains 8 bits (Fig. 20.5). Each byte or groups of bytes is allocated a specific function:

**Fig. 20.5** *Construction of a teletext data line*

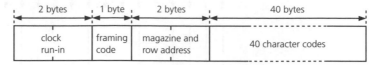

Bytes 1 and 2      *clock run-in pulses for clock synchronisation*
Byte 3                 *framing code to enable byte synchronisation*
Bytes 4 and 5      *magazine and row address*
Bytes 6 to 45     *character bytes representing 40 display characters*

**Fig. 20.6** *Page header*

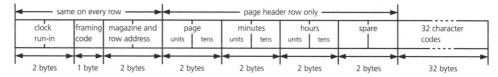

The data line of the first row of a page, the page header, carries extra data, so its construction is different. For this reason, eight of the 40 character bytes listed above are used to carry general information related to the page, such as *page number*, *time code* and *display control information* (Fig. 20.6). For the page header, therefore, only 32 display characters are available.

## Error detection and correction

The data line includes a number of bits that do not carry any message information. These bits are included for the purposes of protecting the integrity of the data against errors due to noise and bad reception. Two levels of error protection are used in teletext broadcasting, one of character codes and another for row and page addresses. For character codes, it is adequate to have a simple parity check which detects a single-bit error in the received data. For the more important parts of the data, such as page and row addresses, where a single-bit error could display the wrong magazine or the wrong page, there is a more effective detection and correction technique using several data bits.

## The clock run-in sequence

The *clock run-in (CR)* comprises a sequence of 16 alternating bits (a total of eight pulses) beginning with logic 1 (Fig. 20.7). The purpose of clock run-in sequence is twofold:

- To indicate the presence of a teletext data line.
- To establish the correct clock pulse rate at the receiving end.

In essence, the clock run-in is a sample of the encoding clock used at the receiving end, very similar to the colour burst in the normal television transmission.

**Fig. 20.7** *Clock run-in*

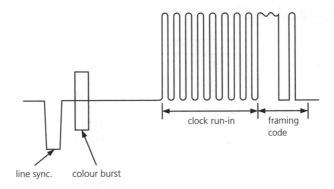

# The framing code

Once the decoder clock has been locked to the transmitter clock by the clock run-in sequence, the next stage is to accurately locate the start of each byte, a process known as *byte synchronisation*. For this purpose, an 8-bit sequence called the *framing code* (*FC*) is inserted immediately following the clock run-in sequence (Fig. 20.7). The bit pattern for the framing code (11100100) is chosen to allow accurate byte synchronisation even if one of the bits is faulty.

Detection of the framing code is carried out by feeding the incoming data stream into a shift register (Fig. 20.8). At the end of the clock run-in sequence, the bits held in the register are the last eight bits of the run-in sequence, 10101010. At the next clock cycle, the first bit of the framing code enters the shift register, shifting all the other bits one place to the left. The new contents of the register are 01010101 (Table 20.1). This is now compared with the framing code and the number of matching bits is noted. In this case, four of the bits in the register match four bits of the framing code, a bit match of 4. The process is repeated for the next clock cycle, and again the bit match is 4. The next clock cycle produces a bit match of 3, and so on. At clock cycle 8, the contents of the register are the framing code itself, 11100100, and a full bit match of 8 is obtained. This indicates the end of the framing code, setting a reference for the beginning of each byte. Byte synchronisation is thus established. Notice that during this process the second highest bit match is 5, which occurs at clock cycle 4. Assume that, due to bad

**Fig. 20.8** *Framing code detection*

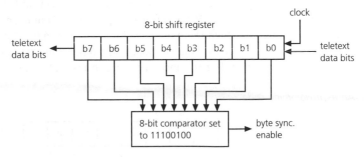

TABLE 20.1

| | Clock no. | Contents of register | No. of matched bits |
|---|---|---|---|
| End of RN | 0 | 10101010 | 4 |
| | 1 | 01010101 | 4 |
| | 2 | 10101011 | 3 |
| | 3 | 01010111 | 3 |
| | 4 | 10101110 | 5 |
| | 5 | 01011100 | 4 |
| | 6 | 10111001 | 3 |
| | 7 | 01110010 | 4 |
| Framing code | 8 | 11100100 | 8 |

reception, one bit of the framing code is wrongly received. The maximum number of matching bits will now drop to 7, and this too will occur only at clock cycle 8. The second highest bit match will be a maximum of 6 and it will occur at clock cycle 4. It therefore follows that a bit match of 7 as well as 8 gives the correct indication of the end of the framing code byte in the received data stream.

## Magazine and row addresses

The fourth and fifth bytes of the teletext data line provide magazine and row address codes. Because of the importance of these codes to the accuracy of the displayed information, *Hamming* error protection is employed, in which alternate bits are used for parity checking. Figure 20.9 shows an 8-bit Hamming code in which M bits 1, 3, 5, and 7 convey message information and P bits 0, 2, 4, and 6 provide odd parity checks. The process of identifying errors in the received data involves four parity tests: A, B, C and D. The following bits are tested by each parity check:

- Test A examines bits 0, 1, 5 and 7.
- Test B examines bits 1, 2, 3 and 7.
- Test C examines bits 1, 3, 4 and 5.
- Test D examines all eight bits.

Notice that, apart from b6, each bit is involved in more than one parity check (Table 20.2).

If all the tests prove to be correct, the data is perfect and it can be passed on for processing by the decoder. If test D proves positive (P) and one or more of the other three tests is failed (F), then more than one bit is faulty and the data is rejected. However, if test D is failed together with one or more of the other tests, the faulty bit can be identified and the error corrected. For instance, a failure in tests D, A and C indicates an error in

**Fig. 20.9** *An 8-bit Hamming code*

| 0 | 1 | 2 | 3 | 4 | 5 | 6 | 7 |
|---|---|---|---|---|---|---|---|
| P | M | P | M | P | M | P | M |

**TABLE 20.3**

*Parity test results*

| A | B | C | D | Conclusion | Action |
|---|---|---|---|---|---|
| P | P | P | P | no errors | message bits accepted |
| P | P | P | F | error parity b6 | message bits accepted |
| F | P | P | F | error parity b1 | message bits accepted |
| P | F | P | F | error parity b2 | message bits accepted |
| F | F | P | F | error message b7 | b7 inverted and data accepted |
| P | P | F | F | error parity b4 | message bits accepted |
| F | P | F | F | error message b5 | b5 inverted and data accepted |
| P | F | F | F | error message b3 | b3 inverted and data accepted |
| F | F | F | F | error message b1 | b1 inverted and data accepted |
| Other combinations | | | | multiple errors | data rejected |

**Fig. 20.10** *Bit pattern for the first five bytes of a teletext data line*

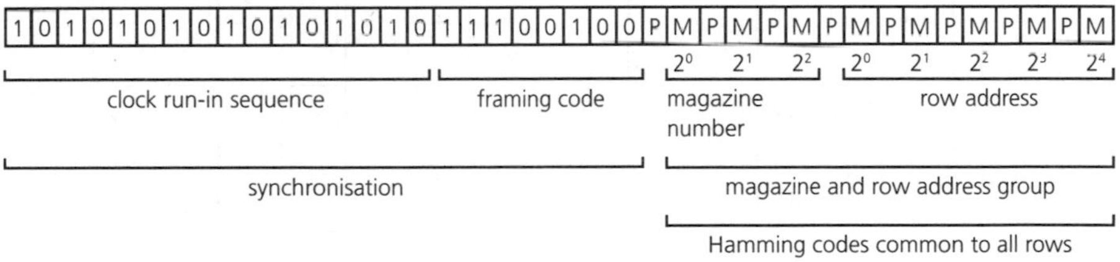

clock run-in sequence — framing code — magazine number — row address

synchronisation — magazine and row address group

Hamming codes common to all rows

b5, which is then inverted; once the data has been corrected, it is accepted for processing. Alternatively, a failure in tests D and A indicates an error in b1; b1 is a parity bit, so it need not be corrected, and the data can immediately be accepted. Table 20.3 shows all possible combinations.

Figure 20.10 shows the bit pattern for the first five bytes of the teletext data line. The first three message bits of the magazine and row address group of bytes are allocated for

# TABLE 20.4 *Character code chart**

| b6→ b5→ b4→ | 0 0 0 | 0 0 1 | 0 1 0 | | 0 1 1 | | 1 0 0 | 1 0 1 | 1 1 0 | | 1 1 1 | |
|---|---|---|---|---|---|---|---|---|---|---|---|---|
| b3 b2 b1 b0 / Col Row | 0 | 1 | 2 | 2a | 3 | 3a | 4 | 5 | 6 | 6a | 7 | 7a |
| 0 0 0 0 — 0 | | | □ | | 0 | | @ | P | — | | p | |
| 0 0 0 1 — 1 | | | ! | | 1 | | A | Q | a | | q | |
| 0 0 1 0 — 2 | | | " | | 2 | | B | R | b | | r | |
| 0 0 1 1 — 3 | | | £ | | 3 | | C | S | c | | s | |
| 0 1 0 0 — 4 | | | $ | | 4 | | D | T | d | | t | |
| 0 1 0 1 — 5 | | | % | | 5 | | E | U | e | | u | |
| 0 1 1 0 — 6 | | | & | | 6 | | F | V | f | | v | |
| 0 1 1 1 — 7 | | | ' | | 7 | | G | W | g | | w | |
| 1 0 0 0 — 8 | | | ( | | 8 | | H | X | h | | x | |
| 1 0 0 1 — 9 | | | ) | | 9 | | I | Y | i | | y | |
| 1 0 1 0 — 10 | | | * | | : | | J | Z | j | | z | |
| 1 0 1 1 — 11 | | | + | | ; | | K | ← | k | | $\frac{1}{4}$ | |
| 1 1 0 0 — 12 | | | , | | < | | L | $\frac{1}{2}$ | l | | ‖ | |
| 1 1 0 1 — 13 | | | – | | = | | M | → | m | | $\frac{3}{4}$ | |
| 1 1 1 0 — 14 | | | . | | > | | N | ↑ | n | | ÷ | |
| 1 1 1 1 — 15 | | | / | | ? | | O | # | o | | ■ | |

\* Columns 0, 1, 2a, 3a, 6a, and 7a are completed in Table 20.5.

the magazine number. With three bits, the maximum number of magazines that can be broadcasted is $2^3 = 8$. The remaining five message bits are allocated for the row address. With five bits, $2^5 = 32$ rows may be addressed, of which only 24 are used (rows 0–23).

## The character bytes

A character byte consists of a 7-bit character code and an odd-parity bit b7. The bits are transmitted in numerical order, b0 followed by b1 and so on. With a 7-bit code, $2^7 = 128$ different codes are available. Of the 128 codes, 96 are allocated for alphanumeric characters, including associated symbols such as %, & and ?. The remaining 32 codes are used for function control and mode selection. The 96 codes and their corresponding alphanumeric symbols are listed in Table 20.4.

## The page header

The first line or row of a page, known as the page header, contains general information such as the *page number* and a *time code*, as well as control functions relating to the display of information such as subtitles and newsflashes. In order to accommodate this extra information, eight additional bytes are included (Fig. 20.11). Although a total of $8 \times 8 = 64$ bits are available for this extra information, only 32 bits are used for messages; the remaining 32 bits are used for parity protection.

**Fig. 20.11**

The first two bytes are used to identify a two-digit page number between 00 and 99. The first of these bytes provides the most significant digit (tens) and the second the least significant digit (units). Two techniques may be employed to increase the number of pages:

- Allocate one page number to a set of pages, usually containing related information such as sport, then send them out in sequence using the same page number. Pages sent out in this fashion are called type B and referred to as rotating pages. Each page will be sent for a period of say one minute, to allow the viewer sufficient time to read the text, then the next page is sent to replace it on the display.
- Send several 'editions' of the same page number. This is achieved by utilising the 'time code' available on the page header. Four bytes are allocated for the time code; they are used to specify a time in hours and minutes. The time code is attached to a page number to provide several versions of the page number. As can be seen from the bit-by-bit construction of the time code (Fig. 20.12), 13 message bits are used to indicate the time code:

**Fig. 20.12** *Time code bit construction*

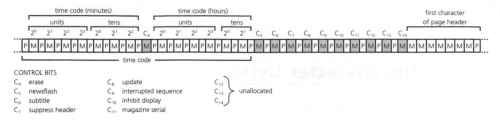

CONTROL BITS
$C_4$    erase
$C_5$    newsflash
$C_6$    subtitle
$C_7$    suppress header
$C_8$    update
$C_9$    interrupted sequence
$C_{10}$    inhibit display
$C_{11}$    magazine serial
$C_{12}$
$C_{13}$ } unallocated
$C_{14}$

*4 bits for the minutes units digit*
*3 bits for the minutes tens digit*
*4 bits for the hours units digit*
*2 bits for the hours tens digit*

The page header contains a further 11 control bits (C4–C14). The functions of these bits are listed in Fig. 20.12.

## Character generation

In the character or text mode, a teletext page comprises 40 character columns × 24 character rows (Fig. 20.13). Each character is allocated a matrix of typically *5 × 7 pixels*. A row of characters thus occupies seven pixel lines. With three lines left blank to provide a space between each row, the total number of lines per row is increased to ten. The process of scanning involves the electron beam sweeping across the first pixel line of all the characters in the first row, energising the appropriate dots along that line scan. This is followed by pixel line 2, and so on. The video signal thus consists of a series of *dot patterns* (also known as *bit patterns*) corresponding to the first pixel line of each successive character, followed by the dot patterns of the second pixel line of the same characters, and so on, up to the seventh line. If, for example, V and F are the first two characters in a row (Fig. 20.14), the monochrome video signals for each of the seven scanlines are as shown in Fig. 20.15. Following three blank scanlines, the process is then repeated for the second row of characters.

These video signals are generated using the arrangement shown in Fig. 20.16. The contents of the selected teletext page are stored in display memory RAM, character code by character code and row by row. These codes are read in the appropriate sequence by the address generator, which places the sequence of addresses on the address bus. The codes appearing on the data bus of display memory are then fed into the character generator. The character generator is a *look-up ROM* chip which contains the dot patterns of each pixel line of every character. When the character code is fed into the ROM character generator, the dot pattern of that character appears at the output of the chip, pixel line by pixel line. The required pixel line is selected by the 3-bit select address.

For instance, assuming that characters V and F in Fig. 20.14 are to be displayed, the character code for the first character, V, is fed to the character generator and

**Fig. 20.13** *Character display: 40 columns × 24 rows*

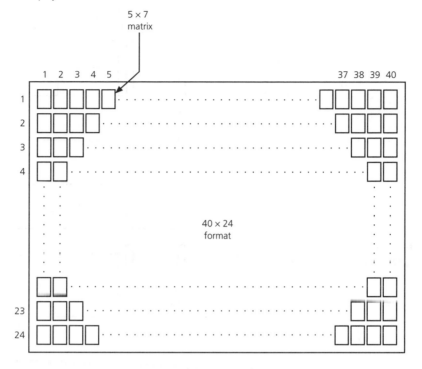

5 × 7
matrix

40 × 24
format

**Fig. 20.14** *Bit pattern of characters V and F*

pixel line

space {

line scan dot pattern
(V)        (F)

1 0 0 0 1 0 1 1 1 1 1
1 0 0 0 1 0 1 0 0 0 0
1 0 0 0 1 0 1 0 0 0 0
1 0 0 0 1 0 1 1 1 0 0
1 0 0 0 1 0 1 0 0 0 0
0 1 0 1 0 0 1 0 0 0 0
0 0 1 0 0 0 1 0 0 0 0

space

simultaneously 000 is placed on the select address to select the first pixel line. The character generator will then produce the dot pattern of the first pixel line of V (100010). A 6-line output is produced, representing the five dots of the character plus one dot for a space between characters. This parallel output is then converted into a serial signal using a simple *parallel-in serial-out* (*PISO*) shift register to form the first part of the first video line. The character code input then switches to the next character, F. With the select address remaining at 000, the dot pattern of the first pixel line of the second character (111110) is produced and applied to the shift register to form the second part of the first video line, and so on, for all other characters along the first row. At the start of the second line scan, the address generator reverts back to the location where

**Fig. 20.15**  *Line waveforms for characters V and F*

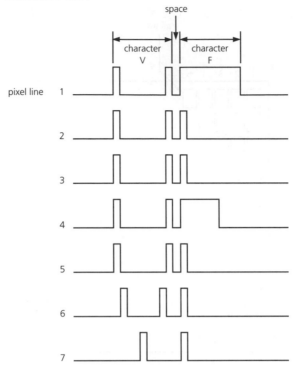

**Fig. 20.16**  *Video character generation*

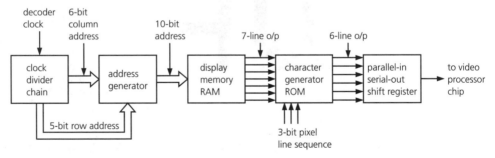

the first character V is stored, but this time the scanline address is incremented to 001 to select the dot pattern of the second pixel line, and so on. The process is then repeated for the next field, and so on.

Modern character generator chips usually combine the functions of character generation and parallel-to-serial conversion into a single i.c. package. For colour applications, four signals are produced by the character generator chip, the three RGB drives and a blanking signal (Fig. 20.17).

---

**Fig. 20.17** *Colour character generation*

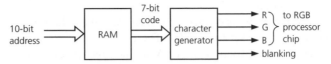

## The dot frequency

The frequency which triggers the shift register determines the total number of dots that may be energised by a single line scan of the screen. The *dot frequency* is therefore determined by the number of characters per row and the pixel format employed. For a given number of characters per row, a higher dot frequency is necessary for a 7 × 9 format than for a 5 × 7 format. Assuming a 5 × 7 format, the dot frequency may be calculated by dividing the number of dots per line scan by the effective duration of a line scan. The total number of dots per scanline may be calculated as follows:

number of horizontal dots per matrix = 5

Assuming a one dot space between characters, then

number of horizontal dots per character = 6
total number of dots per row = 6 × 40 = 240

The total time period of a line scan is 64 $\mu$s, of which 12 $\mu$s are used for synchronisation, etc., leaving 52 $\mu$s. Since the line timebase is normally set to give a slight overscan of a few microseconds, the duration of a line scan is further reduced to about 48 $\mu$s. On the teletext display it is usual to have a small margin on each end, a total of say 8 $\mu$s, which reduces the effective duration to around 40 $\mu$s. This gives a dot frequency of 240/40 = 6 MHz. In order to avoid wavy character edges and unsteady display, the dot frequency is chosen to be a harmonic (a whole multiple) of the line frequency. In this case, 6 MHz (384 × the line frequency) is a suitable frequency to use. A character generator which is set for a different matrix format, say 7 × 9, requires a higher dot frequency. In this case a frequency of 6.9375 MHz will be adequate.

## Address generation

The address of each location consists of two components: a 5-bit row address and a 6-bit column address. Two binary counter/dividers are used: one to increment the row address every 10 lines at one-tenth of the line frequency and one to increment the column address every character at a frequency given by

$$\frac{\text{dot frequency}}{\text{number of dots}} = 6\,\text{MHz}/6 = 1\,\text{MHz}$$

The generated address combines these two components to produce the absolute address of the location.

The memory read operation thus starts with the row address set to 00000 and the column address set to 000000. The code of the first character of the first row is then retrieved from display memory. With the row address remaining unchanged, the column address is incremented to 000001 to retrieve the code of the second character of the first row, and so on, until the column address count reaches 100111 (denary 39) which retrieves the code of the last character in the first row. The column address is then reset by the line sync. and the process is repeated for nine more lines. At the end of the tenth line, the row address is incremented to 00001 and the process is repeated until row address 10111 (denary 23), when the whole process is reset by a field sync. pulse.

This process of address generation is incorporated within the display RAM device (Fig. 20.18).

**Fig. 20.18** *RAM incorporating address generation*

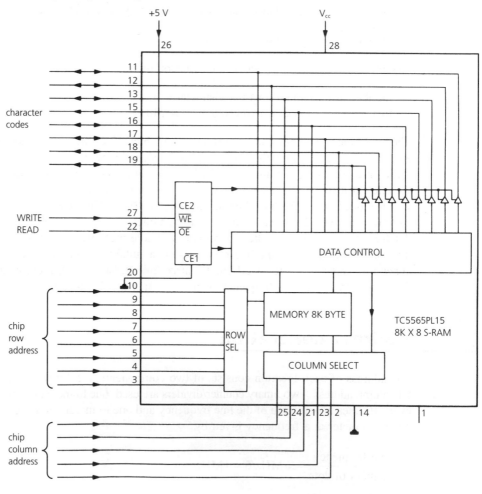

# Graphic representation

For the production of graphics, the space allocated for each character matrix is divided into six segments (Fig. 20.19). Each of the segments in the array can be either on or off. Since there are six segments, it is possible to have $2^6 = 64$ different graphic patterns or symbols. A 6-bit code, one bit for each segment, is used to define each graphic symbol, as illustrated. Thus bit b0 controls the top left-hand segment, bit b3 controls the middle right segment, and so on. Bit b5, which is omitted from the code, is always set to 1 for all graphic codes.

For instance, graphic code 01110110 (b0 = 0, b1 = 1, b2 = 1, b3 = 0, b4 = 1, b6 = 1 b7 = parity) produces the graphic symbol shown in Fig. 20.20 and code 01101110 will display the symbol shown in Fig. 20.21. Recall that the codes 01110110 and 01101110 represent alphanumeric characters as well (v and n respectively). In fact, two-thirds of the alphanumeric codes are shared by graphic codes. In order to distinguish between the two types of display, control codes are used which set the display to the correct mode. Two separate ROM generators (normally incorporated within a single chip) are therefore used: the normal character generator and a *graphic generator* (Fig. 20.22). In the graphic mode, character codes are interpreted as graphic symbols and fed into a graphic generator to produce the appropriate dot patterns. When set to the alphanumeric mode, the character codes are fed into the character generator in the normal way to

**Fig. 20.19**  *Six-segment matrix for graphic generation*

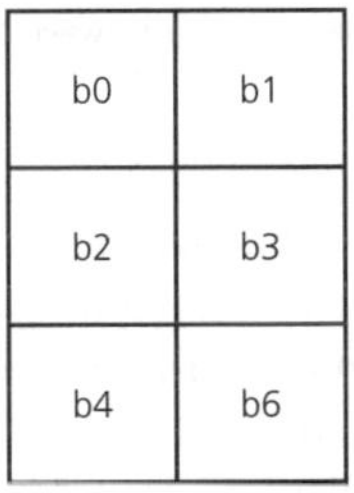

| b0 | b1 |
|----|----|
| b2 | b3 |
| b4 | b6 |

**Fig. 20.20**  *Graphic symbol for code 01110110*

**Fig. 20.21** *Graphic symbol for code 01101110*

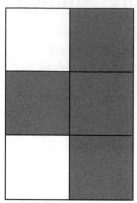

**Fig. 20.22** *Character and graphic generator arrangement*

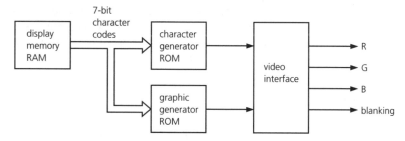

display the appropriate alphanumeric characters. Table 20.5 gives a complete list of the character codes and their corresponding alphanumeric and graphic symbols. Notice that the capital letters in columns 4 and 5 have no graphic symbols associated with them; for these codes, b5 (which is set to 1 in the graphic mode) is set to 0. The logic state of b5 can thus be used to display capital letters within a graphic display without changing the mode of operation. When b5 is set to 0 the character code is diverted to the character generator for an alphanumeric display, regardless of the setting of the display mode.

## Control codes

A set of 32 codes are allocated for control purposes. Table 20.5 lists these codes and their respective functions (columns 0 and 1). Bits b4 and b5 set the display mode, with 10 for the graphic mode and 00 for the alphanumeric mode respectively. Once the mode is set, bits b0, b1 and b2 (the three least significant bits) control the R, G and B colour contents respectively. A total of seven different colours may thus be displayed: red (001), green (010), yellow (011), blue (100), magenta (101), cyan (110) and white (111). The eighth combination, 000, is not used. The remaining control codes cover such display modes as contiguous and separated graphics (Fig. 20.23), flashing and box display.

TABLE 20.5 *Complete character code chart*

| b3 | b2 | b1 | b0 | Row | 0 | 1 | 2 | 2a | 3 | 3a | 4 | 5 | 6 | 6a | 7 | 7a |
|----|----|----|----|-----|---|---|---|----|---|----|---|---|---|----|---|----|
| 0 | 0 | 0 | 0 | 0 | NUL | DLE | | | 0 | | @ | P | — | | p | |
| 0 | 0 | 0 | 1 | 1 | Alpha red | Graphics red | ! | | 1 | | A | Q | a | | q | |
| 0 | 0 | 1 | 0 | 2 | Alpha green | Graphics green | " | | 2 | | B | R | b | | r | |
| 0 | 0 | 1 | 1 | 3 | Alpha yellow | Graphics yellow | £ | | 3 | | C | S | c | | s | |
| 0 | 1 | 0 | 0 | 4 | Alpha blue | Graphics blue | $ | | 4 | | D | T | d | | t | |
| 0 | 1 | 0 | 1 | 5 | Alpha magenta | Graphics magenta | % | | 5 | | E | U | e | | u | |
| 0 | 1 | 1 | 0 | 6 | Alpha cyan | Graphics cyan | & | | 6 | | F | V | f | | v | |
| 0 | 1 | 1 | 1 | 7 | Alpha white | Graphics white | ' | | 7 | | G | W | g | | w | |
| 1 | 0 | 0 | 0 | 8 | Flash | Conceal display | ( | | 8 | | H | X | h | | x | |
| 1 | 0 | 0 | 1 | 9 | Steady | Contiguous graphics | ) | | 9 | | I | Y | i | | y | |
| 1 | 0 | 1 | 0 | 10 | End box | Separated graphics | * | | : | | J | Z | j | | z | |
| 1 | 0 | 1 | 1 | 11 | Start box | ESC | + | | ; | | K | ← | k | | $\frac{1}{4}$ | |
| 1 | 1 | 0 | 0 | 12 | Normal height | Black background | , | | < | | L | $\frac{1}{2}$ | l | | ‖ | |
| 1 | 1 | 0 | 1 | 13 | Double height | New background | – | | = | | M | → | m | | $\frac{3}{4}$ | |
| 1 | 1 | 1 | 0 | 14 | SO | Hold graphics | . | | > | | N | ↑ | n | | ÷ | |
| 1 | 1 | 1 | 1 | 15 | SI | Release graphics | / | | ? | | O | # | o | | ■ | |

Alpha = alphanumeric

**Fig. 20.23** *Graphics character 01111111: (a) contiguous; (b) separated*

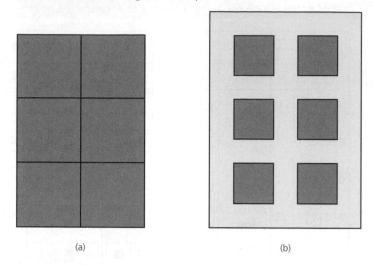

(a)          (b)

**Fig. 20.24** *Character/graphic and character insertion arrangement*

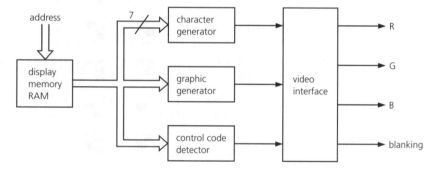

When a control code is recognised, it is fed into a control code detector which sets the colour attributes and the other properties of the display, such as flashing and box insertion (Fig. 20.24).

## Fastext

The minimum size of display memory must be able to store one full page of character codes, so the minimum size is 24 rows × 40 characters = 960 locations of 7 bits each, a total of 6720 bits. In practice a memory of 1K × 8 (1 kbyte) is used. Fastext decoders incorporate a larger memory which can store a number of pages at once. A memory size of 8K × 8 (8 kbyte) for eight pages is the most popular size of display RAM for Fastext decoders. Fastext loads the selected page into memory together with the next seven pages of the magazine. These additional pages are therefore available for display almost instantaneously. Since viewers normally access pages in sequence, the effect is almost instantaneous access to teletext pages. As one page is viewed and discarded, the decoder is prompted to acquire and load the next page in the sequence, and so on. The system is further improved by the teletext broadcaster anticipating the viewer's requirements and sending additional coded instructions to the decoder. These instructions, known as packets, are sent as extra rows of control data in the same way as normal teletext data. They are received, decoded and implemented to offer the viewer additional functions that provide quick access to the pages of information which he or she is most likely to ask for. A packet takes the form of the top or bottom row of the displayed page and contains four coded prompts news, sport, etc. The remote control handset has a corresponding set of four colour-coded buttons which, when pressed, will select and display the appropriate pages of text. A total of eight such additional packets (numbers 24 to 31) may be transmitted to provide other services such as different language support and programme identification for VCRs.

## Background memory

Faster access to a large and random selection of pages requires a large memory capable of storing hundreds of pages. All logically relevant pages may thus be stored into memory, reducing the process of page selection to a search and read operation in a memory chip. Such memory devices are called *background memory*. Cheaper and more compact dynamic RAM (DRAM) devices are used in addition to the existing static RAM. A DRAM chip with a capacity of say 256 kbyte (256 KB) is capable of storing over 200 pages of teletext.

## Access time

With the current practice of using 12 teletext lines per field, it takes 2 fields to transmit a page, and at a field frequency of 50 Hz, each page takes 0.024 s to transmit. This means 25 pages can be transmitted every second. *Access time*, defined as the time between selecting a page and the complete reception of that page, thus depends on the number of pages transmitted. If 100 pages are transmitted, then

max access time = 0.04 × 100 = 4 s

And if the full 800 pages are transmitted, then

access time = 0.04 × 800 = 32 s

a comparatively long time for the viewer to wait. To improve access time, high priority is given to the more highly used pages, such as indices, which are then transmitted more often than other pages. Two other techniques may be used to improve access time: the row-adaptive technique enables blank rows to be omitted from the transmission; the row interleaving technique interleaves the rows of several pages, giving the impression of faster access time. Subtitles on teletext page 888 are given the highest priority.

**Fig. 20.25** *Character rounding arrangement*

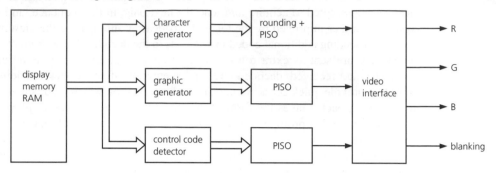

## Character rounding

Characters displayed using a normal character generator have a jagged appearance on their sloping parts. They may be smoothed by arranging for half-dots to be inserted on alternate fields. The process involves comparing the dot patterns of one pixel line of a character with the previous line, then generating half-dots where a slope is identified. Character rounding elements are added to the dot pattern from the character generator (Fig. 20.25).

## Data acquisition

Figure 20.26 shows the basic functional diagram for the data acquisition of a teletext decoder. Composite video is fed into a sync. separator/slicer, which removes the sync. pulses and clips the data signals to produce square pulses. The sync. separator also provides a sync. output for the timebase so the text may be displayed even when there is no video signal being transmitted, known as *after hours display*. Before further processing, the incoming pulses in NRZ format are converted back into an ordinary pulse stream by the bit recognition block. The reconstructed pulses are then fed into an 8-bit serial-in parallel-out shift register to convert the serial data stream into an 8-line parallel output. The shift register is clocked by a 6.9375 MHz clock. The data is then fed byte by byte into a latch; the latch is enabled by the byte sync. clock and synchronised by the framing code detector. The data from the latch is then fed into the *display memory RAM* via a data latch. Before the data is written into the appropriate locations in memory, it is checked for parity and Hamming code errors and corrected where possible. If the data is found to be faulty, the WRITE INHIBIT from the parity/Hamming code decoder is enabled and the data is disregarded. Otherwise, the *WRITE OK (WOK)* control line is enabled by the data accept/reject logic and data is written into the location specified by the row and column addresses. The process is then repeated for the next byte of information, and so on.

The process of reading the memory locations is initiated by a READ command from the user control, which receives instruction via a keypad or more usually via a remote control unit.

**Fig. 20.26** *Data acquisition of a teletext decoder*

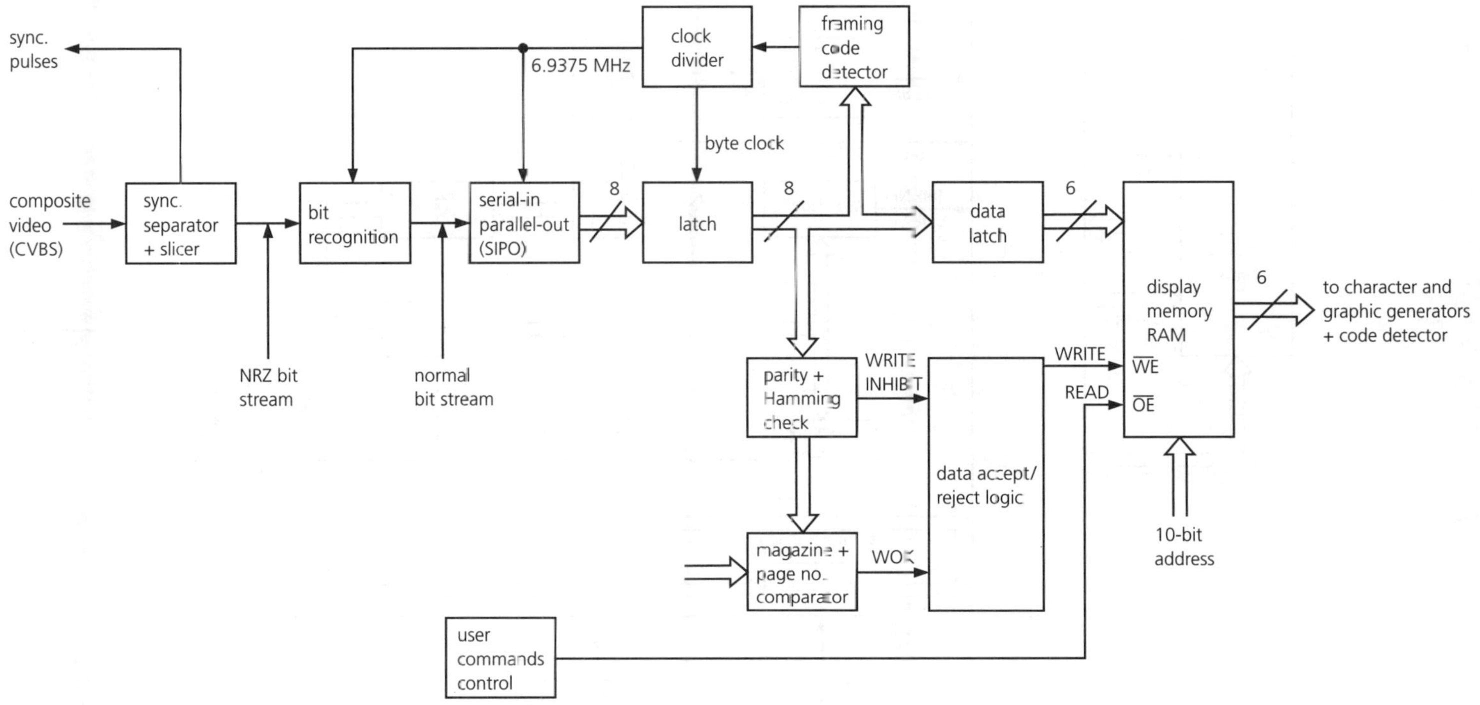

**Fig. 20.27** *Teletext decoder*

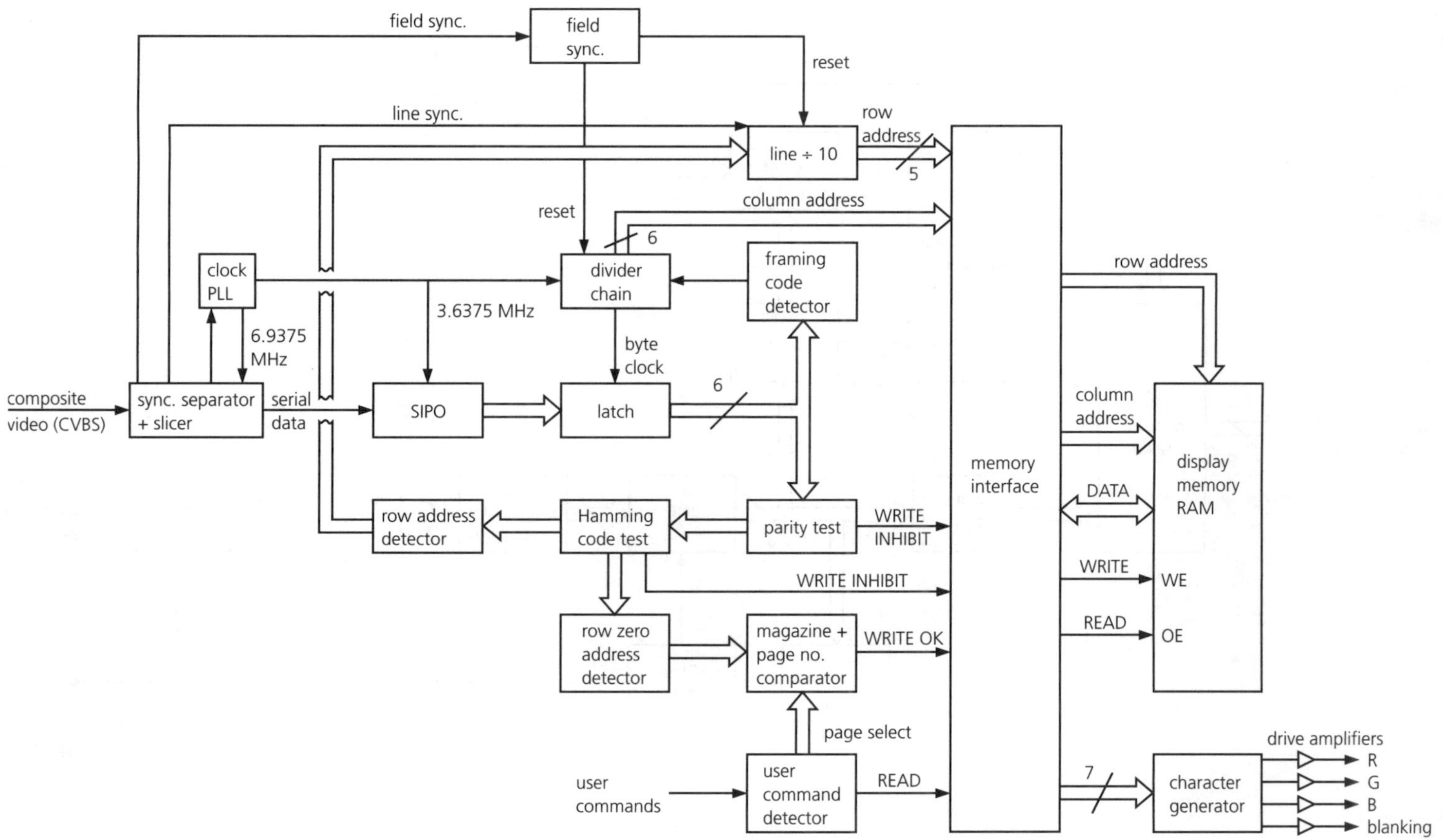

# The teletext decoder

Figure 20.27 shows the main functional elements of a teletext decoder. The memory interface generates the appropriate row and column addresses, controls the READ and WRITE signals to display memory RAM and sends the appropriate sequence of character codes to the character generator. The character generator incorporates the functions of the alphanumeric, rounding and graphic generators as well as the character code detector. Although a total of 11 address lines are shown (5 for the row and 6 for the column addresses), where the memory size is larger than 1 kbyte, the column address would also be larger. With a normal memory of 8K × 8 (8 kbyte) for 8-page memory storage capacity, an 8-bit column address must be used, making a total absolute address of 13 bits.

Data will be written into memory when the memory interface enables the WRITE control line. This takes place only if the following conditions are satisfied:

- The magazine and page number (including the time code if appropriate) matches those requested by the viewer. Matching is carried out by the magazine and page number comparator.
- The data is complete and uncorrupted. The integrity of the data is checked by the parity testing unit, which sends a WRITE INHIBIT if an error is detected.
- The control bits in the page header (row 0) are present and correct. This is carried out by the Hamming code testing block, which sends a WRITE INHIBIT signal to the memory interface if an error that cannot be corrected is detected.

When these conditions are met, data may be written into memory locations as identified by the row and column address lines. The speed or access time of the memory chip must be less than 1 $\mu$s. For this reason, static RAM (SRAM) chips are used which have very short access time.

The data clock is synchronised to the incoming clock run-in sequence by a phase-locked loop (PLL), which produces a 6.9375 MHz pulse that runs in phase with the encoding clock at the transmitter. The line divide-by-10 ($\div$10) unit sets the row address as specified by the row address detector. The sequence of column addresses is produced by the divider chain. Both the line $\div$10 and the divider chain are reset at the start of every field by a field sync. pulse from the sync. separator.

In practice a teletext decoder is designed using one or two specially fabricated ASICs (*application-specific i.c.s*), which carry out the illustrated functions under microprocessor control via the I²C bus or its equivalent.

Figure 20.28 shows a *microcontrolled teletext decoder* circuit used in conjunction with an 8 KB memory chip. The circuit uses two LSI devices: the SDA5231 VIP (*video input processor*) and the SDA5243 CCT (*computer-controlled teletext*). Composite video enters at pin 27 of the video VIP chip, where it takes two paths: one path goes to an *adaptive sync. separator* and the other goes to an *adaptive slicer* via a conditioning circuit. The data pulses eventually emerge at pin 15. The data clock oscillator operates at 13.875 MHz, with the crystal connected to pin 11. This frequency is divided by 2 to produce the required clock frequency of 6.9375 MHz. It is locked to the clock run-in sequence by a phase-locked loop before it appears at pin 14. The 6 MHz dot frequency at pin 7 is produced by an oscillator which is phase locked to the off-air sync. pulses

**Fig. 20.28** Teletext decoder: circuit diagram

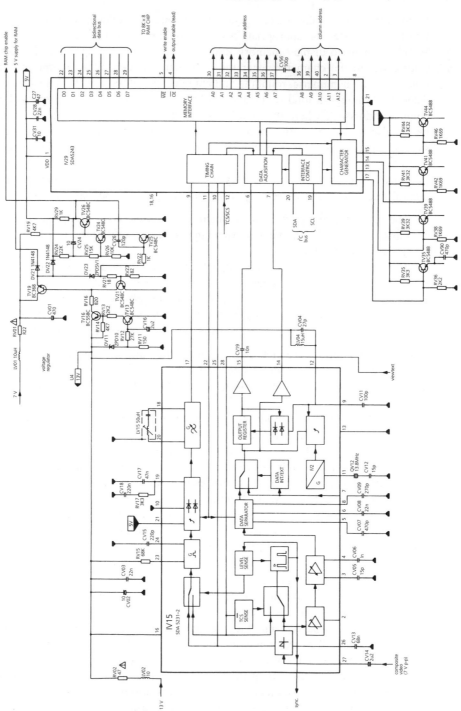

from the sync. separator at the input. The CCT chip receives three signals from the VIP: the serial data at pin 6, the decoder clock pulses at pin 7 and the dot clock pulses at pin 9. The data clock pulses drive a sampling gate incorporated within the data acquisition block; this samples the data and makes logic 1 and 0 decisions, converting the data format from NRZ to a normal bit stream. The reconstituted data bits are then passed to the memory interface. Hamming code error detection and correction along with parity detection are also carried out within the data acquisition block. The interface control section decodes user commands arriving on the $I^2C$ bus. Interfacing with the RAM chip is carried out via 12 address lines and 8 data lines, together with WE (write enable) and OE (output enable) control lines. The CCT device has an onboard character generator containing a complete set of characters; each character occupies $12 \times 10$ bits of storage space of the ROM generator. The VIP chip provides a facility for displaying text from another source for *onscreen display* (*OSD*) or other functions. In this mode the two internal switches are operated, allowing the data on pin 28 to be processed through the CCT device.

## QUESTIONS

1. In relation to teletext, state
   (a) the number of bytes per data line
   (b) the bit rate

2. In relation to teletext, state the purpose of
   (a) the clock run-in
   (b) framing code

3. Describe the digital make-up and purpose of **each** of the following components of a teletext signal:
   (a) clock run-in
   (b) framing code
   (c) magazine and row address group

4. Explain the function of the parity checker and the reason for using odd parity in the teletext system.

# 21 Satellite television

Today satellites form a common part of worldwide communications. The advantages offered by satellite transmission far outweigh the high costs associated with launching a satellite into space. Satellite communication offers flexibility, reliability and reduced running costs when compared with terrestrial communication. For example, a single satellite provides coverage for a very large area using far less power than a terrestrial transmitter, which requires a large number of relay stations and associated undersea and other cabling. In some areas of the world, satellites are the only possible method of communication.

## System basics

A basic satellite communication system consists of two earthbound stations, a transmitter and a receiver together with a satellite in space (Fig. 21.1). The satellite, which is known as a *transponder*, acts as a frequency changer, receiving signals from the transmitting station (*uplink*) on one frequency and transmitting them back (*downlink*) to an earthbound receiving station on another, lower frequency. Both the uplink and the downlink frequencies are well above 100 MHz to pass through the ionosphere surrounding the earth. Frequency modulation is used for both the video and the audio components of the TV signal. A video bandwith of either 5.0, 5.5 or 6.0 MHz is provided by broadcasting companies. Satellite broadcasting provides more than one sound subcarrier (Fig. 21.2). The audio that is associated with the video programme is normally transmitted on the main subcarrier at 6.6 MHz. Stereophonic sound is transmitted using a pair of sub-carriers such as 7.02 and 7.2 MHz. The subcarriers may also be used for multilingual sound broadcasting. Each sound channel is allocated an audio bandwidth of 15 kHz and a subcarrier bandwidth of 200 kHz. Table 21.1 provides a summary of satellite TV broadcasting parameters.

## Geostationary orbit

Satellites rotate in an orbit a few hundred kilometres above the surface of the earth. Apart from an initial thrust to place it in orbit, a satellite continues to circle the earth at

**Fig. 21.1**  *Satellite communication*

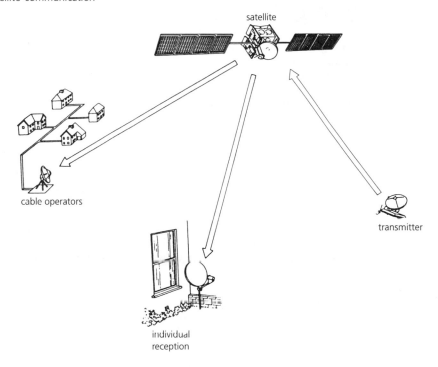

**Fig. 21.2**  *Sound carriers in satellite analogue TV broadcasting*

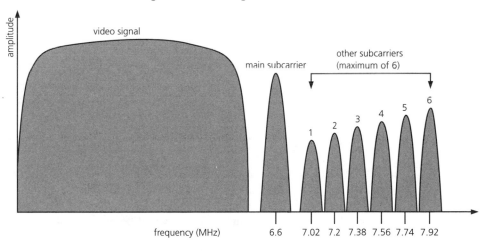

a constant speed completely unaided, provided the *centrifugal force* $F_C$ which attempts to take it away from earth is equal to the earth's *gravitational force* $F_G$ (Fig. 21.3). The closer the satellite to the surface of the earth, the greater the gravitational force $F_G$. To stay in orbit, the centrifugal force must also increase; this is accomplished by increasing the speed of the satellite, making it circle the earth more often. Conversely,

**TABLE 21.1**

**Video signal**

| | |
|---|---|
| Amplitude | 1 V peak-to-peak |
| Video bandwidth | 5.0, 5.5 or 6.0 MHz |

**Main audio signal**

| | |
|---|---|
| Audio bandwidth | 15 kHz |
| Audio pre/de-emphasis | 50 $\mu$s or J17 |
| Audio subcarrier frequency | 6.60 MHz |
| Deviation of subcarrier | |
| Frequency | 170 kHz peak-to-peak at pre-emphasis crossover |
| Subcarrier bandwidth | 200 kHz (in the composite baseband signal) |

**Auxiliary audio signals (analogue)**

| | |
|---|---|
| Companding/pre-emphasis | PANDA I adaptive pre-emphasis (2:1 compression) |
| Audio bandwidth | 15 kHz |
| Max number companded | 6 with an uncompanded (main) sound subcarrier |
| Subcarriers | 8 without an uncompanded sound subcarrier |
| Audio subcarrier frequency | $7.02 + (n - 1) \times 0.18$ MHz |
| Deviation of subcarrier | 100 kHz peak-to-peak at a test tone frequency of 400 Hz |
| Max subcarrier bandwidth | 130 kHz (in composite baseband signal) |

**Composite RF signal**

| | |
|---|---|
| Frequency deviation sensitivity | 16–25 MHz/V |
| Nominal bandwidth | 27–36 MHz |
| Modulation index, uncompanded | |
| Sound subcarrier | 0.26 |
| Modulation index, companded | |
| Sound subcarrier | 0.14 |
| Energy dispersal | 2 MHz peak-to-peak (modulating signal present) |
| | 4 MHz peak-to-peak (modulating signal absent) |
| | 25 Hz triangular waveform, locked to video field rate |

a satellite in an orbit that is further away from the earth's surface will require a lower speed to keep it there, hence it will circle the earth less often.

At a distance of 35 765 km from the earth's surface (and a speed of 3073 km/s), a satellite period of rotation around the earth would be precisely that of the earth itself. The satellite is said to be synchronised to the earth. Such an orbit is known as a *geosynchronous* or *geostationary orbit* (*GSO*). If the orbit were located over the equator (Fig. 21.4) the satellite would appear to be stationary to an earthbound observer. Such an orbit is used for satellite TV transmission. A geostationary orbit is also known as the *Clarke orbit*, after the engineer and writer Arthur C. Clarke, who outlined the requirements for satellite communication in the October 1945 issue of *Wireless World*; he was the first person to do so.

**Fig. 21.3**  *Forces acting on a rotating satellite: gravitational $F_G$ and centrifugal $F_C$*

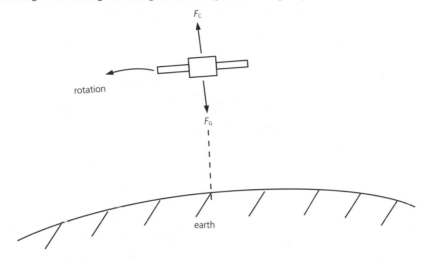

**Fig. 21.4**  *Geostationary orbit*

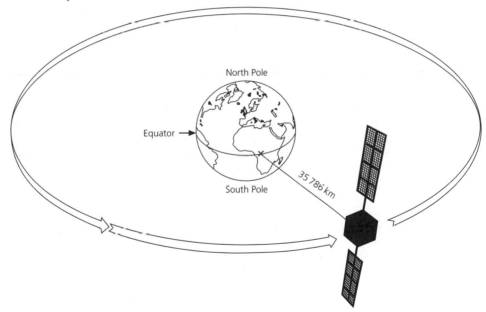

## Satellite footprints

Where satellite transmission is aimed at individual home users, known as *direct to home (DTH)* broadcasting, the radiated power from the satellite is directed not to an individual receiving antenna, but to a specific area known as the *coverage area* or *zone*.

**Fig. 21.5** *Satellite footprints*

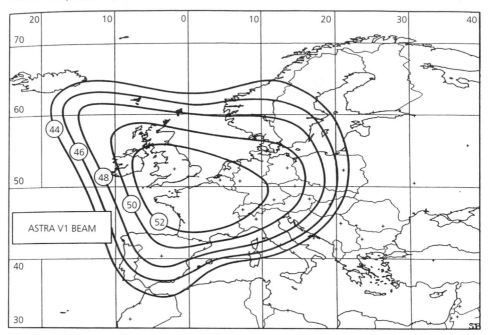

Contour maps are used to represent the distribution of the radiated power of the satellite, with each satellite beam forming its own '*footprints*' on the globe. Figure 21.5 shows a typical set of footprints. The figures associated with each footprint represent radiated power in dBW. High-power satellites capable of radiating more than 100 W are used. The effective radiated power, known as *EIRP* (*effective isotropic radiated power*), may be greatly enhanced by the use of a parabolic reflector onboard the satellite; the reflector magnifies the output by more than 10 000 times, giving EIRP values of 50–60 dBW. These high powers allow individual homes to receive good quality pictures on dishes as small as 30 cm in diameter.

## Satellite frequency bands

Satellite TV transmission is allocated frequencies in the microwave range (Fig. 21.6). Two bands within the super high frequency (SHF) range are used: the *C-band* (3.4–4.7 GHz) and the *Ku-band* (10.95–17.7 GHz). The allocated working range for the European direct *broadcast by satellite* (*DBS*) is 11.7–12.5 GHz. Provision has been made for *high-definition television* (*HDTV*) broadcasting in the Ka-band (21.4–22 GHz).

### Advantages and disadvantages of microwave transmission

Microwaves are used primarily because, unlike UHF signals, they can penetrate the ionosphere to reach the satellite with minimal attenuation. Compared with UHF broadcasting,

**Fig. 21.6**

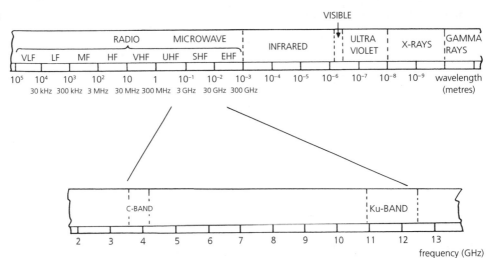

they offer a number of other advantages. Microwaves are less prone to atmospheric noise interference, they provide wider bandwidth and may be aimed with precision to pick out an individual satellite without interfering with neighbouring satellites. A further advantage is the relative newness of microwave broadcasting technology, which makes it amenable to worldwide planning. On the other hand, microwave links suffer from a number of drawbacks which must be overcome when designing a satellite broadcasting system.

Microwave links suffer from attenuation due to oxygen and water vapour in the atmosphere. More serious losses are caused by rainfall. These losses must be compensated by corresponding 'gains' at the transmitting and receiving antennae. Furthermore, at microwave frequencies, standard electric components such as resistors, capacitors and inductors no longer behave in the same way as at lower frequencies. Coaxial cables which are adequate for UHF signals introduce unacceptable losses. They have to be substituted with hollow tubes of rectangular or circular construction, known as *waveguides*. Waveguides may be fashioned to behave as inductors, capacitors or tuned circuits. A waveguide that behaves like a tuned circuit is called a *resonant cavity*.

## Polarisation

Radio transmission is the propagation of electromagnetic waves which have two transverse (right angles) components: the *electric field E* and the *magnetic field H*. The two fields form a plane at right angles to the propagation direction of the wave (Fig. 21.7). The orientation of the *E*-field with respect to the earth's surface is called the *polarisation* of the electromagnetic wave. If the *E*-field is vertical, the transmitted signal is called *vertically polarised*; if the *E*-field is horizontal, the transmitted signal is called *horizontally polarised*. If the *E*-field vector is rotating, the transmitted signal is *circularly polarised*.

**Fig. 21.7** *Electromagnetic propagation*

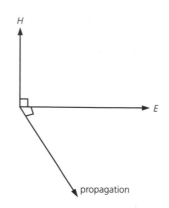

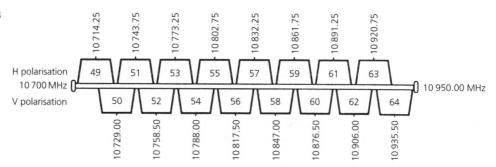

**Fig. 21.8**

Satellite TV broadcasting uses vertical and horizontal polarisation only. This dual-polarisation technique allows for the same frequency to be used by two different channels, one vertically polarised and one horizontally polarised. However, overlapping channel frequencies could cause *cross-talk* and are not normally used. In practice, *non-overlapping* frequency allocation is used, and adjacent channels are transmitted at opposite polarisation to minimise intermodulation between channels. The allocation of channel frequencies (the *frequency plan*) of the ASTRA 1D satellite is shown in Fig. 21.8. The odd channels are horizontally polarised and the even channels are vertically polarised; each channel is limited to an r.f. bandwidth of 28 MHz.

## Modulation and noise

In frequency modulation a constant amplitude carrier changes its frequency in sympathy with the modulating signal. This process produces a carrier and an infinite number of sidebands. Theoretically, an infinite bandwidth is required. However, 99% of the modulating information is contained within a few innermost sidebands, resulting in a more manageable bandwidth requirement. Carson's rule provides a convenient estimation for the bandwidth of an f.m. waveform:

$$\text{bandwidth} = 2(f_m + f_d)$$

where $f_m$ is the highest modulating frequency and $f_d$ is the frequency deviation, i.e. the maximum change in carrier frequency. Satellite TV broadcasting uses a bandwidth of 27–36 MHz.

Compared with amplitude modulation, frequency modulation offers two main advantages: it requires considerably less transmitted power and it is less susceptible to noise. However, noise does exist and has to be minimised.

### F.m. noise

In f.m. systems, noise may take two forms: *thermal noise* and *impulse noise*. Thermal noise is associated with the a.m. system since it mainly takes the form of changes in the amplitude of the carrier. This type of noise may be removed by the inclusion of an a.m. limiter before demodulation. However, thermal noise may also be produced by an interfering frequency which, if close to the carrier, could phase modulate it, producing a high-frequency *noise transient*. The colour subcarrier in the PAL system, being of higher frequency, is more susceptible to this type of noise than the main luminance carrier. This causes *the signal-to-noise ratio* (*S/N*) for the chrominance signal to be worse than S/N for the luminance signal. Impulse noise is caused by poor f.m. reception. On the screen it creates comet-shaped black and white dots, known as *sparklies*.

Thermal noise may be reduced by *pre-emphasis* at the transmitter and a corresponding *de-emphasis* at the receiving end. The S/N ratio may be further improved by increasing the frequency deviation. A change in the deviation from 13.5 MHz to 25 MHz will produce an almost 7 dB improvement in the S/N ratio.

Impulse noise (sparklies) may be reduced by extending what is known as the *threshold* of the f.m. demodulator at the decoder. The threshold is defined as the level of the carrier-to-noise ratio (C/N) at which the f.m. demodulator can no longer distinguish between signal and noise. When that happens, the discriminator generates its own white noise, which is added to the signal to produce sparklies. To reduce the effect of sparklies, an f.m. discriminator with a low threshold should be used. This may be achieved by using a phase-locked loop (PLL), *dynamic tracking filter* (*DTF*) and frequency *feedback loop* (*FFL*). All of these techniques ensure more precise tracking of the carrier, thus reducing instantaneous noise.

An alternative to the threshold extension technique is to reduce the bandwidth of the i.f. stage at the decoder. This improves the S/N ratio at the expense of picture resolution as some of the outermost sidebands are removed. Low picture resolution shows up as a tearing or streaky effect on sharp vertical edges of the picture, an effect known as *truncation*.

## Energy dispersal

The peaky nature of the composite video signal, particularly during the period associated with the line sync. pulse, produces an uneven frequency spectrum of the frequency

**Fig. 21.9** *Dispersal waveform*

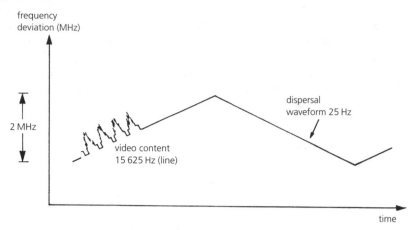

frequency
deviation (MHz)

2 MHz

video content
15 625 Hz (line)

dispersal
waveform 25 Hz

time

modulated signal. The carrier radiated energy is therefore highly concentrated at spot frequencies that represent black levels and sync. tip voltages, and this can cause interference with terrestrial radio and TV services. To minimise signal interference, the spectral energy of the modulated signal is made to spread more evenly by adding a triangular waveform at field or half-field frequency to the composite video signal before modulation (Fig. 21.9). A low-amplitude triangular waveform is chosen to provide a small frequency deviation of 2 MHz peak-to-peak. When the instantaneous frequency deviation produced by the energy dispersal waveform is added to the deviation produced by the composite video, it produces an even frequency spectrum. The frequencies representing the black levels and sync. tip levels are thus gradually changed line by line. This technique is known as *energy dispersal*.

At the receiving end, a d.c. clamping circuit is used to remove the energy dispersal triangular waveform and to restore the black level to each line of the composite video signal. Failure to do so would leave behind an intolerable 25 Hz flicker. A simple circuit employing a capacitor and a zener diode may be used for this purpose. However, a more advanced circuit is normally used to ensure tight and stable clamping.

A typical clamping circuit is shown in Fig. 21.10. Clamping is carried out at the start of every line and sets the charge on capacitor C1 to the right level. During the back porch, field-effect transistor TR1 is turned on by a short-duration clamping pulse. This causes TR2 to conduct, charging capacitor C1 and setting TR3's base voltage to a level determined by reference voltage $V_{ref}$. At the end of the clamping pulse, TR1 is turned off, and because it is a field effect transistor, it acts as a very high resistance which prevents any leakage. Capacitor C1 thus retains its charge, and the composite signal for that scan line is correctly clamped. At the next back porch, a second clamping pulse turns on TR1, which turns on TR2, pumping a further charge into capacitor C1 to clamp the next line on the upward gradient of the triangular wave. The process is repeated with a further incremental charge added for the next line, and so on. During the sloping gradient, C1 is incrementally discharged to set the correct d.c. clamping voltage for each line.

**Fig. 21.10** *D.c. clamping circuit to remove energy dispersal waveform*

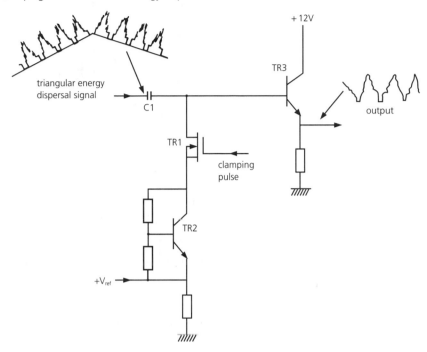

## Satellite system overview

Figure 21.11 shows a block diagram of the essential components for a satellite TV receiving system. The satellite antenna receives the uplink frequency modulated signal (14.25–14.5 GHz) and converts it to a lower frequency (10.7–11.7 GHz and 11.2–11.45 GHz) and sends it back to earth (downlink). The change in frequency is carried out by reducing the uplink frequency by a specific amount such as 3.0 GHz. The downlink f.m. signal is picked up by a domestic satellite dish and passed to the *low noise block* (*LNB*), placed on the receiving dish itself. The basic purpose of the LNB is to convert the super high frequency (SHF) signals into a more manageable range of frequencies, known as the *first i.f.*, of 950–2150 MHz. Following amplification, the converted signal is sent to the satellite receiver via a low-noise coaxial cable. The receiver selects a channel, demodulates the f.m. signal (tuner) and reproduces the original video baseband signal.

The video baseband signal thus reproduced contains two components: a composite video channel and a frequency modulated sound channel. Each component is processed separately.

The main task of the video processing channel is to restore the composite video to its original form before pre-emphasis and energy dispersal. The composite video is thus fed into a de-emphasis circuit followed by an energy dispersal clamping circuit. Where videoCrypt is used, it must be removed by a videoCrypt decoder. The composite video

**Fig. 21.11** *Satellite receiver*

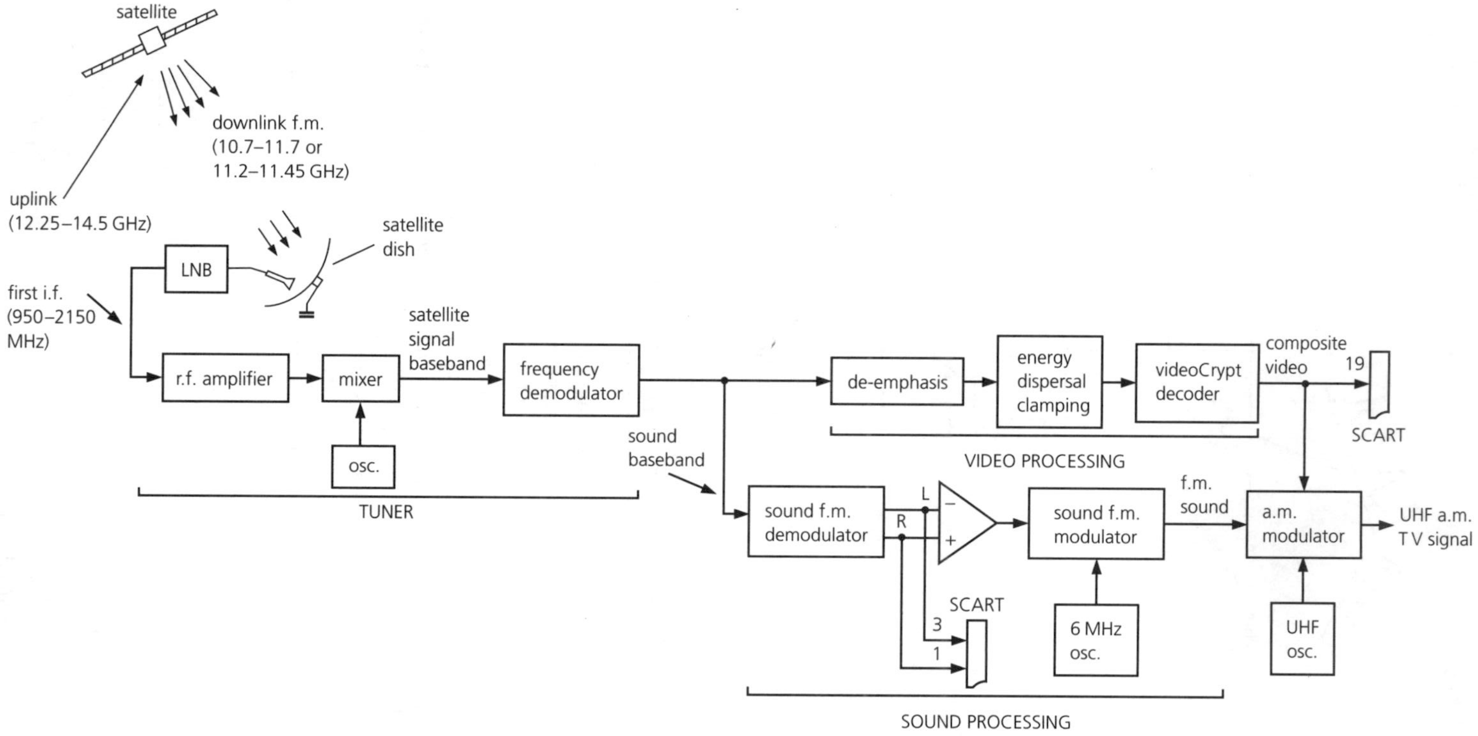

is now in a shape that can be sent to pin 19 (CVBS) of a SCART connector or to a UHF modulator for r.f. output.

Sound processing involves demodulating the f.m. audio channels to produce the original L and R stereo sound, which may then be fed directly into pins 3 and 1 of the SCART connector. Alternatively, the two stereo channels are combined by a summing amplifier into a single mono channel, which is frequency modulated onto a 6 MHz subcarrier before being fed into the UHF modulator for the r.f. output.

The UHF modulator receives the two component parts, composite video and the 6 MHz sound carrier, then amplitude modulates them onto a UHF carrier frequency in the same way as would a UHF transmitter. The UHF modulated composite video may then be fed into the aerial socket of a television receiver for processing into a TV picture display.

## The receiving dish

The SHF signals from the satellite are reflected by the parabolic receiving dish to converge at a focal point $F_P$ (Fig. 21.12). The *feedhorn* is circular in shape to allow both vertically and horizontally polarised signals to be collected. And normally it is flared to reduce its '*beamwidth*' to prevent ground noise pick up. A waveguide feedhorn positioned at the focal point collects the signal and feeds it into the resonant cavity at the input of the low noise block (LNB).

**Fig. 21.12**  *Satellite dish*

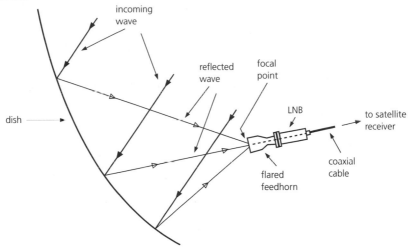

## The low noise block

The low noise block (LNB) receives weak microwave signals from the feedhorn, amplifies them and converts them into an intermediate frequency known as the *first i.f.* Figure 21.13 shows a block diagram for an LNB. To cope with the high microwave

**Fig. 21.13** *Low noise block (LNB): block diagram*

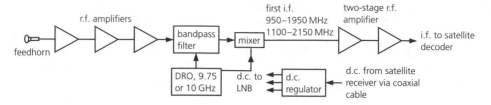

frequencies, *microstrip* and *surface mounted components* are used throughout, together with low-noise field-effect transistors. The signal is picked up by a waveguide cavity, also known as a *probe*, tuned to receive C-band and Ku-band frequencies. This is followed by three or more stages of r.f. amplification using a low-noise *galium arsenate f.e.t. (GaAsfet)*. The amplified microwave signal is then fed via a microstrip bandpass filter to a mixer–oscillator. The oscillator is a *dielectric resonant oscillator (DRO)* set to 9.75 GHz for the low C-band and 10.6 GHz for the high Ku-band. The mixer, normally a microwave diode operating along the non-linear part of its characteristics, produces four different frequencies: the received carrier frequency, the local oscillator frequency, their sum and their difference. The difference frequency is chosen as the new intermediate frequency, the first i.f. This gives an intermediate frequency of 950–1950 MHz for the low band and 1100–2150 MHz for the high band of satellite frequencies. Next come two or three stages of amplification. D.c. power is provided by the satellite receiver itself via the coaxial cable.

## The universal LNB

A universal LNB can receive and process vertically (V) and horizontally (H) polarised satellite signals within either the C-band or the Ku-band, respectively known as the *low band* and the *high band*. The H and V polarised signals are captured by two separate H and V cavity probes mounted at 90° to each other (Fig. 21.14). Switching between the two signals is carried out by the 13/18 V supply line. The LNB supply voltage is

**Fig. 21.14** *Universal LNB*

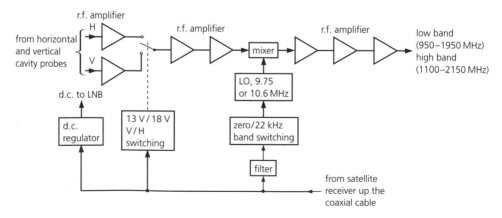

sent along the coaxial cable by the satellite receiver: a low voltage (about 13 V) selects vertically polarised signals whereas a high voltage (about 18 V) selects horizontally polarised signals. Band selection is achieved by a 22 kHz switching tone (amplitude about 600 mV). The tone is also sent up the coaxial cable from the satellite receiver. The absence of the tone, known as a zero signal, switches the LNB to the low C-band, setting the local oscillator (LO) to 9.75 GHz, and the 22 kHz signal switches it to the high Ku-band, setting the LO to 10.6 GHz. Universal LNBs with twin outputs are also available, providing one output for the low band and a second for the high band. The two outputs may then feed two receivers.

## The DiSEqC LNB

The universal LNB described above allows reception from a single satellite for both low and high bands as well as for horizontal and vertical polarisation. The signal must in the first place emanate from a single satellite transponder.

Reception from two different transponders has so far involved the use of motorised dish systems. This is necessary if the satellites involved are some distance away from each other. However, where the satellites are fairly closely spaced, as in the case of the Eutelsat *Hot Bird* and the old Astra satellites, *dual-focus dual-LNB* dishes may be used. The extra LNB requires its own power supply from the receiver and must be able to feed signals to the receiver. Where a receiver has two LNB inputs, simple software switching between the two inputs is used to receive signals from one of the satellites. Where this is not available, switching must take place at the receiving dish itself. The *DiSEqC* (*digital satellite equipment control*) system provides such switching facilities.

Developed by Eutelsat, DiSEqC is an advanced version of the universal LNB switching technique. The 22 kHz band switching frequency is used to carry modulated digital information. Coded information in the form of 0's and 1's are sent up the existing coaxial download to the LNBs (Fig. 21.15).

There are three levels of DiSEqC operation: *simple DiSEqC, version 1.0* and *version 2.0*. The simple version is used for switching between two universal LNBs. One long burst of 22 kHz tone lasting 12.5 ms selects one LNB and thus one satellite, whereas

**Fig. 21.15** *A 22 kHz switching frequency for an LNB*

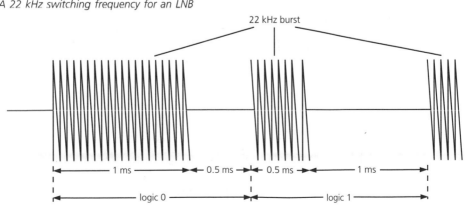

**Fig. 21.16**  *LNB 4-byte data package*

| byte 1 | byte 2 | byte 3 | byte 4 |
|--------|--------|--------|--------|
| 8-bit | 8-bit | 8-bit | 8-bit |
| framing | address | command | data |

a series of nine bits selects the other LNB, i.e. the other satellite. DiSEqC versions 1.0 and 2.0 are entirely digital; information on polarisation, band switching and LNB selection is sent in a 4-byte data packet (Fig. 21.16). Up to four universal LNBs can thus be controlled. Unlike version 1.0, which is unidirectional, version 2.0 provides bidirectional facility for returned messages from the LNB to the receiver. This enables the receiver to check the status of the selected LNB.

# The tuner

The full range of the frequency modulated first i.f. from the LNB enters the tuner section of the satellite receiver via an r.f. BNC connector. The function of the tuner is to select a channel; downconvert the selected carrier to an intermediate frequency of 479.5 MHz; then amplify, filter and demodulate the signals received from the LNB. Modern satellite receivers employ synthesised tuning (Fig. 21.17). Incoming signals from the LBN are first amplified by an r.f. amplifier before being mixed with the frequency from the local oscillator. The local oscillator frequency is controlled by a *phase-locked loop* (*PLL*). Channel selection is carried out under the control of the main micro-controller chip via the $I^2C$ bus. The microcontroller enters the channel 'value' into the programmable counter, the programmable counter sets the frequency for the phase detector, and the phase detector compares it with the frequency of the crystal reference oscillator. The d.c. output of the phase detector is filtered to remove the residual r.f. ripple before it is used to control the local oscillator, in accordance with the 'value' entered into the programmable counter. The 479.5 MHz i.f. produced by the mixer is passed through a *surface acoustic wave* (*SAW*) filter with a nominal bandwidth of 27 MHz and used to drive the frequency demodulator. The baseband video obtained

**Fig. 21.17**  *Synthesised tuning*

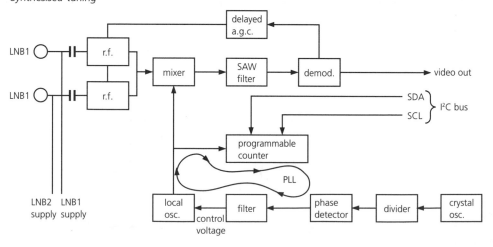

from the demodulator is then fed to the video processing channel and the f.m. sound carriers are fed to the sound processing channel.

# Video processing

Figure 21.18 shows a standard diagram of the video processing section of a satellite receiver. The demodulated video signal from the tuner is fed into a de-emphasis circuit before going into a video source selector, which provides for the demodulated signal to be fed to an external decoder via a SCART connector as shown. Furthermore, the video selector may switch to a different signal source via a SCART connector. The output signal from the video selector is then fed into a controlled amplifier to ensure a minimum amplitude of about 1.5 V. The gain of the amplifier may be controlled manually or by the microcontroller via the bus. A low amplitude would produce unreliable descrambling. Following the sound trap and the removal of the sound subcarrier, the signal is fed into the energy dispersal clamp circuit to remove the energy dispersal waveform and reproduce the CVBS signal. The CVBS signal may take three paths:

- If the signal is scrambled, it is fed into the videoCrypt decoder to produce a descrambled or 'clear' video, which is then fed into the switching unit.
- Unscrambled signals bypass the videoCrypt decoder, going directly to the switching unit. The switching unit makes the selection according to the type of signal received.
- The CVSB signal is also fed to the timing and sync. generator, which produces line and field sync. pulses.

The signal from the switching unit is fed to a character generator on its way to the UHF modulator and pin 19 of the TV SCART connector. The character generator provides a facility for onscreen display (OSD) messages relating to the descrambling operation.

**Fig. 21.18** *Video processing*

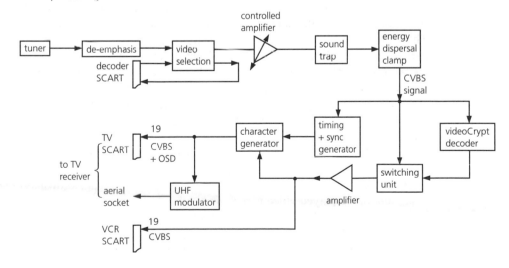

**Fig. 21.19** *STV 0030 video processing chip*

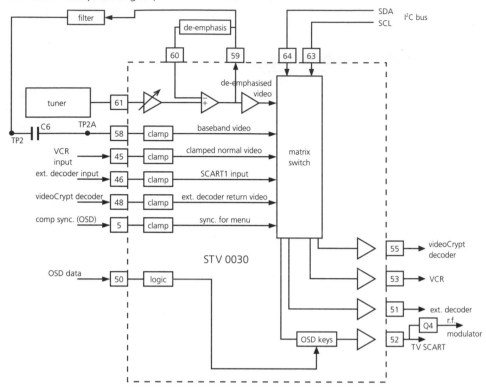

Message-free recording is carried out by the signal going directly to pin 19 of the VCR SCART connector, as shown in the diagram.

In modern satellite receivers, video processing is carried out by a single i.c., such as the STV 0030, which carries out both audio and video processing (Fig. 21.19). Video from the tuner is a.c. coupled to the processor chip at pin 61. The signal is amplified by an in-built controlled amplifier whose gain is set via the I²C bus (pins 63 and 64). The output from the controlled amplifier forms the baseband input to the video-matrix part of the chip. De-emphasis is carried out by the external circuit connected to pins 59 and 60. The de-emphasised signal is filtered externally and used to drive the energy dispersal clamp at pin 58. Figure 21.20 shows the video signals before and after the removal of the energy dispersal waveform observed on TP2 and TP2A respectively. Signals from the videoCrypt decoder, external decoder and VCR are received at pins 48, 46 and 45 respectively. The signals received by the matrix switch are routed to one of the four outputs shown, as determined by the data on the I²C bus coming from the microcontroller.

For OSD without video signal, such as during manual tuning, known as *out-of-hours* operation, the displayed video is provided with the necessary line and field sync. pulses by a locally generated SYNC signal fed into pin 5.

Figure 21.21 shows the video processing part of the STV 0030 circuit diagram. The input from the tuner enters at pin 61. De-emphasis is carried out by the circuit

**Fig. 21.20** *Video signal at field rate, 500 mV per division: (a) before and (b) after removal of energy dispersal signal*

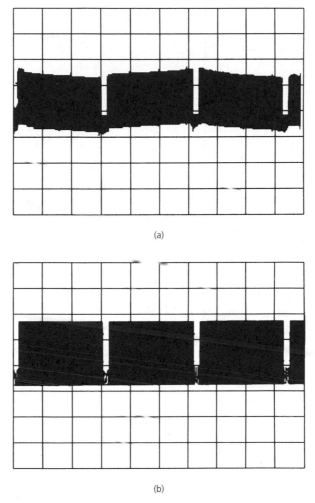

(a)

(b)

containing R6, R7, C8 and associated components. The input to the de-emphasis circuit (from pin 59) is also fed to a filter circuit, which among other things extracts the frequency modulated audio signals and feeds them back to pin 24 of the processor chip. A simple energy dispersal clamping circuit is employed (Q14, Q35 and associated components) with C6 as the clamping capacitor. Three different CVBS outputs are provided at pins 51, 52 and 53 for SCART connectors 3, 2 and 1 respectively. SCART1 is also used to feed into the UHF modulator. Transistors Q3, Q4 and Q5 are emitter follower buffers. Subsidiary composite video blanking and sync. inputs from the SCART connectors are fed into pins 2, 3 and 4. A 4.43 MHz crystal for the colour subcarrier is connected at pins 12 and 14. The chip is under the control of the main microcontroller via the I²C bus at pins 63 and 64. Onscreen display is fed at pin 50 and a satellite baseband output signal is provided at pin 55 for an external decoder.

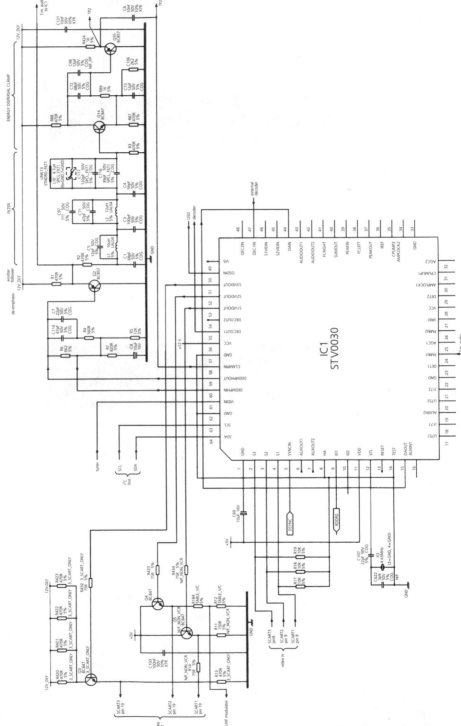

**Fig. 21.21** *The video processing part of the STV 0030*

# Sound processing

Sound signals from the tuner – frequency modulated sound subcarrier in the range 5–10 MHz – are on two separate channels (Fig. 21.22). The audio carriers have their amplitudes fixed by an a.g.c. input to the respective amplifier before they are applied to the f.m. demodulator. The particular sound carrier to be demodulated is selected by $I^2C$ commands from the mircrocontroller. Following demodulation the audio signals are fed into the appropriate de-emphasis network before going into the sound processor chip to produce the left and right sound channels, as well as centre and surround where available. Finally, a switching chip is used to select the required sound output, including audio signals from the SCART connectors. Sound processing is normally carried out by two dedicated chips (Fig. 21.23).

**Fig. 21.22** *Sound processing*

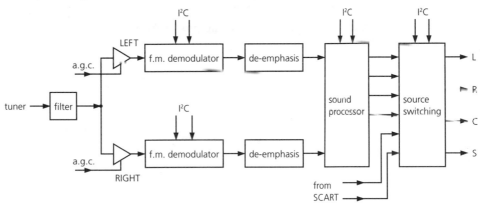

**Fig. 21.23** *Sound processing using two dedicated chips*

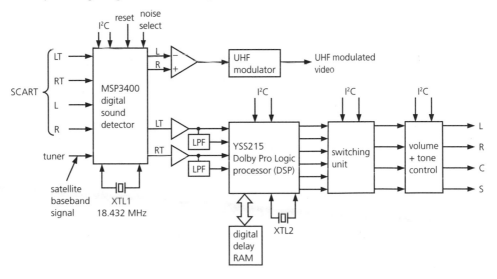

1. Explain what is meant by
   (a) the footprint of a satellite transmission
   (b) a geostationary orbit

2. State three advantages of satellite transmission over terrestrial transmission.

3. For the satellite transmission system used in the UK, state
   (a) the range of frequencies used for the downlink of transmissions
   (b) the channel spacing between copolarised channels
   (c) the channel spacing between oppositely polarised channels
   (d) the type of modulation employed for video signals.

4. Explain the cause of sparklies in satellite reception.

5. State the function of the LNB

6. Describe the types of switching provided by
   (a) universal LNB
   (b) DiSEqc LNB

# 22 Digital TV broadcasting

Digital television transmission involves transmitting moving pictures as well as stereophonic sound. Just as it was necessary to compress the digitised audio data in NICAM, in order to fit the broadcast into the available bandwidth, so it is with digital video signals. Complex techniques are used to ensure the bandwidth is not inhibitive. In Europe, MPEG-2 (Motion Picture Expert Group) compression specifications for digital television broadcasting have been agreed at what is known as Main Profile at Main Level (MP@ML) standards.

Digital TV broadcasting has several advantages over the traditional analogue system:

- very good picture quality
- an increased number of programmes
- lower transmission power requirements
- lower signal-to-noise ratio requirements
- no ghosting

A lower transmission power requirement means there will be less interference between adjacent channels.

## Digitising the TV picture

Digitising involves scanning a picture or frame line by line and sampling its contents. To maintain the quality of the picture, there must be as many samples per line as there are pixels; each sample represents one pixel. The number of pixels in a television picture is determined by the number of lines per picture and its aspect ratio. The British PAL system uses 625 lines, of which 576 are 'active' in that they may be used to carry video information. A figure of 5:4 is chosen for the aspect ratio; this is to accommodate the process of video encoding, which involves organising the picture contents into blocks and microblocks.

If vertical and horizontal resolutions are to be the same, the number of pixels per line may be calculated as

$$576 \times 5/4 = 720$$

Each line will therefore be represented by 720 samples, with each sample representing one pixel: sample 1 represents pixel 1, sample 2 represents pixel 2, and so on (Fig. 22.1).

**Fig. 22.1** *Sampling a picture frame*

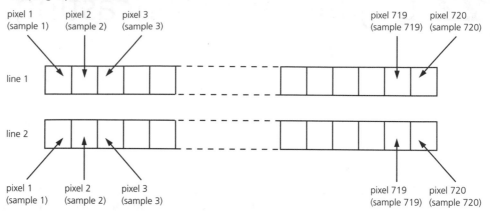

The process is then repeated for the second line, and so on, to the end of the frame, when it is repeated all over again for the next frame, and so on. To ensure the samples are taken at exactly the same point on each line within every frame, the sampling frequency must be locked to the line frequency. For this to happen, the sampling rate must be an exact multiple of the line frequency.

Provided each sample or group of samples is given an identification as to the pixel or group of pixels it represents, it may be reorganised, regrouped or generally manipulated, then processed and reassembled back to its place in the original order.

## The sampling rate

As outlined in Chapter 1, the total period for one line of composite video is 64 $\mu$s. Of this, 12 $\mu$s is used for the sync. pulse, the front porch and the back porch, leaving 52 $\mu$s to carry the video information. With 720 pixels per line, the sampling rate is

$$\frac{\text{number of pixels per line}}{\text{time for video information}} = \frac{720}{52\ \mu s} = 13.8\,\text{MHz}$$

However, since the sampling frequency must also be a whole multiple of the line frequency, a sampling rate of 13.5 MHz (864 × line frequency) is recommended by the CCIR (Comité Consultatif International des Radiocommunications).

Recall from Chapter 15 that, in order to avoid aliasing and other distortions, the sampling frequency must be greater than twice the highest frequency of the analogue input. Thus, for the video signal, which at the studio may have a frequency of up to 6 MHz, a sampling rate in excess of 2 × 6 = 12 MHz is necessary. The selected rate of 13.5 MHz is therefore more than adequate.

# Bandwidth requirements

Colour TV broadcasting involves the transmission of three components: the luminance signal Y and two colour difference signals: Y – R (known as $C_R$) and Y – B (known

**Fig. 22.2** *Digital TV broadcasting: one programme*

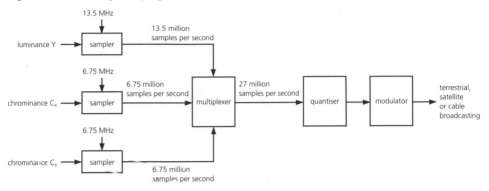

as $C_B$). In the analogue TV system, these components are transmitted directly using amplitude modulation (terrestrial broadcasting) or frequency modulation (satellite broadcasting). In digital television broadcasting, the three components are first converted into digital data streams before modulation and subsequent transmission (Fig. 22.2). For the luminance signal, which contains the highest video frequencies, the full sampling rate of 13.5 MHz is used. For the chrominance components, $C_R$ and $C_B$, which contain lower video frequencies, a lower sampling rate is acceptable. The CCIR recommends half the luminance rate, i.e. $0.5 \times 13.5 = 6.75$ MHz. For the chrominance components, therefore, only half of the pixels (i.e. every other pixel) are sampled. This gives a total number of samples per second of 13.5 million + 6.75 million + 6.75 million = 27 million samples per second. Sampling is following by the quantisation block, where each sample is converted into a multibit code. An 8-bit coding is regarded as the minimum for adequate picture representation to provide $2^8 = 256$ discrete signal levels. This gives a bit rate of

$27 \times 8 = 216$ million bits per second (Mbits/s)

Such a high bit rate requires a very wide bandwidth. The precise bandwidth depends on the type of modulation used at the transmitter. For instance, using pulse code modulation (PCM), a bandwidth of $0.5 \times 216 = 108$ MHz is required. The bandwidth may be reduced by employing complex modulation techniques such as quadrature phase shift keying (QPSK) and quadrature amplitude modulation (QAM). Nonetheless, the bandwidth remains inhibitive for all types of broadcasting media: terrestrial, satellite and cable. Hence the need for data compression and bit reduction techniques. These techniques reduce the number of bits required to define the contents of a picture while maintaining its quality.

## Picture quality

In digital TV broadcasting the quality of the picture is constrained by two main factors: the video bandwidth (or the number of pixels per picture) and the bit rate. Given 720 pixels per line and 576 active lines per picture, the total number of pixels per picture is

$720 \times 576 = 414\ 720$

With every two pixels requiring one complete video cycle, the maximum analogue video frequency is

$$414\,720 \times 25/2 = 5.18\,\text{MHz}$$

which maintains the same picture quality as traditional analogue broadcasting.

The second constraint to picture quality is the bit rate. Given adequate data compression, as defined by MPEG-2, good quality digital TV broadcasting requires a bit rate of up to 15 Mbits/s going into the modulator. A lower bit rate will result in poor picture quality. High-definition TV broadcasting employing 1152 lines requires a bit rate of 60 Mbits/s for 1440 pixels per line and 80 Mbits/s for 1920 pixels per line.

## System overview

Figure 22.3 shows the basic blocks for a digital TV system broadcasting four programmes within a single r.f. channel. A number of programmes containing compressed video and audio together with service data packets are multiplexed into a single bit stream known as the *programme elementary stream* (*PES*). The elementary stream of each programme is further multiplexed into a common bit stream known as the *transport stream*. The programme elementary stream contains identification data, time stamp and programme specific information (PSI); they allow the receiver to unpack the data packets, programme by programme and frame by frame, to reconstitute the original picture. The transport stream is fed into a modulator before transmission along a single 8 MHz r.f. channel.

**Fig 22.3**  *Digital TV broadcasting: multiplex*

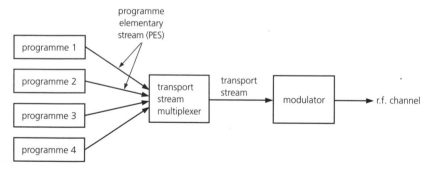

## Programme encoding

The basic elements of programme encoding are shown in Fig. 22.4. Analogue video and audio signals are first sampled at the appropriate rates (13.5 MHz for Y and 6.75 MHz for $C_R$ and $C_B$) before being fed into their respective encoders. The encoders remove non-essential or redundant parts of the picture and sound signals and they perform bit reduction operations to produce individual data packets. Together with the service or housekeeping data packets, the individual data packets are then fed into a multiplexer (MUX) from which the programme elementary stream (PES) is obtained. As there is more video information than either audio or service data, the elementary stream contains

**Fig. 22.4** *Programme encoding*

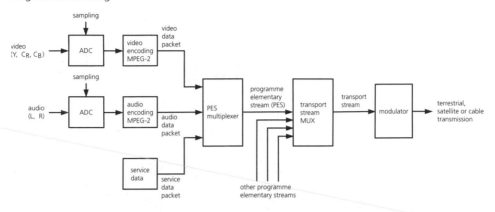

more video data packets than audio or service packets (Fig. 22.5). The elementary stream is then multiplexed with the elementary streams of other programmes before going into the modulator for transmission. The type of modulation employed is determined by the type of broadcasting media: terrestrial, satellite or cable.

**Fig. 22.5**

| video | video | audio | video | video | video | audio | service | video |
|-------|-------|-------|-------|-------|-------|-------|---------|-------|

## Video encoding

Video encoding consists of three major parts: video data preparation, video data compression and quantisation (Fig. 22.6). Video data preparation ensures the raw coded samples of the picture frames are organised in a way that is suitable for data reduction. Video data compression is carried out in accordance with the internationally accepted standards established by the MPEG-2 system. MPEG-2 performs two major data reduction exercises, temporal redundancy removal and spatial redundancy removal. *Temporal redundancy* removal is an *interframe* data reduction which compares two successive picture frames, removes the similarities and produces the differences for processing. *Spatial redundancy* removal, also known as *intraframe* compression, removes unnecessary repetitions of the contents of an individual picture frame. It is carried out using a complex mathematical formula known as the *discrete cosine transform* (*DCT*), hence it is also known as DCT data reduction. Data compression is followed by the quantisation block, which provides further bit reduction. The quantising block translates the DCT coefficients into 8-bit digital codes which form the data bit stream.

**Fig. 22.6** *MPEG-2 video encoding*

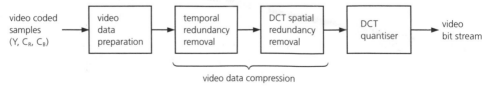

**Fig. 22.7** *Basic blocks: Y, C$_R$ and C$_B$*

| 177 | 178 | 179 | 178 | 179 | 180 | 180 | 180 |
|-----|-----|-----|-----|-----|-----|-----|-----|
| 175 | 176 | 177 | 177 | 177 | 177 | 178 | 179 |
| 176 | 178 | 180 | 178 | 178 | 178 | 178 | 178 |
| 175 | 180 | 181 | 180 | 181 | 179 | 178 | 177 |
| 175 | 179 | 179 | 179 | 177 | 178 | 178 | 177 |
| 176 | 177 | 177 | 177 | 176 | 176 | 169 | 169 |
| 178 | 175 | 176 | 177 | 175 | 175 | 169 | 169 |
| 177 | 176 | 176 | 174 | 174 | 174 | 165 | 165 |

Y

| 67 | 67 | 68 | 69 | 69 | 69 | 70 | 70 |
|----|----|----|----|----|----|----|----|
| 66 | 67 | 67 | 68 | 69 | 69 | 68 | 68 |
| 67 | 68 | 68 | 69 | 69 | 69 | 68 | 68 |
| 69 | 69 | 70 | 70 | 70 | 69 | 68 | 67 |
| 68 | 69 | 69 | 70 | 69 | 67 | 65 | 64 |
| 66 | 67 | 67 | 67 | 66 | 64 | 62 | 60 |
| 65 | 65 | 66 | 66 | 64 | 62 | 59 | 57 |
| 65 | 66 | 66 | 66 | 64 | 61 | 58 | 56 |

C$_R$

| 77 | 77 | 78 | 79 | 79 | 79 | 80 | 80 |
|----|----|----|----|----|----|----|----|
| 76 | 77 | 77 | 78 | 79 | 79 | 78 | 78 |
| 77 | 78 | 78 | 79 | 79 | 79 | 78 | 78 |
| 79 | 79 | 80 | 80 | 80 | 79 | 78 | 77 |
| 78 | 79 | 79 | 80 | 79 | 77 | 75 | 74 |
| 76 | 77 | 77 | 77 | 76 | 74 | 72 | 70 |
| 75 | 75 | 76 | 76 | 74 | 72 | 69 | 67 |
| 75 | 76 | 76 | 76 | 74 | 71 | 68 | 66 |

C$_B$

## Video data preparation

The video information which enters the video encoder is in the form of a series of coded samples of luminance Y, and chrominance C$_R$ and C$_B$; these frames are then regrouped into blocks, macrogroups and slices. The Y, C$_R$ and C$_B$ frames are first divided into a series of blocks of pixels (Fig. 22.7). Each block is an array of $8 \times 8$ digitally coded pixel samples. The three components represented by the basic blocks are brought together to form a macroblock (Table 22.1). Each macroblock consists of four blocks of luminance and two blocks of each of the chrominance components, C$_R$ and C$_B$. This is known as the 4:2:0 format. The macroblocks are then arranged in the order they appear in the picture to form a slice. A slice may consist of one or more macroblocks. At this stage, error detection bits are added to each slice. If an error is detected at the receiver decoding stage, the decoder ignores the information contained in the slice and moves to the next one. The complete picture frame containing all three picture components (Y,C$_{R,(B)}$) is then reconstructed by a series of these slices (Table 22.1) ready for the next stage of the video encoding.

## Temporal redundancy removal

This technique exploits the fact that the difference between two successive picture frames is very slight. Thus it is not necessary to transmit the full contents of every picture frame since most of the current frame is merely a repetition of the previous frame. Temporal compression is carried out on a group of pictures (GOP) composed of 12 non-interlaced frames. The contents of the first frame of the group, known as the I frame, are stored in memory and used as a reference frame for the subsequent 11 frames (Fig. 22.8). The

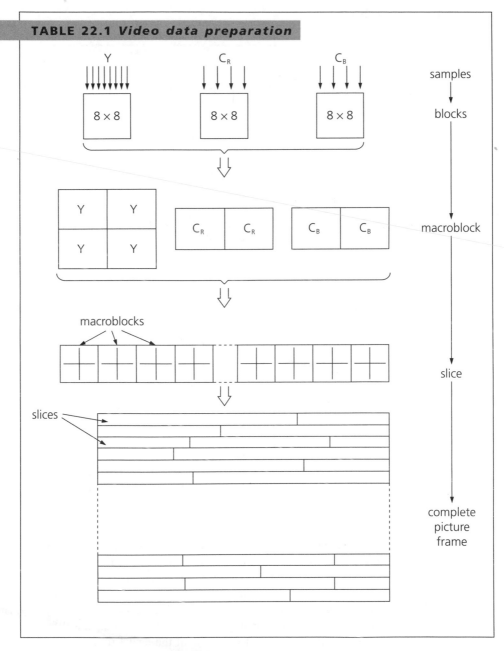

**TABLE 22.1 Video data preparation**

contents of the frame immediately following the reference I frame are compared with the I frame to obtain a difference frame, known as a P frame (P for predicted), which is used for processing. The second frame after the I frame is then compared with the first frame after the I frame, the third frame with the second frame, and so on, until the end of the group of 12 picture frames. A new reference I frame is then produced for the next group of 12 frames, and so on. The amount of compression possible for I frames

**Fig. 22.8** *Group of pictures (GOP)*

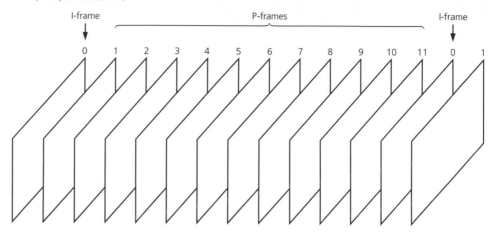

**Fig. 22.9** *I, P and B frames*

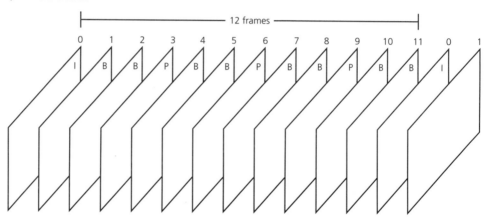

is limited; most of the bit reduction is obtained in the P frames. Increased compression may be achieved by two other techniques: forward prediction and motion prediction.

Forward prediction is a method used for constructing the P frames; it involves predicting the expected difference between macroblocks of consecutive frames and sending the predicted frame for processing. This requires more than one video frame of memory. Forward prediction may use an I frame for reference or a previously reconstructed P frame. The drawback of this technique is that if errors occur in a P frame, they will be carried forward to the subsequent frames until the next I frame arrives.

Motion prediction involves comparing the contents of a previous frame and a subsequent frame to construct the current frame. The constructed frame is known as a B frame, B for backward-predicted (or bidirectional since its contents depend on past frames as well as future frames). Unlike I frames and P frames, B frames cannot be used for reference purposes. They also require two frames of video memory storage. A typical video sequence of 12 frames (0–11) is illustrated in Fig. 22.9.

The MPEG-2 data stream thus contains a continuous series of coded frames consisting of a combination of predicted frames and reference frames. Since predicted frames – P frames and B frames – provide more efficient data compression, it is desirable to ensure that most transmitted frames are predicted frames.

## Motion compensation

Motion compensation is used to correct errors that may occur in predicted frames. Precise calculations to obtain the speed and direction of a moving object are made possible by comparing its position in successive frames. From these calculations, the position of the object in subsequent frames (normally luminance frames) may thus be predicted. A relatively small amount of data is needed to describe the vector for the speed and direction of the motion; it is fed into the P and B frame generators. Once the motion vector has been worked out, it is then used for all three components, Y, $C_R$ and $C_B$.

**Fig. 22.10** *DCT redundancy removal*

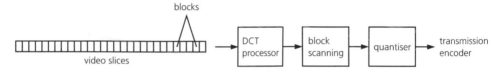

## Spatial DCT redundancy removal

The heart of video encoding is the discrete cosine transform (DCT) processor. The DCT processor receives I, P and B picture frames as a stream of $8 \times 8$ blocks arranged in macroblocks and slices forming a single picture frame. The blocks may be part of a luminance (Y) or chrominance ($C_R$ or $C_B$) frame. Data representing the samples of each block is then fed into the DCT processor (Fig. 22.10), which translates them into an $8 \times 8$ matrix of coefficients representing the video pattern of the block. Before DCT each figure in the $8 \times 8$ block represents the value of the relevant sample, i.e. the brightness of the pixel represented by the sample (Fig. 22.11). The DCT processor examines the spatial frequency components of the block as a whole and translates the time domain array into a frequency domain array. The process involves creating a new set of coefficients in an $8 \times 8$ matrix, starting with the top left-hand cell representing the d.c.,

**Fig. 22.11**

Original block

| 146. | 144. | 149. | 153. | 155. | 155. | 155. | 155. |
|------|------|------|------|------|------|------|------|
| 150. | 151. | 153. | 156. | 159. | 156. | 156. | 156. |
| 155. | 155. | 160. | 163. | 158. | 156. | 156. | 156. |
| 163. | 161. | 162. | 160. | 160. | 159. | 159. | 159. |
| 159. | 160. | 161. | 162. | 162. | 155. | 155. | 155. |
| 161. | 161. | 161. | 161. | 160. | 157. | 157. | 157. |
| 161. | 162. | 161. | 163. | 162. | 157. | 157. | 157. |
| 160. | 162. | 161. | 161. | 163. | 158. | 158. | 158. |

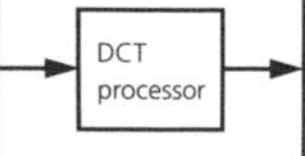

DCT processor →

DCT block

| 314.91 | −0.26 | −3.02 | −1.30 | 0.53 | −0.42 | −0.68 | 0.33 |
|--------|-------|-------|-------|------|-------|-------|------|
| −5.65 | −4.37 | −1.56 | −0.79 | −0.71 | −0.02 | 0.11 | −0.30 |
| −2.74 | −2.32 | −0.39 | 0.38 | 0.05 | −0.24 | −0.14 | −0.02 |
| −1.77 | −0.48 | 0.06 | 0.36 | 0.22 | −0.02 | −0.01 | 0.08 |
| −0.16 | −0.21 | 0.37 | 0.39 | −0.03 | −0.17 | 0.15 | 0.32 |
| 0.44 | −0.05 | 0.41 | −0.09 | −0.19 | 0.37 | 0.26 | −0.25 |
| −0.32 | −0.09 | −0.08 | −0.37 | −0.12 | 0.43 | 0.27 | −0.19 |
| −0.65 | 0.39 | −0.94 | −0.46 | 0.47 | 0.30 | −0.14 | −0.11 |

**Fig. 22.12** *Frequency domain block*

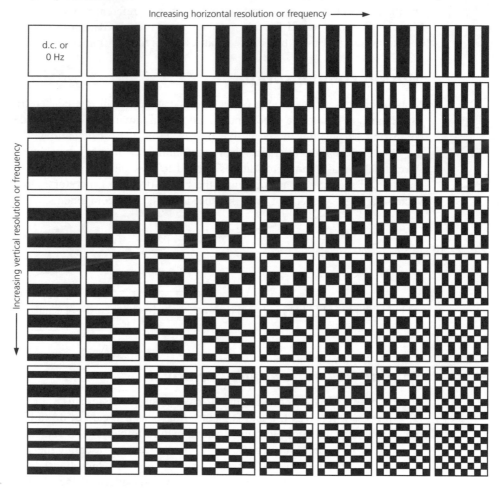

i.e. the 0 Hz frequency component. The coefficient in this cell represents the average brightness of the block. Each of the other cells represents an increasing frequency component of the block (Fig. 22.12). The values of the coefficients in the other cells are determined by the amount of picture detail within the block. Thus, a block containing identical luminance (or chrominance) throughout, e.g. part of a clear sky, will be represented by the d.c. component only; the coefficients in the other cells will be zero. Where a block contains picture detail, it will be represented by non-zero coefficients in the appropriate cells. Coarse picture detail will use low coefficients and only a few cells will have non-zero coefficients; fine picture detail will use higher coefficients and many of the cells will have non-zero coefficients. Fine horizontal picture details (high horizontal frequency) are represented by moving horizontally to the right; greater vertical detail (higher vertical frequency) is represented by moving vertically downwards, as illustrated in Fig. 22.12. The highest possible picture detail, i.e. the highest video frequency, is represented by the bottom right-hand cell of the matrix.

**Fig. 22.13** *Rounded up/down DCT block*

| 315 | 2 | −1 | −2 | 0 | 0 | 0 | 0 |
|---|---|---|---|---|---|---|---|
| −4 | −3 | −2 | 1 | 0 | 0 | 0 | 0 |
| −2 | −3 | 0 | 0 | 0 | 0 | 0 | 0 |
| 0 | 0 | 0 | 0 | 0 | 0 | 0 | 0 |
| 2 | 0 | 0 | 0 | 0 | 0 | 0 | 0 |
| 0 | 0 | 0 | 0 | 0 | 0 | 0 | 0 |
| 0 | 0 | 0 | 0 | 0 | 0 | 0 | 0 |
| 0 | 0 | 0 | 0 | 0 | 0 | 0 | 0 |

As can be seen from Fig. 22.11, which represents an average block DCT matrix, most of the coefficients and therefore the energy is concentrated at and around the top left-hand corner; the bottom right quadrant has very few coefficients of any value. This is not surprising since an 8 × 8 pixel block is unlikely to contain greatly differing picture detail. The DCT coefficients are then rounded up or down to a smaller set of possible values, resulting in a greatly simplified set of coefficients (Fig. 22.13).

## Zigzag scanning of the DCT matrix

Before quantisation, the DCT matrix of each block is reassembled into a serial format by scanning each coefficient in a zigzag pattern, starting at the top left-hand cell (the d.c. component) as shown in Fig. 22.14. For the example in Fig. 22.13, the scanned order is 315, 2, −4, −2, −3, −1, −2, −2, −3, 0, 2, 0, 0 and 1. No further transmissions

**Fig. 22.14** *Zigzag scanning*

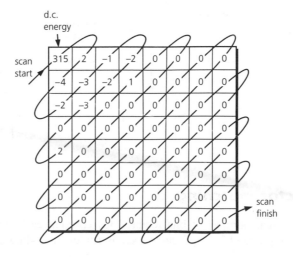

are necessary since the remaining coefficients are zero and thus contain no information. The end of the block is indicated by a special end-of-block (EOB) code which is appended to the end of the scan. Sometimes a significant coefficient may be trapped within a block of zeros, then other special codes are used to indicate a long strings of zeros.

## Variable length quantisation

Quantisation of the DCT coefficients in not carried out linearly. Instead, entropy coding is used. An entropy coding technique gives an individual quantisation level to each coefficient according to its position in the matrix. Each coefficient is given a weighting factor or scale to indicate its relative importance. The highest level of quantisation is given to the top left-hand cell of the matrix, which represents the d.c. component. The d.c. component is encoded to the highest level of accuracy because noise is most visible on low-frequency video information. High-frequency picture information can tolerate larger quantisation errors; it is therefore given lower quantisation levels.

The quantisation weighting scales are further modified to take account of the bit rate emerging from the DCT processor. Where picture detail is coarse, with most of the DCT coefficients at zero or insignificant levels, a short string of coefficients is produced from the DCT processor, placing minimal demands on bit rate and bandwidth. However, a block with highly defined picture details will present a long string of coefficients, hence higher bit rates and wider bandwidth requirements, which may fall outside the specified limits. To avoid this, variable length coding is employed (Fig. 12.15). This technique provides for the weighting quantisation factor to be dynamically changed depending on the bit rate emerging from the DTC processor itself. The quantised bits are first fed into a memory buffer store before being fed out at a constant rate to the transmission encoder. If the bit rate increases and the buffer begins to overflow, the data rate control unit is activated; this causes the quantising level to be reduced, decreasing the data bit rate. In this way, the output bit rate is kept constant.

**Fig. 22.15** *Variable length quantisation*

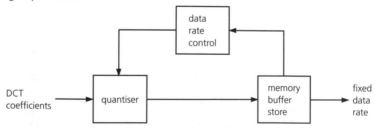

## Vector matching

The actual number of bits required to represent each sample can be further reduced by using coded binary strings such as runlength coding and vector coding. *Runlength coding* replaces a long string of the same number such as 3, 3, 3, 3, 3, 3, with 6, 3 (3 repeated six times). *Vector coding* is a sort of predictive coding in which a quantised group of pixels, such as the array of 8 × 8, is represented by a code vector. The vector, a mathematical representation of the block of pixels, is compared with a set of pre-loaded vectors in a ROM chip. The best match is produced and duly transmitted. At the receiving end, the transmitted vector is converted back into the original block using a look-up table that contains the same set of vectors and their corresponding images.

# The audio encoder

Audio encoding involves splitting the audio baseband into 32 sub-bands of equal band-widths (Fig. 22.16). Audio signal frequencies falling within each sub-band are sampled and converted into multibit codes. Three sampling rates are possible: 32, 44.1 and 48 kHz. Information on the chosen sampling rate is included within the control part of the audio packet. Audio compression is obtained by the use of special algorithms which removes parts of the audio data without affecting the sound quality, a process known as *masking*. Masking exploits two characteristics of human hearing: namely that a quiet sound is made inaudible by a loud sound at a nearby frequency, and quiet high frequencies are masked by louder lower frequencies. Furthermore, human hearing has a finite fre-quency resolution in which certain frequency bands sound alike. Audio masking is carried out by Fourier transformation. The audio signals are fed into a *fast Fourier trans-form (FFT)* processor, which analyses the sound spectrum into frequency coefficients to be used for determining the quantisation levels of the various sub-bands. The quan-tised sub-bands together with the necessary control and error detection information are fed into a multiplexer to obtain the MPEG-2 audio packet shown in Fig. 22.17. The sub-band samples are preceded by a header, CRC, bit allocation and the scaling data. The header carries synchronisation and system information; the CRC bits are used for error correction; the bit allocation data defines the resolution of the samples; and the scaling

**Fig. 22.16** *Audio encoding*

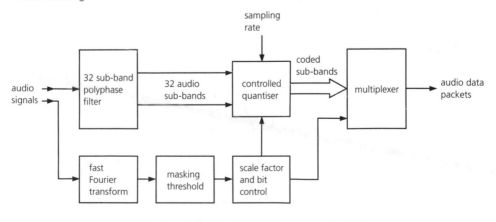

Fig. 22.17  *Audio data packet*

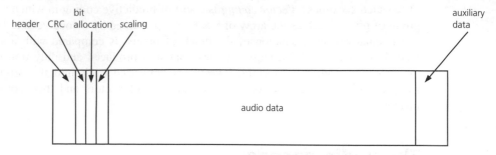

bits provide information on the scaling factor. The auxiliary part provides data on centre and Dolby surround sound. MPEG-2 provides two-channel stereo audio as well as up to five channels per video broadcast, with a bit rate of 64–256 kbits/s.

## The service data packet

The service data packet is the third component of the digital TV programme elementary stream (PES). It contains housekeeping and programme data. Also known as the private or housekeeping data packet, it contains among other control data, the programme-specific information (PSI) required for channel identification and the selection of the corresponding video and audio packets of the selected programme from among the other multiplexed programmes.

## The PES packet structure

The PES packet (Fig. 22.18) consists of a header together with a video, audio or service data packet. The length of the data packet is normally 2 KB (2048 bytes) but may be as large as 64 KB. The header contains the information needed to identify and define the packet, including

Fig. 22.18  *PES data packet*

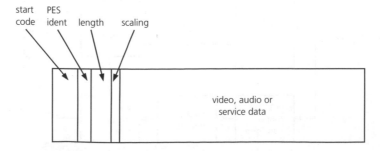

- a start code (24 bits), namely 000001 (hex)
- PES identification (8 bits), e.g. video, audio or private
- packet length (16 bits)
- scrambling control (2 bits)

The header will also include a presentation time stamp (PTS) which provides information on the time at which a frame is to be displayed, an audio piece is to be heard or a caption is to be inserted. These time stamps are essential for audio and video synchronisation.

## Programme multiplexing

Figure 22.19 shows a two-programme multiplexing arrangement. First the video, audio and service data packets of each programme are multiplexed to form elementary programme streams PES1 and PES2. The two elementary streams are then multiplexed to form the transport stream. The transport stream thus contains individual packets of each programme, identified by the programme-specific information (PSI) contained within the service data packet. The transport stream is then fed into the channel encoder, where *forward error correction* (*FEC*) is introduced, before going to the modulator on its way to the transmitter, which may be an uplink for satellite, an aerial for terrestrial or amplifier for cable television. The modulated transport stream is allocated an 8 MHz bandwidth, the same as for a single analogue channel. Up to four different programmes may be contained within a single transport stream without reducing the quality of the picture.

## Transport stream packet structure

The MPEG-2 transport stream consists of a series of 188-byte data packets. Each packet contains a 4-byte header followed by 184 bytes of actual video, audio or service data information known as the *payload* (Fig. 22.20). The header starts with a standard 1-byte sync. word (hex code 47) which provides a run-in clock sequence for the packet. The header (Fig. 22.21) provides the necessary information for unpacking the various programmes and reproducing the selected PES at the receiving end. A list of the header bits and their function is provided in Table 22.2.

As the transport packets are shorter than the PES packets, which are typically 2 KB (2048 bytes), the PES packets have to be divided into data blocks of 184 bytes to fit into the transport stream packet. A single PES packet will be spread across a number of transport stream packets. And since a PES packet is not an exact multiple of 184 bytes, the last transport packet (which carries the residue of the PES packet) will only be partially occupied. The unoccupied part of the transport stream packet is 'stuffed' with an *adaptation field*, the length of which is the difference between 184 bytes and the PES residue (Fig. 22.22). In addition to this making-up or stuffing function, the adaptation field carries the *programme clock reference* (*PCR*), which is used at the receiving end to synchronise a 90 kHz base clock to provide a yardstick for measuring the programme time stamps (PTSs).

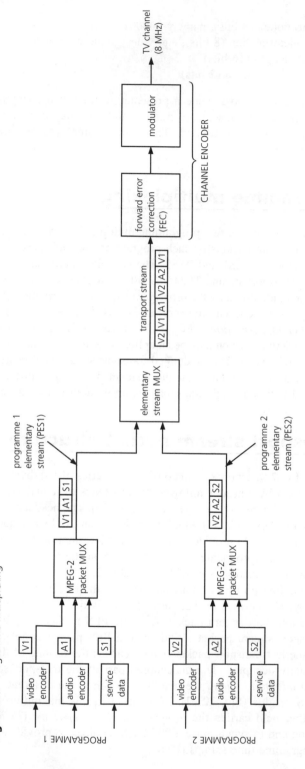

**Fig. 22.19** *Programme multiplexing*

**Fig. 22.20** *Transport stream: data packet*

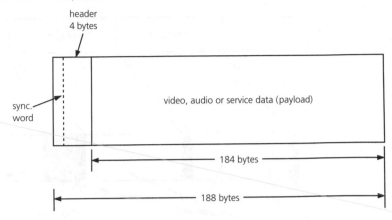

**Fig. 22.21** *Transport stream: header structure*

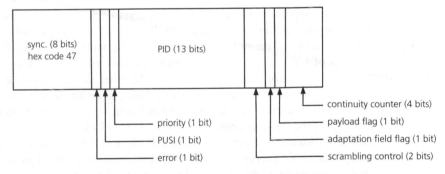

**TABLE 22.2 *Contents of MPEG-2 transport stream header***

| Field | Bits | Function |
|---|---|---|
| Sync. word | 8 | header start sequence, hex code 47 |
| Error indicator | 1 | indicates error in previous stages |
| PUSI | 1 | payload unit start indicator, indicates start of payload |
| Priority | 1 | indicates transport priority |
| PID | 13 | packet identifier, indicates content of packet |
| Scrambling control | 2 | indicates type of scrambling used |
| Adaptation field | 1 | indicates the presence of an adaptation field |
| Payload flag | 1 | indicates the presence of payload data in the packet |
| Continuity counter | 4 | keeps count of truncated PES portions |

**Fig. 22.22** *Packet allocation: PES packets are allocated to transport stream packets*

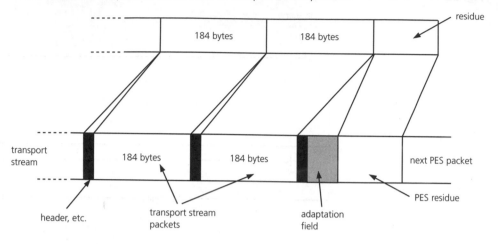

## Forward error correction

Digital signals, especially signals with a high level of data compression, require efficient error detection and error correction. In digital television broadcasting, the *bit error rate (BER)* must be of the order of $10^{-10}$ to $10^{-12}$, equivalent to 0.1 to 10 erroneous bits in one hour of transmission. A transmission channel with such a low bit error rate is known as a *quasi-error-free (QEF)* channel. In order to accommodate such stringent specifications, preventive measures must be taken to ensure that errors introduced by the physical transmission medium are detected and, where possible, corrected. This is the function of the *forward error correction (FEC)* block.

Before forward error correction is applied, the FEC chip carries out what is known as energy dispersal. Energy dispersal involves scrambling the data to obtain evenly distributed energy across the channel. To ensure the data may be descrambled back into its original format, *pseudo-random binary scrambling (PRBS)* is used, similar to the scrambling in NICAM stereo sound transmission. FEC has three stages:

- outer coding (Reed–Solomon)
- convolutional interleaving
- inner coding (convolutional)

Inner coding, the third stage, is not required for cable television. At the receiving end, the three stages are reversed (Fig. 22.23).

Outer coding employs the Reed–Solomon (RS) detection and correction code. This technique does not provide correction for error bursts, i.e. errors in adjacent bits. Interleaving overcomes this; it ensures that adjacent bits are separated before transmission. If the transmission medium introduces a lengthy burst of errors, they are broken down at the receiving end by the deinterleaver before reaching the outer decoder. The RS code chosen for digital TV transmission is 204:188, which adds an extra 16 bytes to the transport stream packet (Fig. 22.24). The RS outer coder can detect and correct 16 bytes of errors in the 204-byte packet.

**Fig. 22.23**

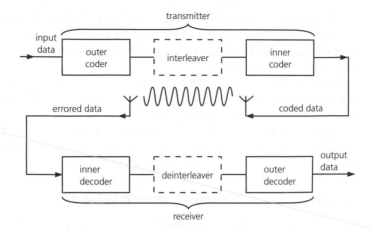

**Fig. 22.24** *Transport packet with FEC error bits*

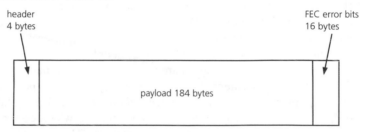

Finally, inner coding already mentioned is another type of convolutional coding which introduces further error correction capabilities. A full (100%) convolutional inner coder produces two simultaneous output bitstreams, X and Y, each replicating the original data stream. Bit streams X and Y are modulated and transmitted. Although this provides a very powerful error correction, it introduces a very high redundancy count, effectively doubling the bandwidth requirement of the channel. Redundancy may be improved by using a technique called *puncturing*, by which only one of the two simultaneous bits of the X and Y bit streams are selected for modulation. Alternate X and Y bits are used within a certain ratio, known as the *puncturing ratio*. A high puncturing ratio improves correction efficiency at the expense of channel capacity. Puncturing ratios of 3/4 or 2/3 may be chosen by the broadcaster, based on the power of the transmitter, the size of the receiving antennae and the desired quality.

## Modulation

The final reduction in the bit rate is provided by the use of advanced modulation techniques. Simple frequency modulation, in which logic 0 and logic 1 are represented by two different frequencies, is highly inefficient in terms of bit rate and bandwidth requirements.

Three types of modulation are used in digital TV broadcasting: *differential quadrature phase shift keying (DQPSK)* for satellite, *quadrature amplitude modulation (QAM)* for cable and *coded orthogonal frequency division multiplexing (COFDM)* for terrestrial digital TV transmission.

Satellite digital TV broadcasting uses differential QPSK, the same modulation technique as NICAM. The four phase settings, 45°, 135°, 225° and 315°, are produced by two equal frequency carriers at right angles to each other. Each phase is used to represent a combination of two bits, as explained in Chapter 16. Pure QPSK requires a reference phase angle. With DQPSK the previous phase is used as the reference phase angle, in which each phase shift represents a 2-bit combination.

## *Quadrature amplitude modulation*

Phase shift keying may be improved by increasing the number of carrier phase angles from 4 in the case of quadrature PSK to 8 or 16 in the case of 8-PSK and 16-PSK respectively. For 8-PSK encoding, the carrier may have one of eight different phase angles (Fig. 22.25) with each phasor representing one of eight 3-bit combinations. Quadrature amplitude modulation (QAM) is an extension of PSK in that the carrier is changed in amplitude as well as phase to provide increased bit representation. For instance, 16-QAM encoding increases the bit width of the modulation to 4, as shown in Fig. 22.26(a). Twelve different carrier phasors are used, four of which have two amplitudes to provide further 4-bit combinations. Figure 22.26(b) depicts all the possible carrier phase angles and amplitudes; it is known as the *constellation map*. Cable digital TV employs a higher order of digital modulation, 64-QAM encoding, in which each carrier phase/amplitude represents one of 64 possible 6-bit combinations. The constellation map for a 64-QAM encoding is illustrated in Fig. 22.27.

**Fig. 22.25**  *Phasor diagram for 8-PSK*

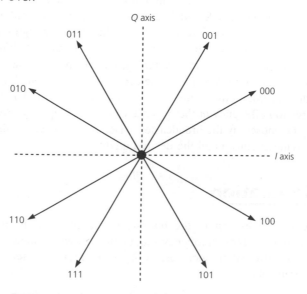

**Fig. 22.26** *(a) Phasor diagram for 16-QAM; (b) constellation diagram for 16-PSK*

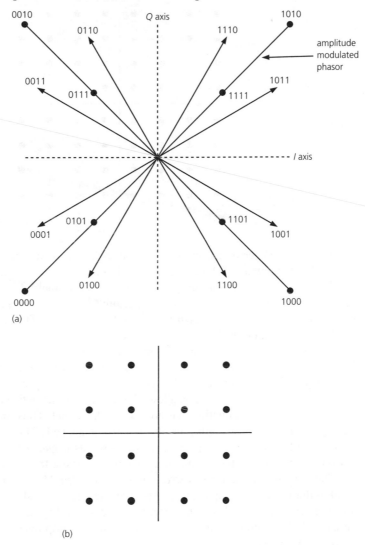

(a)

(b)

## COFDM

Although 64-QAM encoding is very efficient and effective, when used for terrestrial broadcasting it suffers from fading and multiple path interference. In analogue systems, fading and multipath interference causes picture degradation. In digital systems, especially when the reflected path suffers 180° phase shift compared with the direct path, serious picture degradation and complete failure may result. This can be avoided by using a multicarrier modulation technique known as orthogonal frequency division multiplexing (OFDM). As the digital signal is coded for forward error collection (FEC), this modulation process is referred to as coded orthogonal frequency division multiplex (COFDM).

**Fig. 22.27** *Constellation diagram for 64-QAM*

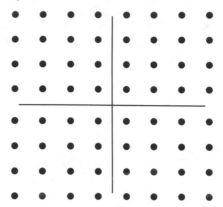

The COFDM technique involves the distribution of the high-rate serial bit stream over a large number of closely spaced individual carriers spread across the available bandwidth, with each carrier only carrying part of the total bit stream. Carriers are processed (or modulated) simultaneously at regular intervals. The set of carriers thus processed at each interval is known as the COFDM symbol. Due to the large number of carriers, the duration of the COFDM symbol is considerably longer than the duration of one bit of the original bit stream. For instance, if 500 modulating bits, each with a symbol duration of 0.1 $\mu$s are used to process (i.e. modulate) 500 carriers to form a COFDM symbol, the length of the COFDM symbol may be crudely calculated to be $0.1 \times 500 = 50\ \mu$s. The long symbol duration allows the receiver to wait until all echoes and reflections have arrived before evaluating and processing the signal. Thus reflected waves arriving during that period will strengthen the direct transmission path. This may be further improved by the addition of a *guard interval* (also known as *a guard band*) before the symbol period, during which the receiver pauses before starting the evaluation of the carriers.

The spacing between the carriers is chosen to be $1/t_s$ where $t_s$ is the modulating symbol duration. The frequency spectrum of each carrier is shown in Fig. 22.28. When all the carriers are included, a flat spectrum is obtained (Fig. 22.29) with parasitic lobes at either end. The introduction of a guard band further improves the frequency spectrum by reducing the secondary lobes.

## Fast fourier transform (FFT)

The set of carriers produced by OFDM is very similar to that produced by a fast fourier transform (FFT) algorithm which analyses a continous waveform into its frequency components. To produce a continuous waveform that may be used to modulate a UHF carrier, the reverse of FFT is performed on the COFDM carriers, a process known as inverse fast fourier transfrom (IFFT). At the receiving end, the original COFDM carriers are obtained by feeding the signal from the UHF demodulator into an FFT chip (Fig. 22.30).

**Fig. 22.28** *A single COFDM carrier: frequency spectrum*

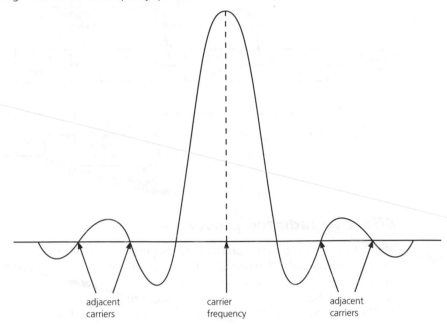

adjacent
carriers

carrier
frequency

adjacent
carriers

**Fig. 22.29** *COFDM carriers: frequency spectrum*

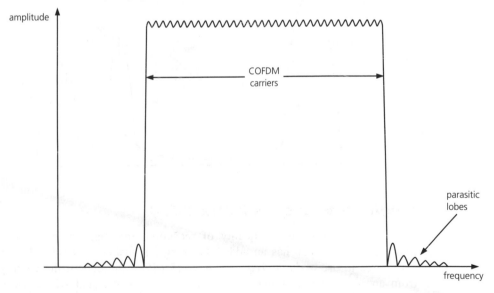

amplitude

COFDM
carriers

parasitic
lobes

frequency

**Fig. 22.30**

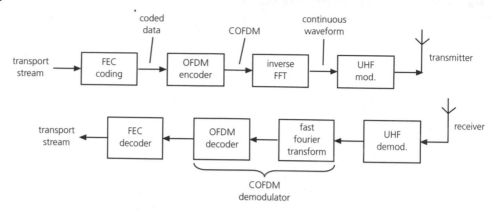

## Effective radiated power

The flat frequency spectrum greatly reduces the effective radiated power (ERP) requirements of the digital TV transmitter by about 20 dBS (100 times) compared with analogue terrestrial broadcasting, in which the carrier power is concentrated in a narrow
band around the vision carrier and the colour, f.m. sound and NICAM subcarriers (Fig.
22.31). COFDM transmitted energy is more efficiently spread across the whole spectrum.

**Fig. 22.31**  *Analogue TV terrestrial broadcasting: energy distribution*

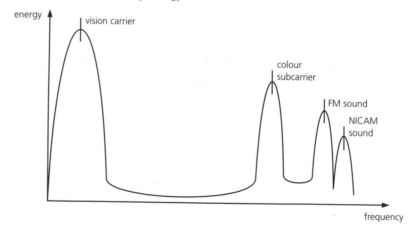

## Single-frequency network (SFN)

Apart from improving the quality of reception, the high tolerance to fading and
multipath interference has an added advantage. It makes it possible for broadcasting
authorities to use a *single-frequency network* (*SFN*) throughout the country. A signal
from an adjacent transmitter broadcasting an identical signal will cause ghosting in
analogue TV transmission. But in terrestrial digital TV broadcasting, they are indistin-
guishable from signals produced by the other transmitter, just like echoes or reflected
waves. And if they arrive during the guard interval, they will be discarded.

## 8K/2K COFDM modes

For terrestrial digital television, the European DVB system is based on COFDM modulation with either 8K (8192) or 2K (2048) carriers with a symbol duration ($t_s$) of 896 $\mu$s and 224 $\mu$s respectively. The effective number of carriers, i.e. the actual number of carriers that may be used for COFDM modulation is 6818 for the 8K mode and 1706 for the 2K mode. The remaining carriers are used for the guard band, and to provide continual pilot carriers and scattered pilot carriers. The continual pilot carriers carry the parameters of the particular transmission and the scattered pilot carriers are used for carrier reference.

# Navigating the digital multiplex

The transport stream multiplex carries two or more programmes, each composed of a number of PESs. Information about the various components of a programme and how they relate to each other are contained in the *programme-specific information* (*PSI*) data bits within the service data packet. PSI consists of a number of tables which contain all the necessary information to tune and select a programme. There are four mandatory tables:

- The *programme allocation table* (*PAT*) establishes the link between the programme and the programme identifier (PID) of the packets carrying the programme map.
- The *programme map table* (*PMT*) indicates the PIDs of the elementary streams making up the programme.
- The *conditional access table* (*CAT*) carries access control data.
- The *private table* carries private information.

Changing programmes or *zapping* is therefore far slower than with conventional analogue TV, as the synchronisation and identification process can take up to one second or more, depending on whether the change is within the same channel or outside the channel.

## QUESTIONS

1. State three advantages of digital TV broadcasting over analogue broadcasting.

2. In the DVB standards, state
   (a) the number of pixels per line
   (b) the number of pixels per picture
   (c) the sampling rate for the analogue video signal

3. State the type of modulation used in
   (a) satellite digital TV broadcasting
   (b) terrestrial digital TV broadcasting

4. State the two types of redundancy removal used in MPEG-2 video encoding.

5. In relation to digital TV broadcasting, what is meant by
   (a) the programme elementary stream (PES)
   (b) the transport stream

6. (a) Explain the reason for using COFDM in terrestrial digital TV broadcasting.
   (b) How many carriers are used in the COFDM 2K mode?

# 23 Digital TV receivers

Digital TV reception is normally provided by an *integrated receiver decoder* (*IRD*) or set-top box. The front end, known as the *channel decoder*, is specific to the broadcasting media: satellite, terrestrial or cable. The remaining part of the circuitry is common to all three broadcasting methods.

## System overview

Figure 23.1 shows the generalised block diagram of a digital TV IRD. The *channel decoder* consists of the tuner, the digital demodulator and the forward error correction (FEC). It receives the modulated r.f. signals and reproduces the original digital transport stream. The transport stream containing packets of four or five multiplexed television programmes is fed into the transport demultiplexer. The *transport demultiplexer* identifies each packet by its programme identifier (PID) and reassembles them to reconstruct the elementary packet of the selected programme. If the packet is scrambled, it is fed into the *conditional access module* (*CAM*). The CAM interrogates the smart card to find out

**Fig. 23.1** *IRD for general digital TV*

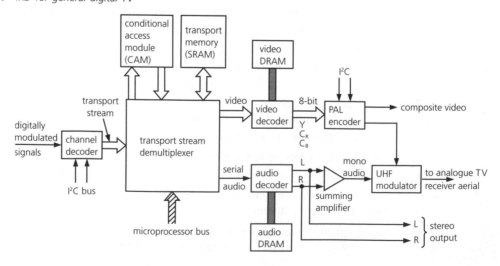

whether the user has a current subscription to the selected programme. If not, the CAM informs the system's microcontroller, which prevents the signals from proceeding any further. Alternatively, the signals are routed back into the transport demultiplxer for further processing by the video and audio decoders. Two sets of digital signals are produced by the transport demultiplexer and fed to their respective decoders: a video output 1 byte (8 bits) wide and a serial audio output. A fast 8K × 8 SRAM memory store is used to buffer the video and audio data so that onward transmission to the video and audio decoders can take place in bursts.

The video section consists of an MPEG-2 *video decoder* which decompresses the video data stream and converts it back into its original components: Y (luminance) and $C_R$ and $C_B$ (chrominance). The picture is reconstructed from the I, P and B frames. This requires the simultaneous storage of these frames, hence the need for a large video memory in the form of a DRAM buffer. The three components are sent to the *PAL decoder*, which converts the digital video into an analogue PAL composite video, before going into the UHF modulator.

The audio section consists of an MPEG *audio decoder*, where the audio packet is decoded by following the same rules as for encoding at the transmitter; this produces left and right analogue audio signals. Audio buffering is provided by an audio DRAM memory chip which, among other things, provides a 1 s delay to ensure audio and video synchronisation. This delay is necessary, given that video processing takes longer than audio processing. The left and right analogue audio signals from the audio decoder are fed into a summing amplifier to produce mono sound for the UHF modulator. A separate stereo (L and R) output is also provided.

## System control

Channel selection and signal processing require normal microcontroller supervision as well as a microprocessor unit, which carries out the necessary software operations. Figure 23.2 shows a control system employing a microcontroller M1 and a microprocessor M2. The *microprocessor* controls the audio and video decoders, including synchronisation,

**Fig. 23.2** *IRD control system*

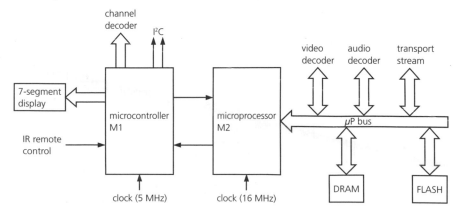

the transport stream mutiplexer and the conditional access module; it also runs the receiver menu system. The microprocessor has its own clock, its own address, data and control bus structure and its own DRAM and flash memory stores. The non-volatile flash type RAM chip is used to store software programmes used in the decoding process. The software may be upgraded off-air by loading the new software into the DRAM chips then transferring it to the non-volatile flash memory or via the telephone system using a modem. Off-air upgrading and reprogramming are performed by the broadcaster; no action needs to be taken by the user. Upgrading through the modem involves the broadcaster sending an off-air message to the digital TV receiver to tell it to 'dial-up' a preloaded telephone number. Provided the telephone line at the subscriber home is free, upgrading Flash memory can take place using the normal telephone line. Such communication can also be used for credit control, billing and other purposes. The *microcontroller* monitors the power supply and decodes the infrared information from the remote control handset and front panel buttons; it controls the channel decoder, including the tuner and the digital demodulator, it controls the PAL decoder and it communicates with the microprocessor.

## The channel decoder

The channel decoder (also known as the front end) is the one part of the receiver; which is specific to the type of reception. The basic component of a channel decoder for a satellite receiver is shown in Fig. 23.3. The *tuner* mixes the incoming analogue modulated r.f. signal with a local oscillator signal to produce a modulated i.f. Before the modulated i.f. is fed into the digital demodulator, it has to be converted into a digital form by the *analogue-to-digital converter (ADC)*. The ADC uses a sampling rate which is at least twice the baud rate of the transmitted signal. The *demodulator* is a digital signal processor chip which may be a QPSK demodulator for satellite reception or COFDM demodulator for terrestrial reception. It is fully controlled and programmed by the system's microcontroller via a data/address bus and control lines. The demodulator estimates

**Fig. 23.3** *Channel decoder*

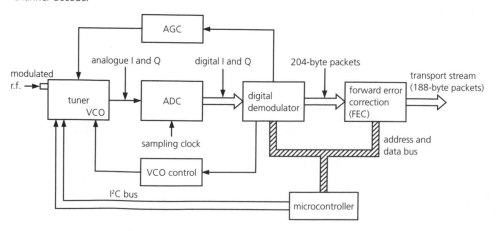

the input signal power and provides an a.g.c. signal to the tuner. The demodulator also provides a synchronising signal for the tuner local oscillator via the VCO control line. The output data is in the same format as the original transport stream, consisting of 204-byte packets (video, audio or programme service) including the 16-byte checksum. The transport stream is then fed into the *forward error correction* (*FEC*) unit.

The FEC unit uses the checksum bytes to make what is known as 'hard' decisions to define the received data. It determines whether the received data packets contain any errors. If they do, the FEC unit will attempt to correct them. Failing that, the FEC will flag the packets that contain errors so they may not be used in subsequent processing. At the end of the process, a transport stream is produced which consists of a number of 188-byte (204−16) multiplexed packets belonging to up to four or five different television programmes. Before the picture is reconstructed, the packets belonging to the selected programme must be identified and placed in the correct order. This is carried out by the demultiplexer.

## A practical channel decoder

The circuit diagram for a PACE satellite channel decoder is shown in Fig. 23.4. The tuner (not shown) is a self-contained isolated unit which receives the modulated signals from the LNB, downconverts the carrier to a second i.f. and produces two analogue signals: in-phase I and quadrature Q. These two signals are fed into pins 12 and 16 (AIN-I and AIN-Q) of the dual-ADC U100, which converts them into two 6-bit digital signals DI0–DI5 and DQ0–DQ5. The sampling rate is set by a control signal from the QPSK demodulator to pin 21 (VCOINP) which controls an in-built voltage-controlled oscillator (VCO). The sampling rate is set to twice the symbol rate (i.e. phase rate) or baud rate. If the baud rate is high, the sampling rate is set to twice the baud rate; if the baud rate is low, the sampling rate is set to three times the baud rate. The I and Q outputs from the dual-ADC (pins 29–34 and 37–42) are fed into the QPSK demodulator U102. The demodulator receives the I and Q samples at pins 55–58, 61, 62 (RI5–RI0) and pins 73, 70, 69, 68, 67 and 64 (RQ5–RQ0). Bits RI0 and RQ6 are tied high at 3.3 V.

As explained in Chapter 16, demodulating a QPSK signal involves two distinct stages: phase recovery and data recovery. Phase recovery requires a reference carrier signal, which is recovered from the reference clock contained within the transmitted packet. In the satellite IRD, phase detection is carried out by the QPSK demodulator chip, whereas data recovery is performed by the subsequent FEC block, where the reference clock is established.

The demodulator carries out the phase detection of the I and Q signals, samples the result at twice the baud rate and quantises it into 3-bit codes. The I and Q phase data are thus represented by 3-bit coded digital signals which appear on pins 19, 20 and 21 (DEMI0–DEMI2) and pins 23, 24 and 25 (DEMQ0–DEMQ2). Tuner local oscillator control is carried out by carrier synchronisation pins 39–42 (CAR-VCO2N, CAR-VCO1N, CAR-VCO2P and CAR-VCO1P), which send the appropriate signal to the tuner via op-amp U105. A.g.c. is provided by PWRN signal at pin 43, which is fed to the tuner via integrating op-amp U104A. The ADC sampling rate is set by signals at pins 37 and 38 (CLK-VCOP and CLK-VCON), which are fed to op-amp U106B before

**Fig. 23.4** *Part of a PACE channel decoder*

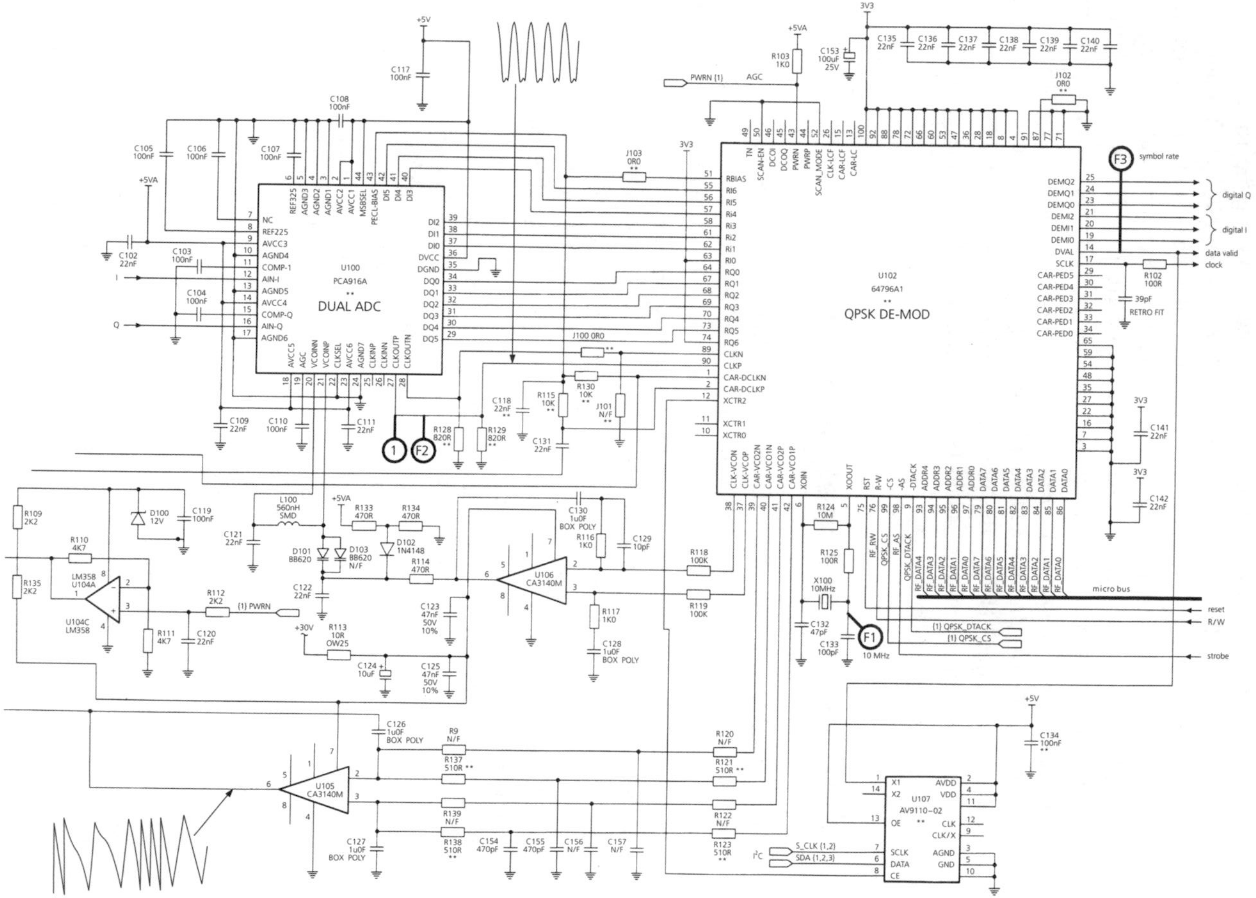

going into pin 21 of the dual-ADC chip. Reference 100 MHz × 100 frequency at pins 5 and 6 is used to synchronise and control the sampling rate of the ADC and the local oscillator at the tuner. Interfacing with the system microcontroller is carried out by an 8-bit data bus (DATA0–DATA7 on pins 79 to 86), a 5-bit address bus (ADDR0–ADDR4 on pins 93 to 97) and control lines: read/write (R–W on pin 76), chip select (QPSK-CS on pin 99) and data acknowledge (QPSK-DTACK on pin 9). The five address lines ADDR0–ADDR4 are multiplexed with the first five data lines DATA0–DATA4 on the microcomputer chip. For this reason, an address strobe control line AS (pin 98) is used to latch the address lines into the demodulator.

## The transport demultiplexer

The input to the demultiplexer (Fig. 23.5) is an 8-bit wide transport stream consisting of video, audio or service information belonging to up to four or five different programmes organised into 188-byte packets. Each packet contains a sync. word for clock run-in, packet ID, programme clock reference, presentation time stamp and video, audio or service data bits. Where scrambling or encryption is used, the input data is sent to the conditional access module (CAM) along a dedicated 8-bit bus (CAM DATAOUT) and if access is approved, the data is sent back along a separate 8-bit data bus (CAM DATAIN). Access is controlled by the microprocessor, which interrogates the smart card to see whether the user has a valid subscription to the selected programme.

**Fig. 23.5**  *A transport demultiplexer*

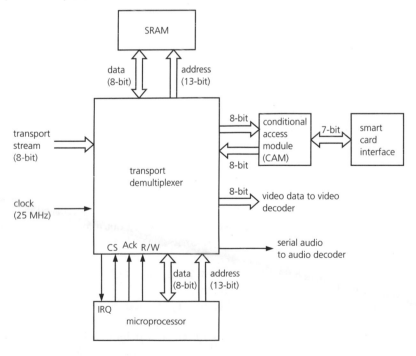

The central task of the demutiplexer is to identify the packets that belong to the selected programme, and using their presentation time stamp together with the 27 MHz programme reference clock, it ensures the selected packets are synchronised with each other. It then filters out the video packets and sends them to the video decoder along a byte-wide video data bus. In the case of the audio packets, the demultiplexer converts them into a serial format and sends them out to the audio decoder. This process is performed under the direction of the microprocessor which, among other things, checks the service packets of the requested programme and sends the appropriate control and processing instructions to the demultiplexer. Fast SRAM (access time 20–25 $\mu$s) is used to store selected video and audio packets for later release in a burst. A dedicated RAM address and data buses are provided for that purpose. The transport demultiplexer is interfaced to the main microprocessor via an 8-bit data bus, a 13-bit address bus and a number of control lines: read/write (R/W), acknowledge (ACK) and chip select (CS) as well as one or more interrupt request (IRQ) lines. The transport demultiplexer uses its interrupt requests to signal the occurrence of certain events such as a full SRAM buffer. Upon receipt of an interrupt request, the microprocessor performs an interrupt service routine associated with the nature of the request. The demultiplexer operates on a 25 MHz system clock (distinct from the 27 MHz programme reference clock for the data stream).

## A typical demultiplexer chip

A typical transport demultiplexer chip is shown in Fig. 23.6 together with an SRAM buffer chip (MT5C6408–20). The speed in microseconds of the memory chip is indicated by the number immediately to the right of the dash –. In this case the speed is 20 $\mu$s. The 8-bit wide transport stream (PARDATA0–PARDATA7) enters the demultiplexer chip at pin 159 and pins 2–8. Three control signals are used to control the transport stream (Oscillogram, Fig. 23.7):

- PARSTART (pin 12) goes high for the first byte of each data packet of the transport stream.
- PARFAIL (pin 11) goes low if the data received has errors which have not been corrected by the FEC.
- PAR-CLK (pin 157) is a data clock.

A register in the demultiplexer determines whether the received data needs to be passed to the CAM. If so, the received data is clocked out into the CAM on pins 26–29, 31–34 and arrives back on pins 15–23. Once again three control signals are employed:

- CA_PKT_START_IN (pin 35) goes high to indicate the first byte of each data packet in the transport stream.
- CA_DATA_VALID_IN (pin 42) goes high if the data received from the FEC has no errors in it.
- CA_BCLKEN (pin 158) is for clock synchronisation.

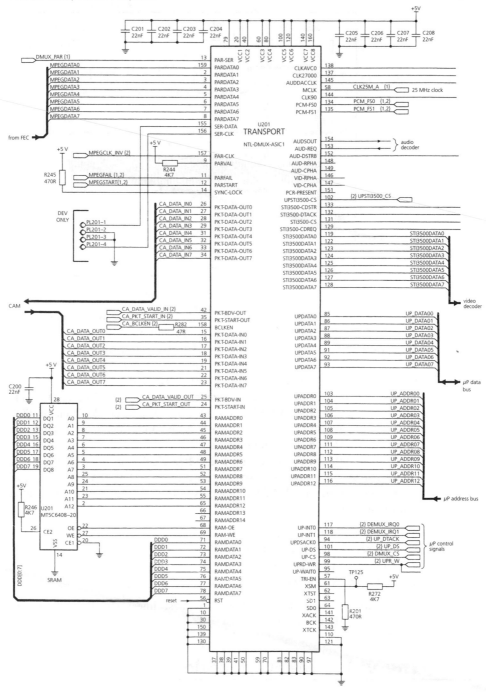

**Fig. 23.6** *A PACE transport demultiplexer*

**Fig. 23.7** *Oscillogram*

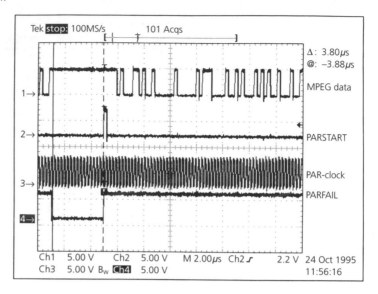

The demultiplexer filters off the video data and sends it to the video decoder along an 8-bit wide bus (ST13500DATA) on pin 119 and pins 122–128. Access to the SRAM chip is via an 8-bit data bus (RAMDATA0–RAMDATA7) on pins 71–78 and a 13-bit address bus (RAMDR0–RAMDR12) on pins 43–49, 51–55, 65 with OUTPUT or READ ENABLE (RAM-OE) on pin 68 and WRITE ENABLE (RAM-WE) on pin 69.

Audio data is sent to the audio decoder serially on pin 154 (AUDSOUT) with a data strobe control line (AUD-DSTB) on pin 152. Data is clocked out on the falling edge of the clock; when the data is received by the audio decoder, it sends an AUD-REQ signal to pin 153 on the demultiplexer.

The demultiplexer operates on a 25 MHz clock into pin 58. It is programmed and controlled by the microprocessor via an 8-bit bus (UP_DATA on pins 85–93) and a 13-bit address bus (UP_ADDR on pins 103–109 and 111–116) together with four control signals:

- read/write (UPR_W) on pin 99
- chip select (DMUX_CS) on pin 98
- data strobe (UP_DS) on pin 101
- data acknowledge (UP_DTACK) on pin 94

The demultiplexer has two interrupt request lines: DEMUX_IRQ0 (pin 117) and DEMUX_IRQ1 (pin 118).

## The video MPEG decoder

The function of the MPEG video decoder (Fig. 23.8) is to restore the picture to its original form. This involves data decompression, inverse DCT, reconstructing the picture

**Fig. 23.8** *MPEG-2 video decoding*

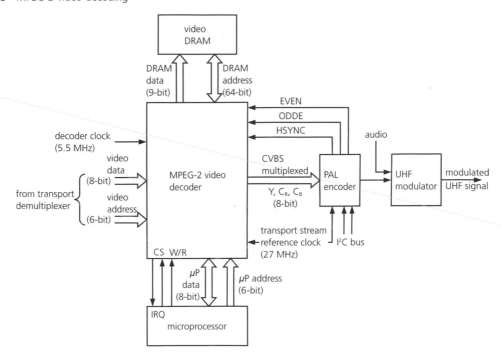

from the I, P and B frames and reproducing the original luminance (Y) and chrominance ($C_R$ and $C_B$) components of each frame. Reconstruction of the picture requires the simultaneous storage of the transmitted frames, making the necessary comparisons to rebuild the complete frames. A large memory store is therefore necessary. The memory required for storing the different frames is provided by a 64-bit wide DRAM buffer. In some receivers, the B frames are not used and the picture is rebuilt from the I and P frames only. This procedure requires smaller DRAM memory store, hence it reduces the cost.

Having reconstructed the picture, the luminance (Y) and chrominance ($C_R$ and $C_B$) pixel data is sent to the PAL encoder along an 8-bit multiplexed YC data bus. The PAL encoder produces a standard television signal, 625 lines per picture and 25 pictures per second. The start of each scan line is indicated by an HSYNC control signal from the PAL encoder to the video decoder. Odd and even fields are indicated by ODDE and EVEN. The composite video (CVBS) signal is fed into a UHF modulator (Fig. 23.14). Both the video decoder and the PAL encoder are clocked by the 27 MHz programme reference clock to synchronise the bit acquisition. The video decoder itself operates on its own clock of around 5.5 MHz.

The decoder is programmed and controlled by the microprocessor via an 8-bit data bus and a 6-bit address bus together with a number of control signals, including R/W and CS. One IRQ is provided for the video decoder to request processing routines such as the start of a field. The PAL encoder is controlled by the microcontroller via the I$^2$C bus, which sets the operational parameters of the encoder.

**Fig. 23.9** *A PACE video decoder*

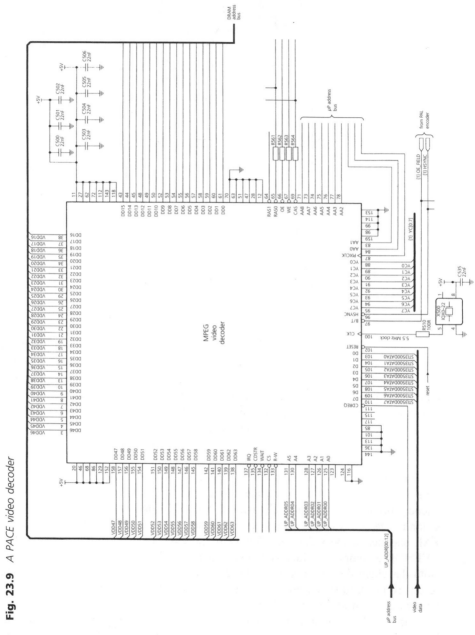

### A practical video decoder chip

A typical video decoder chip is shown in Fig. 23.9. The 8-bit wide video data (ST13500DATA) from the transport demultiplexer is received on pins 103–110. The microprocessor accesses the decoder via the same video data bus (ST13500DATA) and a 6-bit address bus (UP_ADDR) on pins 125–128 and 130, 131. It allows the microprocessor to select one of $2^6 = 64$ registers inside the video decoder. Data inside these registers determines the parameters (type of compression, error correction, etc.) used in the decoding process. The 16 Mbit video DRAM is accessed via a 64-bit data bus (DD0–DD63) and an 8-bit address bus (AA0–AA7). The output to the PAL encoder is carried along an 8-bit bus (YC0–YC7) on pins 88–95, which carries multiplexed pixel data (Y, $C_R$ and $C_B$). Synchronisation between the video decoder and the PAL encoder is obtained by a common pixel 27 MHz clock (PIXCLK on pin 87), horizontal sync. (HSYNC pin 96) and vertical sync. called bottom/top (B/T) on pin 97. The decoder operates on a 5.5 MHz clock generated by a crystal oscillator and fed into pin 100.

## The MPEG audio decoder

The audio decoder (Fig. 23.10) is a digital signal processing chip which receives serial coded audio data from the transport demultiplexer and carries out the necessary decoding to produce two serial pulse code modulated (PCM) left and right audio channels. The audio decoder can give an audio output at three different sampling rates: 32, 44.1 and 48 kHz. Information on the actual sampling rate used by the transmitter is provided

**Fig. 23.10** *An MPEG-2 audio decoder*

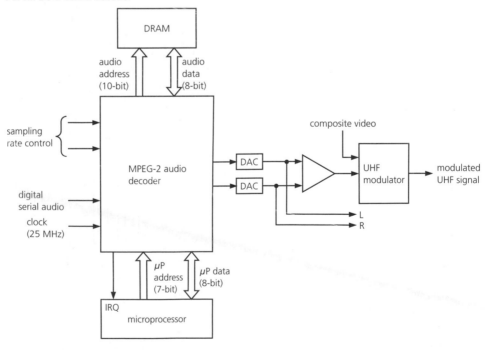

**Fig. 23.11** A PACE video decoder

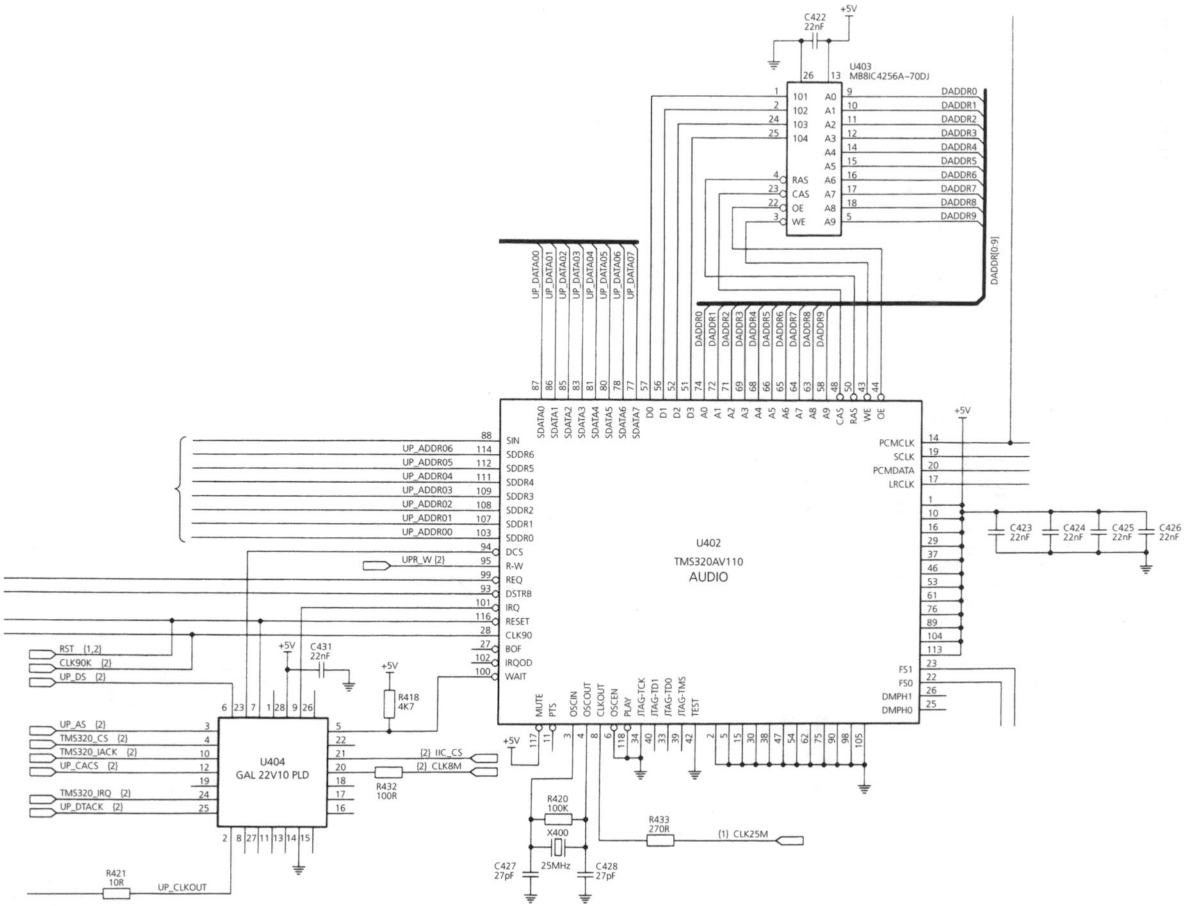

by the transport demultiplexer, which extracts the information from the incoming transport stream. Audio buffering is provided by the DRAM store, which also provides the necessary 1 s delay. The decoder is fully programmed and controlled by the microprocessor with a 7-bit address bus, an 8-bit data bus and control lines R/W and CS. An interrupt request is provided so the decoder can inform the microprocessor when certain events have occurred and make requests for services.

## A practical audio decoder chip

A typical MPEG audio decoder chip is shown in Fig. 23.11. When the audio decoder is ready to receive data, it sends a request (REQ) signal to the demultiplexer on pin 99. Digitally modulated serial data (SIN) is then sent to the decoder on pin 88 together with a data strobe (DSTRB) on pin 93. The sampling rate is indicated by the logic states of signals FS0 and FS1 at pins 22 and 23 (Table 23.1).

The audio output to the audio digital-to-analogue converters is PCMDATA on pin 20. The sampling clock labelled SCK (pin 19) for the DACs is produced by dividing the PCM clock (PCMCLK pin 14). The signal on pin 17 (LRCLK) indicates whether the data is left or right channel (Oscillogram, Fig. 23.12). The interface to the microprocessor

### TABLE 23.1

| FS0 | FS1 | Sampling rate (kHz) |
|-----|-----|---------------------|
| 0   | 0   | 44.1                |
| 0   | 1   | 32                  |
| 1   | 0   | 48                  |

**Fig. 23.12**  *Oscillogram*

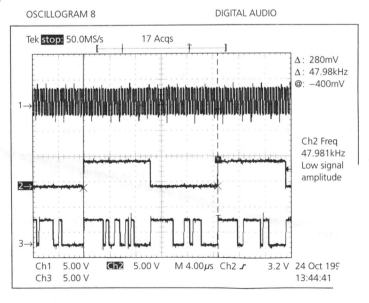

comprises an 8-bit data bus and a 7-bit address bus, which allows the microprocessor access to the decoder's registers. Read/write (R/W pin 95) is connected directly to the processor. Other control signals pass through a programmable logic device (PLD) which translates them into a format the decoder can understand. Apart from the 25 MHz sync. lock (pin 8), the decoder also receives a 90 kHz clock from the transport demultiplexer on pin 28. This is used with a counter within the audio decoder to prevent false locking of the onboard generated clock to the transmitted clock. The memory chip provides the required 1 s delay to ensure audio and video synchronisation. It uses a DRAM chip with 70 ns access time and control lines RAS, CAS, OE and WE.

## A practical PAL encoder chip

A typical PAL encoder is shown in Fig. 23.13. It receives the multiplexed luminance (Y) and chrominance ($C_R$ and $C_B$) signals on an 8-bit bus YC0–YC7 (pins 36–43) and converts them into a standard PAL composite video CVBS (pin 2). The encoder also provides RGB and SVHS outputs. The PAL encoder requires two clocks, 25 MHz (pin 48) and 13.5 MHz (pin 49), both obtained from the system clock generator. Control is carried out by the microcontroller via the $I^2C$ bus (pins 45 and 46). When the receiver is in standby mode, the microcontroller turns off transistor Q500 and puts the decoder in sleep mode (pin 44); this disables the encoder, resulting in a black screen. The composite video passes through a video reconstruction circuit comprising L503, L504 and associated components. The circuit which removes digital noise has traps at 25 MHz and 13.5 MHz.

## UHF modulator

The circuit consists of two sections: a synthesised UHF modulator and a loop-through amplifier. The modulator (see Figure 23.14) comprises of:

- an $I^2C$ controlled phase locked loop frequency synthesiser
- an amplitude modulator
- an audio oscillator for the sound sub-carrier
- a video clamp to ensure correct modulation index.

Tuning of the modulator is carried out by a DC voltage (0–24 V) derived from the PLL frequency synthesiser. The audio signal is used to frequency modulate the sound sub-carrier. The modulated sound carrier is then added to the clamped video signal and used to amplitude modulate a tuned UHF carrier. The modulated UHF is then fed into the loop-through amplifier which mixes the signal with that from a second antenna.

When the digital decoder is on standby, the UHF synthesised oscillator is switched off by a command from the microcontroller via the $I^2C$ bus. The loop-through amplifier, however, operates normally so that the r.f. signal from a the second antenna may loop through to the r.f. output socket.

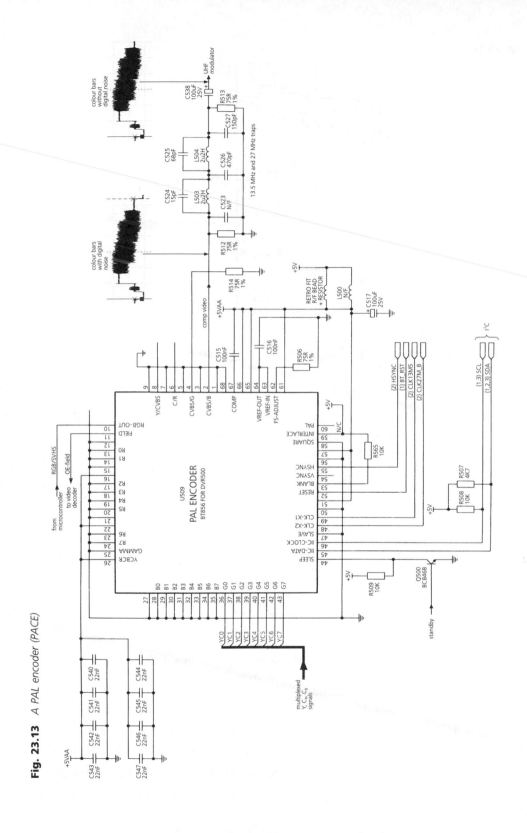

**Fig. 23.13** A PAL encoder (PACE)

**Fig. 23.14**

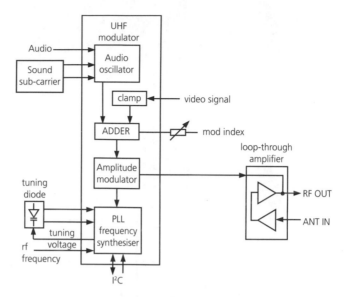

## A satellite digital TV receiver

Figure 23.15 shows a complete block diagram of a PACE satellite digital TV decoder. The receiver provides an RS-232 serial port for direct connection with a personal computer and a high-speed data port for connection to external devices. The smart card is fed into a standard PCMCIA connector for authentication. Two types of clocks are used in the receiver. Device-own clocks are 16 MHz for the microprocessor, 55 MHz for the MPEG video decoder and transport demultiplexer, 25 MHz for the MPEG audio decoder and 4.9 MHz for the microcontroller. Data synchronisation clocks are the 27 MHz clock for the bit rate and the various video and audio sampling rates.

### QUESTIONS

1. Give a brief explanation of the following:
   (a) programme identifier (PID)
   (b) integrated receiver decoder (IRD)
   (c) forward error correction (FEC)

2. Name the three main components of a channel decoder.

3. State the function of the following blocks in a digital TV receiver:
   (a) the transport demultiplexer
   (b) the video decoder
   (c) the PAL decoder

4. State the reason for including an RS-232 port in a digital IRD.

5. What is the purpose of the conditional access (CA) unit in a digital TV receiver?

6. What is the purpose of energy dispersal in digital TV broadcasting?

**Fig. 23.15** *Full IRD block diagram*

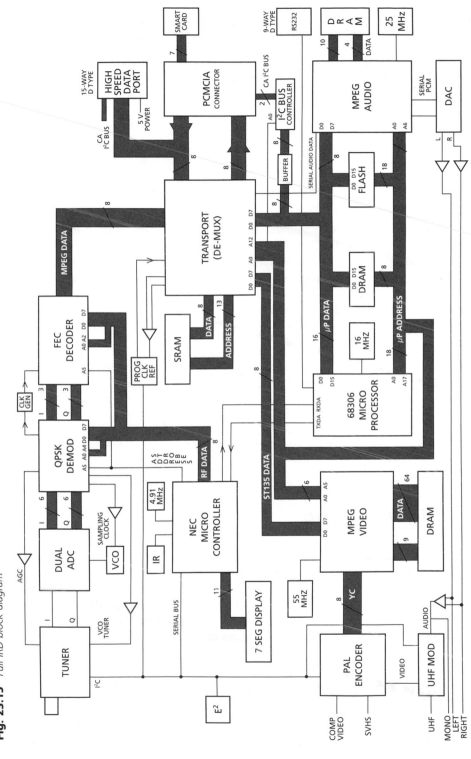

# 24 Testing a digital TV receiver/decoder

For the purposes of servicing, a digital TV receiver/decoder (IRD) may divided into two main parts (Fig. 24.1): an *analogue part* comprising the tuner at the front end and the PAL decoder and UHF modulator at the back end; and a *digital part* comprising digital decoding and processing, software programming and control, and other devices such as analogue-to-digital and digital-to-analogue converters (ADCs and DACs).

**Fig. 24.1**  *A digital TV receiver/decoder has an analogue part and a digital part*

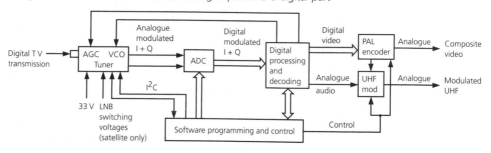

## Signal tracing

Tracing the path of the signal can provide a good indication of the faulty stage. It involves testing both analogue signals and digital data streams at regular points along the signal path. Absence of a signal such as I, Q or a data stream will produce a *total failure* of the picture, the sound or both. *Partial failure*, such as a single data or address line fault, may cause a variety of impairments to the picture, the sound or both, namely a total absence of picture or sound, intermittent picture or sound, or broken-up picture or sound.

## Use of the logic probe

The flow of a digital bit stream may be traced using a logic probe (minimum frequency 100 MHz) (Fig. 24.2). The logic probe is a logic state test instrument which indicates

**Fig. 24.2** *Logic probe*

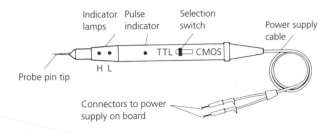

the logic state of a test node. It can indicate a logic 1, a logic 0, an open circuit and a data stream pulse. Two LEDs are used to indicate a high and a low. An open circuit or indeterminate logic state is indicated by no light. The presence of digital activity in the form of a bit stream is indicated by a flickering light or a special indicator. By the use of a pulse stretcher, pulses as narrow as 10 ns may be detected. Although the actual waveform cannot be examined by the logic probe, it nonetheless provides a fast and simple method of testing for digital activity at various points along the signal path. Where real-time waveforms have to be examined, an oscilloscope must be used.

# Use of the cathode-ray oscilloscope

The oscilloscope is used to display both analogue and digital waveforms from which amplitude, frequency and time measurements can then be made. Analogue signals in a digital TV decoder box are somewhat similar to those found on an analogue TV receiver and may be displayed using a normal analogue oscilloscope. However, examining a data bit stream requires a digital storage oscilloscope with a minimum bandwidth of 100 MHz and a sampling rate of at least 500 million samples per second. Such an oscilloscope may be also used to display and examine analogue signals.

The storage oscilloscope captures a part of the data bit stream, stores it and then displays the waveform on the screen for examination and measurements. This is then repeated for another part of the data stream, and so on. Unlike analogue systems – where a test point has its own unique signal in terms of waveshape, frequency and amplitude – a sequence of 1s and 0s in a data stream have the same general waveshape and amplitude regardless of the test point. A typical digital data oscillogram is shown in Fig. 24.3.

**Fig. 24.3** *Typical digital data waveform*

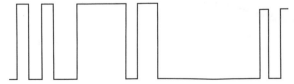

**Fig. 24.4** *Multi-trace display*

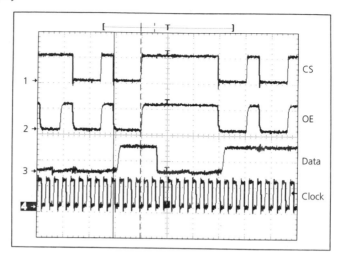

## Timing and control

In microprocessor-based systems, failure may occur due to a wrong or missing clock, or incorrect timing of data and control signals. A fault in timing or synchronisation can result in partial or total failure of the system. The timing and synchronisation of the various data and control bit streams may be examined using a multi-trace oscilloscope which displays two or more signals simultaneously. Figure 24.4 shows a typical multi-trace display and it illustrates the time relationship between chip select (CS), read or output enable (OE), a data line and the clock of a memory chip.

## Testing the tuner

The function of the tuner is to downconvert, amplify, filter and demodulate the transmitted signal to reproduce the digitally modulated carrier (Fig. 24.5). The local oscillator

**Fig. 24.5** *The tuner produces the digitally modulated carrier*

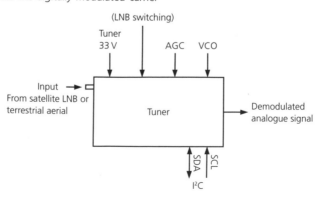

**Fig. 24.6** *Tuner search waveform*

is controlled by the voltage-controlled oscillator (VCO) d.c. voltage from the digital demodulator. When the decoder is turned on, it attempts to lock to the default frequently known as the *home channel*. If it fails to do so, a search sawtooth waveform is fed into the VCO input pin (Fig. 24.6). At the same time, the a.g.c. input is set to provide maximum tuner gain. LNB supply switching voltages may be tested using a DVM.

## Testing the channel decoder

The modulated carrier entering the channel decoder may take the form of two carriers in quadrature: I and Q in the case of satellite transmission or COFDM carriers in the case of terrestrial transmission. The function of the channel decoder is first to digitise the incoming carriers into a multi-bit stream, demodulate it and reproduce error-free MPEG packets. Testing a channel decoder (Fig. 24.7) involves checking the output MPEG packets by examining the MPEG data lines, line by line, using a logic probe or an oscilloscope. The sampling clock is normally set to twice the symbol (modulation) rate. With a symbol rate of 27 500K symbols per second, the sampling clock is then set to $2 \times 27\,500\text{K} = 55\,\text{MHz}$. This clock pulse should be present even when no signal is being received by the IRD. With no MPEG packets present, the digitised I and Q data streams should be checked, and this should be followed by an examination of the reference and sampling clocks, the control signals and the data and address microprocessor communication buses. Some of the control signals and their properties are as follows:

**Fig. 24.7** *The channel decoder reproduces error-free MPEG packets*

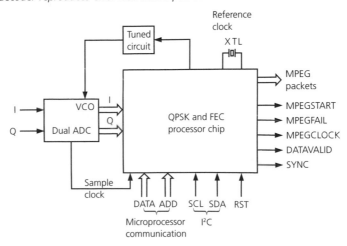

- MPEGSTART: high for the first byte of each packet
- MPEGFAIL: low if the packet contains errors
- DATAVLID: high for 188 bytes of the packet and low for the following 16 bytes
- SYNC: high if synchronisation is correct
- RST: the reset is permanently high if active low, and vice versa
- SCL (clock line of the I²C control bus): a clock pulse
- SDA (data line of the I²C control bus): data stream pulse

# Testing the transport demux and decoders

The general procedure for testing a processing/decoding chip is to check four items:

- The input and output data streams
- The processing and other clocks
- Microprocessor control and communications signals
- Memory data/address bus and control signals

The input to the transport demultiplexer (Fig. 24.8) consists of MPEG data packets together with the control signals MPEGSTART, MPEGFAIL and MPEG clock from the

**Fig. 24.8** *The transport demultiplexer descrambles the MPEG packets*

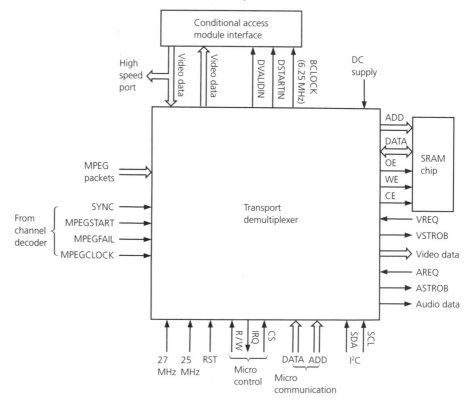

channel decoder. The function of the demultiplexer is to descramble the MPEG packets if appropriate, and reassemble the video and audio packets for the selected programme in the correct order.

If any of the video or audio output input data are missing, the control signals from the channel decoder should be checked followed by the 25 MHz clock for the demultiplexer and the 27 MHz clock for the pixel system. A faulty clock pulse should be traced back through the clock-generating circuitry to ascertain the faulty component.

The conditional access module (CAM) should be tested next. This involves testing the input and output data streams to and from the CAM interface, first with an encrypted programme and then with a 'free' or unencrypted programme. CAM control signals should also be tested using a logic probe or an oscilloscope:

- BCLOCK: the byte clock provides the delay for unscrambling the data
- DSTARTIN: goes high for the first byte of data packet
- DVALIDIN: goes high if the data is free from errors

Demux control signals should be tested for digital activity:

- Chip select (CS)
- Read/write (RW)
- Interrupt request (IRQ)
- Reset (RST)
- SCL (clock line of $I^2C$ bus)
- SDA (data line of $I^2C$ bus)

Absence of the video output may be a result of a faulty video strobe (VSTROB) or a faulty video request (VREQ) signal from the video decoder. Data transfer between the demultiplexer and the video decoder only takes place when the decoder asserts VREQ to indicate its readiness to receive compressed data from the demultiplexer. Compressed data is then strobed through by VSTROB. The same goes for the audio signal, where audio request AREQ is asserted by the audio decoder and the data is strobed by audio strobe ASTROB.

An SRAM memory chip may be tested by checking for digital activity on its data and address lines as well as on the read or output enable (OE) and write enable (WE) control lines. Chip enable (CE) is active low, so it should be low permanently. A faulty SRAM memory chip will cause total video and audio failure.

# The video decoder and the PAL encoder

The function of the video decoder is to convert the MPEG video packets from the demultiplexer into multiplexed pixel data: Y, $C_R$ and $C_B$ suitable for the PAL encoder. The other function of the video decoder is the processing and displaying of on-screen messages including the menu driven activities. These are generated by a software routine which resides in the Flash memory.

**Fig. 24.9** *The video decoder converts MPEG video data into multiplexed pixel data*

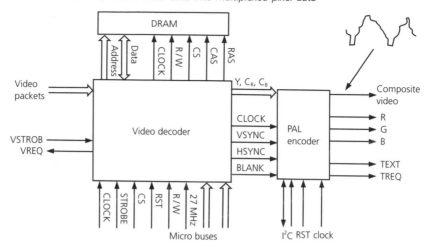

## The 'menu test'

The most effective way of testing the video decoder is to enter the set-up menu through the remote control handset (or directly using front panel keys where available). A positive result indicates good video decoder and video memory chips. The input and output data streams may be tested using a logic probe to check for digital activity, or an oscilloscope to display real-time waveforms (Fig. 24.9). The next items to check are the 27 MHz clock of the pixel system and the decoder processing clock, followed by VREQ and VSTROB, the two control signals that regulate the transfer of data between the demultiplexer and the decoder.

DRAM memory may tested by checking the chip's data, address and control lines and microprocessor communication buses and control signal can be tested in the normal way. A faulty DRAM chip will cause total video collapse. Partial memory failure, such as a single data or address line failure (open circuit, permanently high or permanently low), could result in no picture, an intermittent picture or a broken-up picture.

Testing the PAL encoder involves checking the input and output signals. A typical list of these signals and their types is as follows.

| Signal | Type |
| --- | --- |
| Composite video signal | analogue |
| R, G and B colour signals | analogue |
| TEXT data line | digital |
| Text request (TREQ) which allows text through | digital |
| Video clock from the decoder | analogue/digital |
| 27 MHz pixel system clock | analogue/digital |
| Vertical sync (VSYNC) | analogue |
| Horizontal sync (HSYNC) | analogue |
| Blanking pulse (BLANK) | analogue |

**Fig. 24.10** *The audio decoder takes PCM serial input and produces left and right audio output*

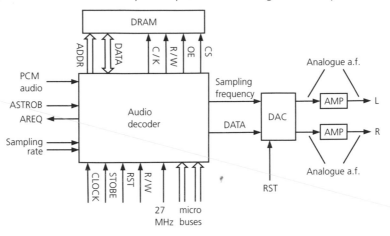

## Audio decoding

The function of audio decoding is to process the pulse code modulated (PCM) serial input into separate left and right stereo audio output (Fig. 24.10). Two control signals – audio request (AREQ) and audio strobe (ASTROB) – regulate the transfer of compressed audio data from the demultiplexer to the decoder. The input to the ADC consists of two parts: a sampling clock and a data line. Both may be checked by a logic probe for digital activity, or an oscilloscope for real-time waveforms. The output of the ADC may be checked with a simple analogue oscilloscope. DRAM memory chips may be tested in the normal way. Faulty sound memory chips normally cause total sound collapse with partial faults causing loss of sync., intermittent sound or broken-up sound.

## The IRD start up sequence

Normally, an IRD is never switched off. When not in use, it remains in the standby mode. Its microprocessor, microcontroller and all other processing chips remain set and ready to receive and process data.

However, when an IRD is switched on from cold, it goes through a comparatively lengthy process of setting, initialising, configuring and programming the processor and decoder chips. This requires the downloading of software routines from flash memory to the microprocessor DRAM memory and other chips. This process is known as the start-up or boot-up sequence.

The main components of the start-up sequence are as follows:

- The power supply voltages builds up to the required levels.
- The microprocessor and the microcontroller are set by taking their respective RST control lines to 5 V.
- The other processing chips are set by taking their RST control lines to 5 V.

- Software is downloaded in two stages: in the microprocessor and the microcontroller, digital activity must be present on the micro bus and control lines only; then the channel decoder and the transport demultiplexer are initialised and programmed, along with the video and audio decoders.
- The sampling frequency, which is normally 2 × the modulating symbol rate, is set by the demodulator.
- The channel decoder begins to search for the default channel, known as the home channel. If the signal is detected, the channel decoder locks to it and data is received, decoded and processed. Picture and sound are produced.
- If the home channel could not be detected, the channel decoder searches for other default channels. Failure to lock to any incoming signal is indicated by a 'no signal' message on the screen.

The start-up sequence takes up to two minutes to complete; it is transparent to the user with coded symbols on the receiver's LED display and messages on the screen.

## Programming and control

A malfunction in the programming and control section, i.e. the microprocessor, micro-controller, and associated memory chips and components, will normally cause the start-up routine to be halted resulting in a fatal video and audio fault.

A quick and a reliable test of the programming and control section is the 'menu test'. The display of a menu on the screen is a positive indication that this section is functioning normally.

Testing the chips for malfunction involves testing for digital activities on the address, data and control lines during, and after, the start-up sequence. No activity at all suggests a d.c. supply failure or reset malfunction.

Apart from a malfunctioning microprocessor/microcontroller system, certain faults in the demultiplexer and decoder chips can also cause the start-up sequence to be halted.

## Examining a suspect chip

The absence of an output from a processing chip does not necessarily mean the chip is faulty. In a microprocessor-controlled system, IC output failure may be caused by one of several malfunctions, including a clock pulse, a control signal, a software routine, data or address bus lines, a memory chip, a d.c. supply line as well as a faulty IC itself. In general, before a suspect IC is replaced, the following items should be checked:

- It is receiving its d.c. supply voltage at the appropriate pins; if not, then the d.c. line should be traced back to the power supply to ascertain the fault.
- It is operating at the correct frequency; this may be checked by a suitable oscilloscope. Where more than one clock pulse is fed into the chip, they must all be checked in the same way.
- It is receiving the correct instructions from the microprocessor, the microcomputer or both. This involves checking the control signals and the data and address lines.
- The necessary software is valid and up to date.

The d.c. check may be carried out using a digital voltmeter (DVM) to measure the d.c. voltage at the appropriate pins of the chip. The clock signal may be checked using an oscilloscope with adequate frequency range. The clock pulse must have

- The correct frequency
- A fast-rising square shape
- An amplitude of 3.5–5.25 V, normally around 4.8 V

The frequency may be calculated from the periodic time of the waveform.

Checking hardware control lines involves using a logic probe to check for digital activity or a storage oscilloscope to examine the actual waveforms of the various control signals such as chip select (CS), acknowledge (Ack), read, write, strobe and interrupt. The reset (RST) may be tested by a logic probe or a DVM. It is normally active low, ($\overline{RST}$ or _RST) i.e. it resets the chip when it is taken low and restarts when taken high. If it is active low, then the RST line should be permanently high. The signals at the data and address lines may be examined using a logic probe or an oscilloscope to check for digital activity.

Faulty software – out-of-date, deleted or corrupted programs – may also cause an interruption of the bit stream. Software processing routines are mainly stored in flash memory chips. Rewriting or upgrading the software can be carried out by using a personal computer that communicates with the decoder via a serial port using an RS232 cable. The PC is then used to program the flash memory with up-to-date software routines. Alternatively, the flash memory chip may be replaced with a fully programmed memory.

## Anti-static precautions

When servicing digital TV decoders, care must be taken to avoid damaging the integrated circuits in the form of electrical damage caused by static discharge. This may be avoided by implementing anti-static precautions.

The movement and contact of the human body can accumulate energy in the form of electro-static charge. When an electric contact with another object, say an integrated circuit, is established, electro-static discharge, ESD takes place in the form of a very brief flow of current. Although the discharge current is very small, the voltage could be in the region of few thousand volts. Such an *electro-static discharge* can cause damage to MOS based integrated circuits such as CMOS and NMOS processing, decoding and memory chips. The damage caused by ESD may be instant failure. However, it is more likely to weaken the chip thus shortening its life span.

Anti-static precautions involve taking the following steps:

- the use of grounded wrist strap designed for static discharge or alternatively, touching a grounded metal when handling components;
- handling the printed circuit boards by their edges only;
- when handling components, touching the pins must be avoided;
- the use of anti-static work surface or mat which ensure that all parts of its surface are kept at the same potential;

- all boards as well as components must be kept in anti-static bags when not in use;
- touching a grounded metal object when removing a board or a component from their anti-static bag;

# Testing logic devices

A digital TV receiver includes a large variety of logic devices such as gates, flip-flops, counters and inverters. Such devices may be tested for *stuck-at faults* (*stuck-at-one* when a pin is shorted to the supply line and *stuck-at-zero* when a pin is shorted to the zero-volt line) and open-circuit (o/c) faults. These faults are normally caused by a failure within the IC itself. Stuck-at and o/c faults may be detected using a logic pulser in conjunction with a logic probe.

The logic pulser changes the logic state of an IC pin or test point and then changes it back again, i.e. it drives a low node high and then low, and a high node low and then high. Testing a logic device, such as a gate, takes the form of stimulating the inputs using the pulser and observing the effect on the output using the logic probe. For example, consider the simple steering network used to set the read (RD) and write (WR) control signals for a modem. A quad AND gate 7400 package is used as shown Fig. 24.11. The AND gate U511A may tested by placing the pulser at pin 2 and monitoring the output at pin 3 with a logic probe. Since the second input to the gate (pin 1) is at +5 V, changing the state of pin 2 will produce a change in the output at pin 3, indicated by a flicker on the probe. No indication on the probe suggests a faulty pin 2 or pin 3. The process can be repeated for the other gates until the faulty pin is identified. The stuck-at fault may be confirmed by placing both the pulser and probe on the suspect pin. The fault is confirmed if the operation of the pulser does not produce a flicker on the probe.

**Fig. 24.11** *A quad AND gate may be tested with a logic pulser (PACE 500)*

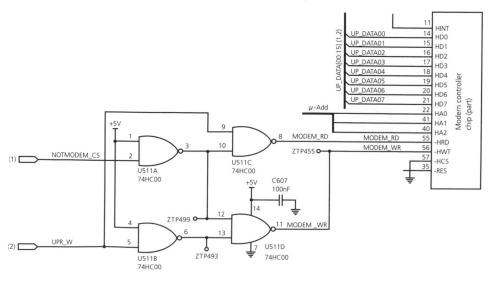

Once an IC pin is identified as stuck-at-zero or stuck-at-one, the source of the short circuit must be established; the short circuit is to earth in the first case and to the d.c. supply voltage in the second case. If in Fig. 24.11, pin 8 is found to be stuck-at-zero, then either pin 8 of the 7400 chip is shorted to earth or pin 55 (MODEM_RD) of the modem controller is shorted to earth. Identifying the cause of the fault requires a very sensitive ohmmeter or a *current tracer* in conjunction with a pulser. The current tracer senses the magnetic field created by the flow of fast-rising pulses and indicates their presence by a light or constant-tone sound. Before a current tracer is used, power must be switched off. The pulser is then placed at one of the suspect pins, say pin 8 of the AND gate in Fig. 24.11. If the actual short circuit is at pin 55 of the modem controller chip, pulses from the pulser will flow from pin 8 to pin 55. The tracer can then be used to detect the presence of the current at pin 55. However, if the short circuit is at pin 8, the tracer will not detect the presence of a current at pin 55.

# Appendix 1
# List of functional symbols

| | | | | | |
|---|---|---|---|---|---|
| | Changer, general | | Band-stop filter | | Amplifier, general |
| | Interference separator | | Band-pass filter | | Stand-by |
| | Synchronisation separator | | Pulse-width modulator | | On/off |
| | Divider | | 90° phase shifter | | Output stage |
| | Rectifier | | Electronic switch | | Controlled amplifier |
| | Automatic Gain Control | | Variable impedance | | Differential amplifier |
| | Flip-flop on half line frequency | | Display | | Amplifier with limiter |
| | Square wave generator | | Delay element | | Positive peak clipper |
| | Sawtooth generator | | Detector | | Black level restorer |
| | Sinewave generator | | Phase detector | | Coaxial aerial input |
| | Adjustable sinewave generator | | Voltage stabilizer | | RC network (integrator) |
| | Rejection filter | | FM detector | | Decoding matrix |
| | Low pass filter | | Phase discriminator | | Infra red transmitter |
| | High-pass filter | | Colour killer | | Infra red receiver |
| | Sound mute | | Search control | | Multi-function switch |
| | VCR switch | | Band selection | | Modulator |
| | Mixer stage | | Constant level | | Mono I or II sound |
| | Emitter follower | | Variable level | | Stereo sound |
| | Tuning control | | Input-control | | Spatial stereo |
| | A.F.C. function | | De-emphasis | | Schmitt trigger |
| | A.F.C. control | | Shaper | | Volume control |
| | General operating command | | AND gate | | Balance control |
| | Search function | | OR gate | | Bass and treble control |

# Appendix 2
# Terrestrial analogue TV: main transmitting stations

| | BBC1 | BBC2 | ITV | CH4 | CH5 | Power (kW) |
|---|---|---|---|---|---|---|
| BORDERS AND NORTHERN ENGLAND | | | | | | |
| Belmont | 22 | 28 | 25 | 32 | – | 500 |
| Belmont | – | – | – | – | 56 | 50 |
| Bilsdale (W. Moor) | 33 | 26 | 29 | 23 | – | 500 |
| Bilsdale (W. Moor) | – | – | – | – | 35 | 250 |
| Burnhope | – | – | – | – | 68 | 50 |
| Caldbeck | 30 | 34 | 28 | 32 | – | 500 |
| Caldbeck | – | – | – | – | 56 | 10 |
| Chatton | 39 | 45 | 49 | 42 | – | 100 |
| Emley Moor | 44 | 51 | 47 | 41 | – | 870 |
| Emley Moor | – | – | – | – | 37 | 870 |
| Pontop Pike | 58 | 64 | 61 | 64 | – | 500 |
| Sandale | 22 | – | – | – | – | 500 |
| Sandale | – | 67 | – | – | – | 20 |
| Selkirk | 55 | 62 | 59 | 65 | – | 50 |
| Selkirk | – | – | – | – | 52 | 50 |
| Winter Hill | 55 | 62 | 59 | 65 | – | 500 |
| Winter Hill | – | – | – | – | 48 | 12.5 |
| MIDLANDS AND EASTERN ENGLAND | | | | | | |
| Lichfield | – | – | – | – | 37 | 1000 |
| Nottingham | – | – | – | – | 34 | 2 |
| Ridge Hill | 22 | 28 | 25 | 32 | – | 100 |
| Ridge Hill | – | – | – | – | 35 | 100 |
| Sandy Heath | 31 | 27 | 24 | 21 | – | 1000 |
| Sandy Heath | – | – | – | – | 39 | 10 |
| Sutton Coldfield | 46 | 40 | 43 | 50 | – | 1000 |
| Tacolneston | 62 | 55 | 59 | 65 | – | 250 |
| Tacolneston | – | – | – | – | 52 | 4 |
| Waltham | 58 | 64 | 61 | 54 | – | 250 |
| Waltham | – | – | – | – | 35 | 250 |

|  | BBC1 | BBC2 | ITV | CH4 | CH5 | Power (kW) |
|---|---|---|---|---|---|---|
| | | | SOUTHERN ENGLAND | | | |
| Alexandra Palace | 58 | 64 | 61 | 54 | – | 70 |
| Beacon Hill | 57 | 63 | 60 | 53 | – | 100 |
| Bluebell Hill | 40 | 46 | 43 | 65 | – | 30 |
| Caradon Hill | 22 | 28 | 25 | 32 | – | 500 |
| Croydon | – | – | – | – | 37 | 250 |
| Crystal Palace | 26 | 33 | 23 | 30 | – | 1000 |
| Dover | 50 | 56 | 66 | 53 | – | 100 |
| Fawley | – | – | – | – | 34 | 1 |
| Fremont Point | 51 | 44 | 41 | 47 | – | 20 |
| Hannington | 39 | 45 | 42 | 66 | – | 250 |
| Hannington | – | – | – | – | 35 | 60 |
| Heathfield | 49 | 52 | 64 | 67 | – | 100 |
| Huntshaw Cross | 55 | 62 | 59 | 65 | – | 100 |
| Huntshaw Cross | – | – | – | – | 67 | 2 |
| Mendip | 58 | 64 | 61 | 54 | – | 500 |
| Mendip | – | – | – | – | 37 | 126 |
| Midhurst | 61 | 55 | 58 | 68 | – | 100 |
| Plympton | – | – | – | – | 30 | 2 |
| Redruth | 51 | 44 | 41 | 47 | – | 100 |
| Redruth | – | – | – | – | 37 | 3 |
| Rowridge | 31 | 24 | 27 | 21 | – | 500 |
| | | | SCOTLAND | | | |
| Angus | 57 | 63 | 60 | 53 | – | 100 |
| Black Hill | 40 | 46 | 43 | 50 | – | 500 |
| Black Hill | – | – | – | – | 37 | 500 |
| Bressay | 22 | 28 | 25 | 32 | – | 10 |
| Craigkelly | 31 | 27 | 24 | 21 | – | 100 |
| Craigkelly | – | – | – | – | 48 | 4 |
| Darvel | 33 | 26 | 23 | 29 | – | 100 |
| Darvel | – | – | – | – | 35 | 100 |
| Durris | 22 | 28 | 25 | 32 | – | 500 |
| Durris | – | – | – | – | 67 | 100 |
| Eitshal (Lewis) | 33 | 26 | 23 | 29 | – | 100 |
| Keelylang Hill | 40 | 46 | 43 | 50 | – | 100 |
| Knock More | 33 | 26 | 23 | 29 | – | 100 |
| Mounteagle | – | – | – | – | 67 | 100 |
| Perth | – | – | – | – | 55 | 2 |
| Rosemarkie | 39 | 45 | 49 | 42 | – | 100 |
| Rumster Forest | 31 | 27 | 24 | 21 | – | 100 |
| Tay Bridge | – | – | – | – | 34 | 4 |

|                  | BBC1 | BBC2 | ITV | CH4 | CH5 | Power (kW) |
|------------------|------|------|-----|-----|-----|------------|
| **NORTHERN IRELAND** | | | | | | |
| Black Mountain   | –    | –    | –   | –   | 37  | 50  |
| Brougher Mountain| 22   | 28   | 25  | 32  | –   | 100 |
| Divis            | 31   | 27   | 24  | 21  | –   | 500 |
| Limavady         | 55   | 62   | 59  | 65  | –   | 100 |
| Londonderry      | –    | –    | –   | –   | 31  | 10  |
| **WALES**        | | | | | | |
| Blaen Plwyf      | 31   | 27   | 24  | 21  | –   | 100 |
| Blaen Plwyf      | –    | –    | –   | –   | 56  | 4   |
| Carmel           | 57   | 63   | 60  | 53  | –   | 100 |
| Kilvey Hill      | –    | –    | –   | –   | 35  | 10  |
| Llanddona        | 57   | 63   | 60  | 53  | –   | 100 |
| Moel–y–Parc      | 52   | 45   | 49  | 42  | –   | 100 |
| Presely          | 46   | 40   | 43  | 50  | –   | 100 |
| Presely          | –    | –    | –   | –   | 37  | 100 |
| Wenvoe           | 44   | 51   | 41  | 47  | –   | 500 |

# Appendix 3
# Terrestrial digital TV: main transmitting stations

The UK has a total of six multiplexes, each offering a data capacity of 24.13 Mbit/s. Each multiplex contains between four and eight programmes. Five of the total number of programmes are digital versions of existing analogue programmes, leaving a capacity of about 30 new programmes. Two multiplexes are allocated for BBC and ITV. The remaining four (Mux A to Mux D) are for new broadcasters. The following table lists the main transmitting stations for terrestrial digital TV.

| | D1 BBC | | D2 ITV + 4 | | D3 Mux A | | D4 Mux B | | D5 Mux C | | D6 Mux D | |
| | Ch | Pwr (kW) | Ch | Pwr (kW) | Ch | Pwr (kW) | Ch | Pwr (kW) | Ch | Pwr (kW) | Ch | Pwr (kW) |
|---|---|---|---|---|---|---|---|---|---|---|---|---|
| **ENGLAND** | | | | | | | | | | | | |
| Beacon Hill | 52 | 1.000 | 61 | 1.000 | 58 | 1.000 | 54 | 1.000 | 56 | 1.000 | 64 | 1.000 |
| Belmont | 30 | 5.000 | 48 | 10.000 | 68 | 10.000 | 66 | 10.000 | 60 | 4.000 | 57 | 4.000 |
| Bilsdale | 34 | 2.400 | 21 | 3.000 | 31 | 3.000 | 24 | 3.000 | 27 | 3.000 | 42 | 0.400 |
| Bluebell Hill | 59 | 3.000 | 24 | 2.000 | 27 | 2.000 | 45 | 3.000 | 42 | 3.000 | 39 | 3.000 |
| Caldbeck | 25 | 5.000 | 23 | 7.500 | 26 | 7.500 | 39 | 1.600 | 45 | 1.600 | 42 | 1.600 |
| Caradon Hill | 34 | 4.000 | 31 | 2.000 | 48 | 2.000 | 21 | 2.000 | 24 | 2.000 | 27 | 2.000 |
| Chatton | 40 | 3.000 | 50 | 3.000 | 43 | 3.000 | 46 | 1.000 | 47 | 1.000 | 51 | 1.000 |
| Crystal Palace | 25 | 6.500 | 22 | 6.500 | 32 | 6.500 | 28 | 6.500 | 34 | 1.000 | 29 | 1.000 |
| Dover | 61 | 1.000 | 68 | 1.000 | 55 | 1.000 | 58 | 1.000 | 57 | 1.000 | 60 | 0.500 |
| Emley Moor | 52 | 5.000 | 40 | 5.000 | 43 | 5.000 | 46 | 5.000 | 50 | 5.000 | 49 | 2.000 |
| Hannington | 50 | 10.000 | 43 | 5.000 | 40 | 5.000 | 46 | 5.000 | 29 | 1.300 | 48 | 0.850 |
| Heathfield | 34 | 1.600 | 29 | 1.600 | 48 | 2.500 | 47 | 1.000 | 54 | 1.000 | 51 | 1.000 |
| Huntshaw Cross | 54 | 2.000 | 58 | 2.000 | 61 | 2.000 | 64 | 2.000 | 53 | 2.000 | 57 | 2.000 |
| Huntshaw Cross Fille | | | | | | | | | 51 | 0.040 | 47 | 0.040 |
| Mendip | 59 | 3.000 | 55 | 3.000 | 62 | 3.000 | 65 | 3.000 | 52 | 0.250 | 48 | 0.250 |
| Midhurst | 56 | 10.000 | 65 | 10.000 | 62 | 2.500 | 59 | 2.500 | 64 | 1.000 | 60 | 1.000 |
| Oxford | 34 | 10.000 | 68 | 2.800 | 56 | 1.250 | 52 | 3.000 | 48 | 1.000 | 67 | 1.000 |
| Plympton | 52 | 0.100 | 67 | 0.100 | 66 | 0.100 | 60 | 0.100 | 63 | 0.100 | 56 | 0.100 |
| Pontop Pike | 48 | 10.000 | 55 | 10.000 | 59 | 10.000 | 62 | 10.000 | 65 | 10.000 | 53 | 2.000 |
| Redruth | 39 | 10.000 | 42 | 10.000 | 45 | 10.000 | 49 | 10.000 | 43 | 1.000 | 50 | 1.000 |
| Ridge Hill | 34 | 5.000 | 30 | 5.000 | 52 | 1.000 | 39 | 1.000 | 42 | 1.000 | 45 | 1.000 |
| Rowridge | 67 | 10.000 | 52 | 10.000 | 30 | 10.000 | 32 | 3.000 | 25 | 3.000 | 28 | 3.000 |
| Sandy Heath | 29 | 10.000 | 45 | 10.000 | 42 | 10.000 | 67 | 10.000 | 40 | 2.500 | 46 | 2.500 |
| Stockland Hill | 22 | 2.500 | 28 | 2.500 | 25 | 2.500 | 32 | 2.500 | 30 | 2.500 | 34 | 2.500 |
| Sudbury | 49 | 7.000 | 68 | 8.100 | 48 | 5.000 | 39 | 7.500 | 54 | 1.500 | 50 | 0.625 |
| Sutton Coldfield | 41 | 4.000 | 44 | 4.000 | 47 | 4.000 | 51 | 4.000 | 52 | 2.000 | 55 | 2.000 |
| Tacolneston | 63 | 10.000 | 60 | 10.000 | 64 | 10.000 | 57 | 10.000 | 43 | 10.000 | 46 | 5.000 |
| The Wrekin East | 39 | 1.000 | 49 | 1.000 | 42 | 1.000 | 45 | 1.000 | | | | |
| The Wrekin West | 21 | 1.000 | 31 | 1.000 | 24 | 1.000 | 27 | 1.000 | 53 | 1.000 | 57 | 1.000 |
| Waltham | 49 | 5.000 | 23 | 10.000 | 26 | 10.000 | 33 | 10.000 | 45 | 5.000 | 42 | 5.000 |
| Winter Hill | 56 | 5.000 | 66 | 5.000 | 68 | 5.000 | 50 | 2.000 | 60 | 2.000 | 63 | 2.000 |

| | D1 BBC | | D2 ITV + 4 | | D3 Mux A | | D4 Mux B | | D5 Mux C | | D6 Mux D | |
| --- | --- | --- | --- | --- | --- | --- | --- | --- | --- | --- | --- | --- |
| | Ch | Pwr (kW) | Ch | Pwr (kW) | Ch | Pwr (kW) | Ch | Pwr (kW) | Ch | Pwr (kW) | Ch | Pwr (kW) |
| CHANNEL ISLES | | | | | | | | | | | | |
| Fremont Point | 38 | 0.200 | 43 | 0.200 | 49 | 0.200 | 32 | 0.200 | 66 | 0.200 | 68 | 0.200 |
| SCOTLAND | | | | | | | | | | | | |
| Angus | 68 | 2.000 | 66 | 2.000 | 59 | 2.000 | 62 | 2.000 | 56 | 2.000 | 65 | 2.000 |
| Black Hill | 41 | 10.000 | 47 | 10.000 | 44 | 10.000 | 51 | 10.000 | 55 | 10.000 | 65 | 10.000 |
| Craigkelly | 33 | 1.000 | 29 | 1.000 | 23 | 1.000 | 26 | 1.000 | 42 | 2.000 | 39 | 2.000 |
| Darvel | 22 | 2.000 | 25 | 2.000 | 32 | 2.000 | 28 | 2.000 | 30 | 2.000 | 34 | 2.000 |
| Durris | 30 | 10.000 | 34 | 10.000 | 52 | 10.000 | 51 | 5.000 | 41 | 5.000 | 44 | 5.000 |
| Eitshal | 34 | 0.800 | 30 | 0.800 | 22 | 0.800 | 25 | 0.800 | 28 | 0.800 | 32 | 0.800 |
| Keelylang Hill | 48 | 1.000 | 52 | 1.000 | 41 | 0.630 | 44 | 0.630 | 47 | 0.630 | 51 | 0.630 |
| Knock More | 34 | 1.000 | 30 | 1.000 | 53 | 1.000 | 57 | 1.000 | 60 | 1.000 | 56 | 1.000 |
| Rosemarkie | 47 | 10.000 | 51 | 10.000 | 41 | 10.000 | 44 | 10.000 | 46 | 4.000 | 50 | 4.000 |
| Rumster Forest | 28 | 1.000 | 22 | 1.000 | 25 | 1.000 | 32 | 1.000 | 62 | 2.000 | 59 | 2.000 |
| Selkirk | 53 | 3.000 | 57 | 3.000 | 60 | 3.000 | 63 | 3.000 | 66 | 0.500 | 56 | 0.500 |
| NORTHERN IRELAND | | | | | | | | | | | | |
| Brougher Mountain | 30 | 0.500 | 34 | 0.500 | 23 | 0.500 | 26 | 0.500 | 29 | 0.500 | 33 | 0.500 |
| Divis | 29 | 8.900 | 33 | 8.900 | 23 | 8.900 | 26 | 8.900 | 48 | 1.800 | 34 | 0.800 |
| Limavady | 67 | 0.800 | 58 | 0.800 | 53 | 0.800 | 57 | 0.800 | 60 | 0.800 | 63 | 0.800 |
| WALES | | | | | | | | | | | | |
| Blaen Plwyf | 28 | 2.000 | 22 | 2.000 | 25 | 2.000 | 32 | 2.000 | 29 | 1.000 | 33 | 1.000 |
| Carmel | 55 | 2.500 | 65 | 2.500 | 59 | 2.500 | 62 | 2.500 | 68 | 1.000 | 66 | 1.000 |
| Llanddonna | 67 | 1.000 | 54 | 1.000 | 58 | 1.000 | 61 | 1.000 | 64 | 1.000 | 46 | 0.500 |
| Moel y Parc | 54 | 0.500 | 58 | 0.500 | 61 | 0.500 | 64 | 0.500 | 30 | 0.250 | 34 | 0.250 |
| Presely | 47 | 0.500 | 51 | 0.500 | 39 | 1.000 | 42 | 1.000 | 45 | 1.000 | 49 | 1.000 |
| Wenvoe | 30 | 10.000 | 34 | 4.000 | 56 | 6.000 | 67 | 10.000 | | | | |

# Revision questions

1. Figure RQ1 is a block diagram of a colour television receiver.
   (a) Name **each** of the following blocks:
      - (i) 6     (iv) 34
      - (ii) 14     (v) 31
      - (iii) 15     (vi) 26
   (b) State which block feeds the input to block 7 (marked Y).
   (c) Name the type of demodulator used in blocks 18 and 19.

**Fig. RQ1**

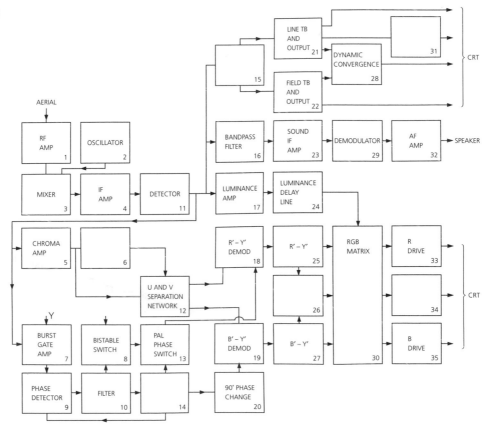

2. Figure RQ1 is a block diagram of a colour television receiver.
   (a) Name the signal and state the frequency at **each** of the following points:
      (i) block 7 output
      (ii) block 8 output
      (iii) block 24 output
      (iv) block 26 output
   (b) Describe the function of the RGB matrix.

3. Figure RQ1 is a block diagram of a colour television receiver. The receiver is displaying a standard colour-bar signal. Sketch the signal voltage waveform at the output of **each** of the following blocks:
   (a) 17
   (b) 18
   (c) 35

4. Explain the functions of the following items in a television receiver:
   (a) the RGB matrix
   (b) the delay line

5. Figure RQ2 is a block diagram of a television remote control receiver.
   (a) State the function of **each** of the following:
      (i) the binary output
      (ii) the integrating circuits to which A, B and C are connected
      (iii) the external circuit connected to point Z
      (iv) the shift register
   (b) Explain the operation of the standby function.
   (c) State the number of channels that can be selected using this receiver.

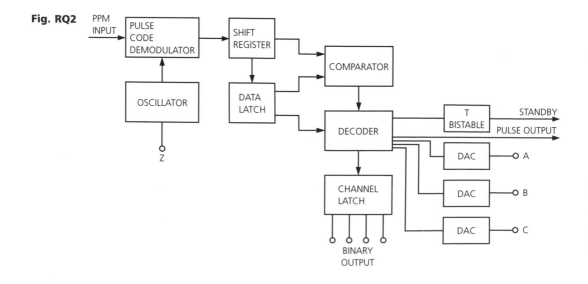

**Fig. RQ2**

**Fig. RQ3**

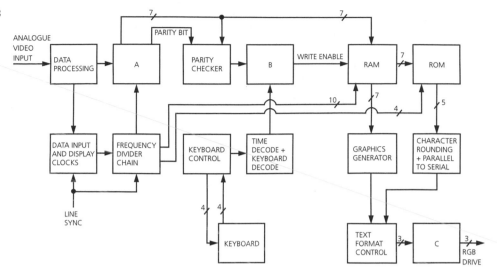

6. Figure RQ3 is a block diagram of a teletext decoder.
   (a) State the frequency of the display clock.
   (b) Name the following blocks and explain briefly why **each** is required in a teletext system:
       (i) A
       (ii) B
       (iii) C

7. Consider the British PAL colour TV system.
   (a) State **each** of the following frequencies:
       (i) vision i.f.
       (ii) chrominance subcarrier i.f.
       (iii) main sound i.f.
       (iv) adjacent channel vision i.f.
   (b) State **two** functions of colour burst.
   (c) State the type of modulation used for
       (i) sound
       (ii) chrominance

8. For the British PAL UHF television system, state
   (a) the line frequency
   (b) the field frequency
   (c) the maximum video modulating frequency
   (d) the frequency spacing between sound and vision carriers on channel 55
   (e) the frequency spacing between vision carriers on adjacent channels
   (f) the time period of the front porch
   (g) the time period of the back porch
   (h) the duration of the line blanking period
   (i) the number of cycles transmitted in **each** colour burst

9. Consider a PAL television receiver.
   (a) State the bandwidth required for **each** of the following:
       (i) the luminance amplifier
       (ii) the green output amplifier
   (b) State the luminance delay time in nanoseconds.
   (c) If the luminance signal is not delayed, what effect will it have on the colour picture displayed by a receiver with a 59 cm (24 inch) c.r.t.? Explain why this effect occurs.

10. Figure RQ4 shows part of a block diagram for a satellite digital TV transmitter.
    (a) State the function of
        (i) block 1
        (ii) block 2
    (b) Name block A and state its function.
    (c) Name the bitstream at the output of
        (i) block 1
        (ii) block 2

**Fig. RQ4**

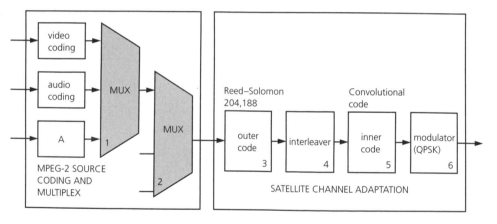

11. Figure RQ5 shows a digital TV receiver.
    (a) Name the following blocks:
        (i) X
        (ii) Y
        (iii) Z
    (b) State the type of signal at
        (i) A
        (ii) B

12. In a digital TV receiver, what happens when **each** of these failures occurs? Give reasons for your answers.
    (a) a failure in the FEC block
    (b) a failure in the conditional access module (CAM)

**Fig. RQ5**

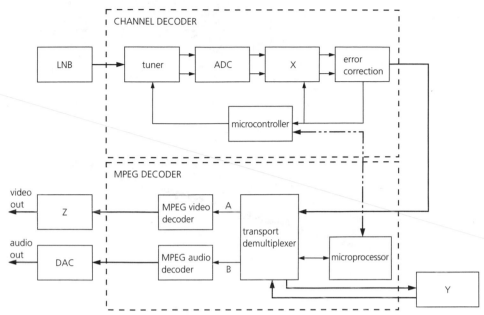

13. Refer to Fig. RQ6.
    (a) Describe the function of the circuit that includes transistors TR61, TR62 and TR67.
    (b) State the functions of transistor TR60.
    (c) The receiver is displaying incorrect colours. The highlights of the picture appear too red and the lowlights too blue.
       (i) Name and identify the controls to be adjusted.
       (ii) Describe how to reset the controls in part (i) so the picture becomes normal.

14. Refer to Fig. RQ7.
    (a) State **two** circuit function of IC1.
    (b) Name SF1 and state its circuit function.
    (c) Consider IC2.
       (i) Name block X.
       (ii) Explain why tuned circuit L3/C25 is required
       (iii) Explain the function of the input to pin 6.

**Fig. RQ6**

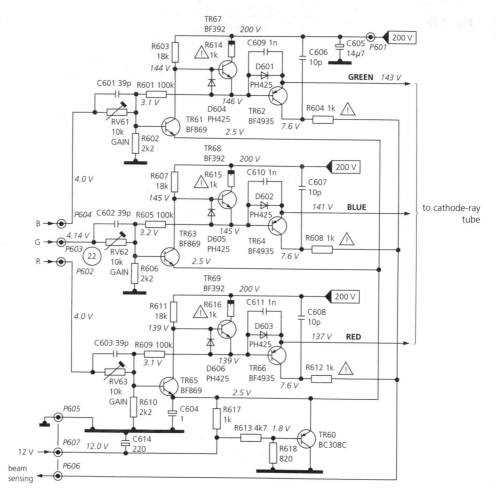

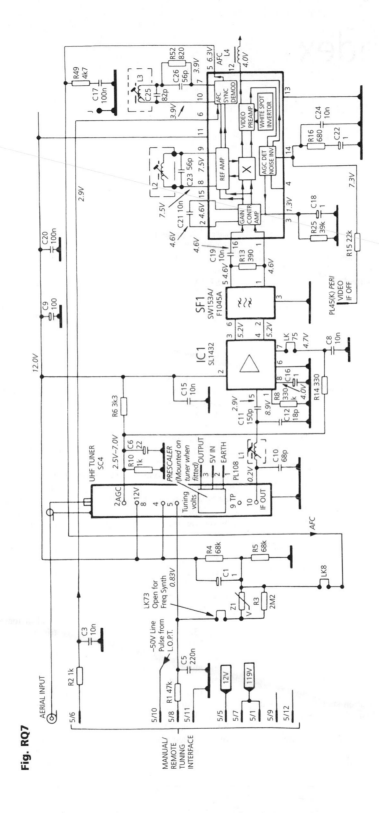

**Fig. RQ7**

# Index

---